Radon Transforms and Tomography

(*Photo courtesy of Professor Steven H. Izen.*)

Conference Participants

Selected Titles in This Series

278 Eric Todd Quinto, Leon Ehrenpreis, Adel Faridani, Fulton Gonzalez, and Eric Grinberg, Editors, Radon transforms and tomography, 2001

277 Luca Capogna and Loredana Lanzani, Editors, Harmonic analysis and boundary value problems, 2001

276 Emma Previato, Editor, Advances in algebraic geometry motivated by physics, 2001

275 Alfred G. Noël, Earl Barnes, and Sonya A. F. Stephens, Editors, Council for African American researchers in the mathematical sciences: Volume III, 2001

274 Ken-ichi Maruyama and John W. Rutter, Editors, Groups of homotopy self-equivalences and related topics, 2001

273 A. V. Kelarev, R. Göbel, K. M. Rangaswamy, P. Schultz, and C. Vinsonhaler, Editors, Abelian groups, rings and modules, 2001

272 Eva Bayer-Fluckiger, David Lewis, and Andrew Ranicki, Editors, Quadratic forms and their applications, 2000

271 J. P. C. Greenlees, Robert R. Bruner, and Nicholas Kuhn, Editors, Homotopy methods in algebraic topology, 2001

270 Jan Denef, Leonard Lipschitz, Thanases Pheidas, and Jan Van Geel, Editors, Hilbert's tenth problem: Relations with arithmetic and algebraic geometry, 2000

269 Mikhail Lyubich, John W. Milnor, and Yair N. Minsky, Editors, Laminations and foliations in dynamics, geometry and topology, 2001

268 Robert Gulliver, Walter Littman, and Roberto Triggiani, Editors, Differential geometric methods in the control of partial differential equations, 2000

267 Nicolás Andruskiewitsch, Walter Ricardo Ferrer Santos, and Hans-Jürgen Schneider, Editors, New trends in Hopf algebra theory, 2000

266 Caroline Grant Melles and Ruth I. Michler, Editors, Singularities in algebraic and analytic geometry, 2000

265 Dominique Arlettaz and Kathryn Hess, Editors, Une dégustation topologique: Homotopy theory in the Swiss Alps, 2000

264 Kai Yuen Chan, Alexander A. Mikhalev, Man-Keung Siu, Jie-Tai Yu, and Efim I. Zelmanov, Editors, Combinatorial and computational algebra, 2000

263 Yan Guo, Editor, Nonlinear wave equations, 2000

262 Paul Igodt, Herbert Abels, Yves Félix, and Fritz Grunewald, Editors, Crystallographic groups and their generalizations, 2000

261 Gregory Budzban, Philip Feinsilver, and Arun Mukherjea, Editors, Probability on algebraic structures, 2000

260 Salvador Pérez-Esteva and Carlos Villegas-Blas, Editors, First summer school in analysis and mathematical physics: Quantization, the Segal-Bargmann transform and semiclassical analysis, 2000

259 D. V. Huynh, S. K. Jain, and S. R. López-Permouth, Editors, Algebra and its applications, 2000

258 Karsten Grove, Ib Henning Madsen, and Erik Kjær Pedersen, Editors, Geometry and topology: Aarhus, 2000

257 Peter A. Cholak, Steffen Lempp, Manuel Lerman, and Richard A. Shore, Editors, Computability theory and its applications: Current trends and open problems, 2000

256 Irwin Kra and Bernard Maskit, Editors, In the tradition of Ahlfors and Bers: Proceedings of the first Ahlfors-Bers colloquium, 2000

255 Jerry Bona, Katarzyna Saxton, and Ralph Saxton, Editors, Nonlinear PDE's, dynamics and continuum physics, 2000

254 Mourad E. H. Ismail and Dennis W. Stanton, Editors, q-series from a contemporary perspective, 2000

(*Continued in the back of this publication*)

CONTEMPORARY MATHEMATICS

278

Radon Transforms and Tomography

2000 AMS-IMS-SIAM Joint Summer Research Conference
on Radon Transforms and Tomography
Mount Holyoke College, South Hadley, Massachusetts
June 18–22, 2000

Eric Todd Quinto
Leon Ehrenpreis
Adel Faridani
Fulton Gonzalez
Eric Grinberg
Editors

American Mathematical Society
Providence, Rhode Island

Editorial Board

Dennis DeTurck, managing editor

Andreas Blass   Andy R. Magid   Michael Vogelius

The 2000 AMS-IMS-SIAM Joint Summer Research Conference on "Radon Transforms and Tomography" was held at Mount Holyoke College, South Hadley, Massachusetts, June 18–22, 2000, with support from the National Science Foundation, grant DMS-9973450.

2000 *Mathematics Subject Classification.* Primary 44A12, 92C55, 92F05, 65T60, 35R30; Secondary 43–02, 42A38, 30E05, 43A77, 32L25.

Any opinions, findings, and conclusions or recommendations expressed in this material are those of the authors and do not necessarily reflect the views of the National Science Foundation.

Library of Congress Cataloging-in-Publication Data
AMS-IMS-SIAM Joint Summer Research Conference on Radon Transforms and Tomography (2000: Mount Holyoke College)
  Radon transforms and tomography: 2000 AMS-IMS-SIAM Joint Summer Research Conference on Radon Transforms and Tomography, Mount Holyoke College, South Hadley, Massachusetts, June 18–22, 2000 /Eric Todd Quinto...[et al.], editors.
    p. cm.— (Contemporary Mathematics, ISSN 0271-4132; 278)
    ISBN 0-8218-2135-0
    1. Radon transforms–Congresses. 2. Tomography–Congresses. I. Quinto, Eric Todd, 1951–. II. Title. III. Contemporary mathematics (American Mathematical Society); v. 278.

QA649.A495 2000
515′.723—dc21   2001033762

Copying and reprinting. Material in this book may be reproduced by any means for educational and scientific purposes without fee or permission with the exception of reproduction by services that collect fees for delivery of documents and provided that the customary acknowledgment of the source is given. This consent does not extend to other kinds of copying for general distribution, for advertising or promotional purposes, or for resale. Requests for permission for commercial use of material should be addressed to the Assistant to the Publisher, American Mathematical Society, P. O. Box 6248, Providence, Rhode Island 02940-6248. Requests can also be made by e-mail to reprint-permission@ams.org.

Excluded from these provisions is material in articles for which the author holds copyright. In such cases, requests for permission to use or reprint should be addressed directly to the author(s). (Copyright ownership is indicated in the notice in the lower right-hand corner of the first page of each article.)

© 2001 by the American Mathematical Society. All rights reserved.
The American Mathematical Society retains all rights
except those granted to the United States Government.
Printed in the United States of America.

∞ The paper used in this book is acid-free and falls within the guidelines
established to ensure permanence and durability.
Visit the AMS home page at URL: http://www.ams.org/

10 9 8 7 6 5 4 3 2 1   06 05 04 03 02 01

Contents

Preface ... ix

I. Expository Papers

Local tomography and related problems
    CARLOS A. BERENSTEIN ... 3

Tomography problems arising in synthetic aperture radar
    MARGARET CHENEY ... 15

Introduction to local tomography
    A FARIDANI, K. A. BUGLIONE, P. HUABSOMBOON, O. D. IANCU, AND J. MCGRATH ... 29

Algorithms in ultrasound tomography
    FRANK NATTERER ... 49

Radon transforms, differential equations, and microlocal analysis
    ERIC TODD QUINTO ... 57

Supplementary bibliography to "A bibliographic survey of the Pompeiu problem"
    LAWRENCE ZALCMAN ... 69

II. Research Papers

Twistor results for integral transforms
    TOBY BAILEY AND MICHAEL EASTWOOD ... 77

Injectivity for a weighted vectorial Radon transform
    JAN BOMAN ... 87

Shape reconstruction in 2d from limited-view multifrequency electromagnetic data
    OLIVER DORN, ERIC L. MILLER, AND CAREY M. RAPPAPORT ... 97

Three problems at Mount Holyoke
    LEON EHRENPREIS ... 123

A Paley-Wiener theorem for central functions on compact Lie groups
    FULTON B. GONZALEZ ... 131

Inversion of the spherical Radon transform by a Poisson type formula
  ERIC L. GRINBERG AND ISAAC PESENSON　　　　　　　　　　　　137

Application of the Radon transform to calibration of the NASA-Glenn icing research wind tunnel
  STEVEN H. IZEN AND TIMOTHY J. BENCIC　　　　　　　　　　　147

Range theorems for the Radon transform and its dual
  ALEXANDER KATSEVICH　　　　　　　　　　　　　　　　　　　167

Moment conditions *indirectly* improve image quality
  S. K. PATCH　　　　　　　　　　　　　　　　　　　　　　　　　193

Principles of reconstruction filter design in 2D-computerized tomography
  ANDREAS RIEDER　　　　　　　　　　　　　　　　　　　　　　207

The k-dimensional Radon transform on the n-sphere and related wavelet transforms
  BORIS RUBIN AND DMITRY RYABOGIN　　　　　　　　　　　　227

Reconstruction of high contrast 2-D conductivities by the algorithm of A. Nachman
  SAMULI SILTANEN, JENNIFER L. MUELLER, AND DAVID ISAACSON　　241

Integral geometry problem with incomplete data for tensor fields in a complex space
  L. B. VERTGEIM　　　　　　　　　　　　　　　　　　　　　　　255

Preface

One of the most exciting features of the fields of Radon Transforms and Tomography is the strong relationship between high level pure mathematics and applications to areas such as medical imaging, remote sensing, and industrial nondestructive evaluation. These proceedings bring together fundamental research articles in the major areas of tomography and Radon transforms.

The proceedings include six expository papers that we hope will be valuable to beginners as well as advanced researchers. Local tomography is an extremely useful new type of tomography in which local data are used to reconstruct the singularities or shapes of the objects to be reconstructed. Fundamental articles are included on local tomography and wavelets [Berenstein] as well as on Lambda tomography and related methods [Faridani, Buglione, Huabsomboon, Iancu, McGrath]. Margaret Cheney gives a mathematical tutorial of Synthetic Aperture RADAR and shows how tomographic methods come up naturally in this setting. Ultrasound tomography is based on an inverse problem for the Helmholz equation. Frank Natterer discusses the general problem and linear approximations and he provides a survey to iterative inversion methods. One article shows how microlocal analysis is used to prove support theorems for the hyperplane transform and how such theorems for a spherical transform characterize stationary sets for the wave equation [Quinto]. The Pompeiu problem is to find sets that determine functions from integrals over rigid motions of the given set. Larry Zalcman updates his comprehensive bibliography on this problem from 1992 and provides valuable commentary.

The major themes in Radon transforms and tomography are represented among the research articles. Bailey and Eastwood use spectral sequences to formulate and examine the X-ray and related transforms on Grassmmann manifolds, obtaining useful results about null spaces and ranges. Vector tomography models integration of vector (or tensor) fields over curves and surfaces. Support theorems and injectivity are proven for vectorial transforms on divergence free vector fields integrating over lines in R^n in real analytic measures, and counterexamples are provided if the measure is only C^∞ [Boman]. Injectivity theorems are proven for the solenoidal part of tensor fields from integrals over complex lines intersecting a given complex curve in C^n [Vertgeim]. The paper by Ehrenpreis discusses three problems posed during the conference. These involve the recovery of both test data and defining measures from Radon transforms values, the uniqueness properties of a transform that integrates over spheres tangent to a surface, and a variation on Morera's theorem for detecting analyticity properties of functions from integrals. Gonzalez proves a Paley-Wiener Theorem for the Fourier coefficients of central functions on compact Lie groups as a consequence of a corresponding Paley-Wiener Theorem for ordinary

Fourier series. Range theorems are proven for the Radon transform and its dual on spaces that do not decay rapidly at infinity [Katsevich].

Spherical Radon transforms (transforms integrating over spheres) and their applications appear in several articles (including the expository articles of Cheney and of Quinto as well as Ehrenpreis' research article). Grinberg and Pesenson provide a Poisson type formula for recovering an approximation of a function on a sphere from a finite number of integrals over great spheres. Rubin and Ryabogin use wavelet transforms to invert k-plane transforms on the sphere.

The applied articles employ high-quality pure mathematics and numerical analysis to solve important practical problems. Algorithms for limited data electromagnetic tomography are developed and tested [Dorn, Miller, and Rappaport]. A uniqueness proof of Nachman is used to develop an algorithm for electrical impedance imaging [Siltanen, Mueller, and Isaacson]. The algorithm is tested on high-contrast simulations and then analyzed. Izen and Bencic test several scanning geometries for tomography in a wind tunnel to detect icing on wings. They analyze the singular values and functions and test their predictions on the data. Patch proposes using range theorems to detect when tomographic scanning equipment is failing. Rieder investigates the principles for designing the convolution kernel in the popular filtered backprojection algorithm, and he suggests alternatives to the kernels currently in use.

All articles have been carefully refereed and are in final form.

We thank James Maxwell, Associate Director, AMS, for his assistance and support throughout this process. The co-organizers and participants are indebted to Wayne Drady, AMS Conference Coordinator for the conference. He was always cheerful and helpful. We thank AMS Acquisitions Assistant Christine Thivierge for her cooperation with the editors and her able and thorough job putting the proceedings together. The organizers could not have arranged the conference as easily and as pleasantly without the help of everyone at the AMS. Finally we thank the National Science Foundation for their generous support.

Todd Quinto for the committee

Leon Ehrenpreis	Temple University
Adel Faridani	Oregon State University
Fulton Gonzalez	Tufts University
Eric Grinberg	Temple University
Eric Todd Quinto, Chair	Tufts University

I. Expository Papers

Local tomography and related problems

Carlos A. Berenstein

Medical tomography is one of the most visible recent contributions of Mathematics to the general well-being. CT and MRI are now almost routine diagnostic tools. These two imaging methods and other tomographic instrumentation have also significant scientific and industrial applications, quite often unexpected. The problems discussed below were inspired by work on a prototype instrumentation to measure space plasma originally conceived with M. Coplan and J. Moore (see [ZCMB] for a 2-d version, the full 3-d case was done jointly with M. Shahshahani.)

Let us start by recalling the definition of the Radon transform we use and its relationship to the wave operator. (The basic references for the Radon transform are [He1, Na, KS], each of them has a different, complementary, point of view.) First, consider the original case studied by Radon, the Radon transform in the Euclidean plane. (Although in his original paper [R], he also briefly discusses the inversion formula for the 3-d case.) Let $w = (\cos\theta, \sin\theta) \in S^1, p \in \mathbb{R}$, thus the equation $x \cdot w = p$ represents the equation of a line ℓ perpendicular to w and of signed distance to the origin equal to p. For a function f sufficiently nice, for instance, continuous of compact support, one defines the Radon transform of f evaluated in this line as given by

$$R_w f(p) := Rf(w,p) := \int f(x)ds = \int_{-\infty}^{\infty} f(x_0 + tw^\perp)dt,$$

where x_0 is any fixed point in ℓ, $w^\perp = (-\sin\theta, \cos\theta)$ is the rotate of w by $\frac{\pi}{2}$. Clearly, Rf is defined in the family of all lines in the plane and depends only on the line ℓ and not in its equation. In particular, it satisfies $Rf(-\omega, -p) = Rf(w,p)$. One can see that the definition of the Radon transform can be easily extended to other classes of functions, e.g. $\mathcal{S}(\mathbb{R}^2), L^1(\mathbb{R}^2)$, and, more interestingly for the usual applications like CT, f bounded and compactly supported, sufficiently smooth except along some curves, where f has jump singularities. The case of 3-d, applicable to MRI, is similar with the exception that the same equation $x \cdot w = p$ represents a 2-d plane in the space and ds has to be interpreted as the area measure along this hyperplane. Although the discussion is more general, we are considering $n = 2$ or 3 in what

---

2000 *Mathematics Subject Classification.* Primary 44A12, Secondary 92C55.

This lecture reflects past and ongoing research with a number of collaborators, Enrico Casadio Tarabusi, Roger Gay, Boris Rubin, and David Walnut, among them. The author's research has been partly supported by NSF grants DMS-9622249 and DMS-0070044.

follows. In fact, mostly $n = 2$. We assume this implicitly anytime the dimension n appears. Furthermore, throughout this lecture we will call "image plane", the plane of the object, what we denoted by x coordinates above, and "Radon plane" or "data plane" the plane of the data acquired via R, whose points were denoted by coordinates (w, p) above. Later on we will briefly discuss tomography in the hyperbolic plane, but we refer the reader to [He1, He2] for an excellent introduction to tomography in more general spaces.

It is convenient to extend the domain transform from $S^1 \times \mathbb{R}$ (or $S^2 \times \mathbb{R}$ in the 3-d case) to $(\mathbb{R}^2 \setminus \{0\}) \times \mathbb{R}$ (resp. $(\mathbb{R}^3 \setminus \{0\}) \times \mathbb{R}$) by making R homogeneous of degree -1 on the spatial variable. Namely, for $\xi \neq 0$ we set

$$Rf(\xi, p) := \frac{1}{|\xi|} Rf(\frac{\xi}{|\xi|}, p)$$

It is now easy to see that

$$\frac{\partial}{\partial \xi_j} Rf(\xi, p) = -\frac{\partial}{\partial p} R(x_j f)(\xi, p),$$

while, assuming f is smooth

$$R(\frac{\partial}{\partial x_j} f)(\xi, p) = \xi_j \frac{\partial}{\partial p} Rf(\xi, p).$$

Thus, for $n = 2$ or 3,

$$R(\Delta f)(\xi, p) = (\xi_1^2 + \cdots + \xi_n^2) \frac{\partial^2}{\partial p^2} Rf(\xi, p),$$

so that for $w \in S^{n-1}$,

$$R(\Delta f)(w, p) = \frac{\partial^2}{\partial p^2}(Rf)(w, p),$$

Note that if f depends on additional variables, for instance, time t, we can conclude that

$$R(\Delta f - \frac{\partial^2}{\partial t^2} f)(w, p, t) = (\frac{\partial^2}{\partial p^2} - \frac{\partial^2}{\partial t^2}) Rf(w, p, t),$$

where, as usual, Δ is the Laplacian in $\mathbb{R}^n$. Fixing $w \in S^{n-1}$, one can express this identity by saying that the Radon transform intertwines the wave operator $\Box_n = \Delta - \frac{\partial^2}{\partial t^2}$ in n-dimensions with the wave operator $\Box_1 = \frac{\partial^2}{\partial p^2} - \frac{\partial^2}{\partial t^2}$ in 1-space dimension. We denote by $R^\#$ the adjoint of R, usually called the backprojection operator, since if $R_w^\#$ is the adjoint of R_w, $R_w^\# g(x) = g(x \cdot w)$, then

$$R^\# g(x) = \int R_w^\# g(x) dw$$

so that we have the classical backprojection inversion formula

$$(*) \qquad f(x) = \frac{1}{2} \int_{\mathbb{R}^2} e^{2\pi i x \cdot \zeta} |\zeta| (R^\# Rf)^\wedge(\zeta) d\zeta = \wedge R^\# Rf(x),$$

It is a well known fact, even to small children, that dropping a small stone in the water causes ripples that last "forever", while a scream cannot be heard after a child stops crying. In other words, the wave equation localizes the initial data for odd dimensions and does not localize it for even dimensions. Because R (the codimension one Radon transform in n dimensions) intertwines $\Box_n$ with $\Box_1$, it follows that it cannot be localized in even dimensions. This is the reason that the true CT image of a head has "artifacts" surrounding the skull. (They are suppressed in the image given to the doctor and patient, so the doctors and patients do not worry about them.) Nevertheless, there are many reasons why it would be convenient to localize the Radon transform in dimension 2, for instance, if the object is very large, or if there is some area that should receive significantly less radiation, etc. In fact, one could argue that in general, to get a very good approximation of a very nice function f, even of compact support, say in the disk $B(0,r)$ one may need to know the values of Rf for all lines that traverse the disk $B(0,4r)$ (or such "large" multiples of r) [Na]. The key point is the comparison of a Gaussian to its Hilbert transform as explained following the statement of Proposition 2.

The reason is that the Hilbert transform is an essential component of the inversion formula of R in dimension 2. On the other hand, one can also see that very oscillatory functions (many vanishing moments) have rather rapidly decreasing Radon and Hilbert transforms [BW2, Appendix]. This leads naturally to the idea that up to an additive constant one should be able to localize "fairly well" the Radon transform in dimension two if one represents any function in terms of oscillatory functions. What one really wants is a solution that given an priori error bound, one can reconstruct the unknown function f in a disk using only lines that cross a slightly larger concentric disk. One typical example of such functions are the compactly suggested wavelets. This is a natural idea that appeared first in the report [W] of the ongoing joint work of David Walnut and myself. One should also point out other approaches to localization of the Radon transfom, for instance, in the work of Faradani et al. [FRK], as well as other references given in [BW2]. There have been since then ideas in the same spirit in [Ho] and others. The algorithm presented in [FLBW] works incredibly well, as the comparison of Figures 2 and 3 indicate (There are a priori good estimates of the error and margin of security, the algorithm proposed in this paper received a US patent last year.)

There are many excellent books on the subject of wavelets, at all levels of sophistication and different points of view, the following is a very partial list [M], [D], [Ka]. There are actually two different, albeit related concepts, the *continuous wavelet transform* (CWT) (easier to understand) and the *discrete wavelet transform* (DWT) (easier to work with), for the sake of completeness we recall here the definition of the CWT. A *wavelet* is an L^2 function ψ which is sufficiently oscillatory, and if we want to consider the behavior of a function $f \in L^2(\mathbb{R})$ at different scales, we define its CWT by

$$W_\psi f(a,b) := \int_{-\infty}^{\infty} f(t)\bar{\psi}(\frac{t-b}{a})dt = <f, D_a\tau_b\psi> = (f * \overline{D_a\check{\psi}})(b)$$

where for $0 < a < \infty, b \in \mathbf{R}, \tau_b\psi(x) = \psi(x-b), D_a\psi(x) = a^{1/2}\psi(x/a), \bar{\psi}$ denotes the complex conjugate of ψ, and $<,>$ denotes the L^2-scalar product. The assumption that the wavelet is "oscillatory", is given by the condition

$$c_\psi := \int_{-\infty}^{\infty} \frac{|\hat{\psi}(\xi)|^2}{|\xi|}d\xi < \infty.$$

where $\hat{\psi}$ denotes the Fourier transform of ψ, namely

$$\hat{\psi}(\zeta) = \int_{-\infty}^{\infty} \psi(x)e^{-2\pi ix\zeta}dx.$$

This condition implies that $\int_{-\infty}^{\infty} \psi(x)dx = 0$. (For instance, when $\hat{\psi}$ is continuous at $\xi = 0$, which occurs if $\psi \in L^1(\mathbf{R}) \cap L^2(\mathbf{R})$.) In fact, we will be interested here in wavelets with many vanishing moments

$$\int_{-\infty}^{\infty} x^k \psi(x)dx = 0, \quad 0 \leq k \leq N.$$

A typical wavelet is the Haar wavelet

$$\psi = \chi_{[0,1/2]} - \chi_{[1/2,1/1]})$$

so that

$$D_{1/2}\psi = \sqrt{2}(\chi_{[0,1/4]} - \chi_{[1/4,1/2]})$$

which shows that for $k \to \infty$, $D_2^{-k}\psi$ "analyzes" smaller and smaller details of the "signal" f, but it has only one vanishing movement. One can see that $W_\psi f$ determines f from the following relation valid for any pair $f, g \in L^2(\mathbf{R})$

$$\int_{-\infty}^{\infty}\int_{-\infty}^{\infty} W_\psi f(a,b)\overline{W_\psi g(a,b)}\frac{dadb}{a^2} = c_\psi <f,g> \|\psi\|_2^2,$$

usually called Calderon's identity. If $\|\psi\|_2 = 1$ one also has the L^2-approximation property

$$\|f - \frac{1}{c_\psi}\int_{A_1 \leq |a| \leq A_2, |b| \leq B} W_\psi f(a,b) D_a T_b \psi \frac{dadb}{a^2}\|_2 \to 0$$

as $A_1 \to 0^+, A_2 \to +\infty, B \to +\infty$.

The generalization to $\mathbf{R}^n$ is easy. A function $\psi \in L^2(\mathbf{R}^n)$ is a wavelet if

$$\int_{\mathbf{R}^n} \frac{|\hat{\psi}(\xi)|^2}{|\xi|}d\xi < \infty,$$

where $\hat{\psi} = \mathcal{F}\psi$ is the n-dimensional Fourier transform of ψ. For a radial wavelet $\psi \in L^2(\mathbf{R}^n)$ and $f \in L^2(\mathbf{R}^n)$ we define the CWT by

$$W_\psi f(a,b) = f * D_2\bar{\psi}^{\vee}(b) \text{ for } a \in \mathbf{R}\setminus\{0\}, b \in \mathbf{R}^n,$$

where this time, $D_a\psi(x) = |a|^{-n/2}\psi(x/a)$.

The interest of CWT for tomography lies in the following two propositions from [BW2].

Proposition 1. Let $\rho \in L^2(\mathbf{R})$ be real valued, even, and satisfying

$$\int_0^{\infty} \frac{|\hat{\rho}(r)|^2}{r^3}dr < \infty$$

Define a radial function ψ in $\mathbf{R}^2$ by $\mathcal{F}_2\psi(\xi) = 2\hat{\rho}(|\xi|)/|\xi|$, then ψ is a wavelet and

$$W_\psi f(a,b) = a^{-1/2}\int_{S^1} (W_\rho R_w f)(a, b \cdot w)dw$$

A similar relation between the Radon transform and the CWT can be found using "separable" wavelets in $\mathbf{R}^2$.

Proposition 2. Given a separable 2-dimensional wavelet of the form
$$\psi(x) = \psi^1(x_1)\psi^2(x_2)$$
where each $\psi^i(t)$ satisfies $|\hat{\psi}(\gamma)| \leq C_1(1+|\gamma|)^{-1}$ for all $\gamma \in \mathbf{R}$, define the family of one-dimensional functions $\{\rho_w\}_{w \in S^1}$ by
$$\hat{\rho}_w(\gamma) = \frac{1}{2}|\gamma|\hat{\psi}^1(\gamma w_1)\hat{\psi}^2(\gamma w_2)$$
where $w = (w_1, w_2) \in S^1$. Then, for every $f \in L^1(\mathbf{R}^2) \cap L^2(\mathbf{R}^2)$,
$$(W_\psi f)(a, x) = a^{-1/2} \int_{S_1} (W_{\rho_w} R_w f)(a, x \cdot w) dw.$$

The point of Proposition 2 is the observation that the wavelet transform of a function $f(x)$ with any mother wavelet and at any scale and location can be obtained by backprojecting the wavelet transform of the Radon transform of f using wavelets that vary with each angle, but which are admissible for each angle.

So far we have not yet shown that the inversion formulas of the Radon transform based on wavelets do a good localization job. Using Proposition 1 the problem is clear, find a function ρ such that ρ has small support and simultaneously ψ has small support. From the inversion formula $(*)$ for R given earlier we see that we have to balance out the Riesz operator of order -1, which is the composition of differentiation and the Hilbert transform. The difficulty is with the Hilbert transform, but if we choose ρ with many vanishing moments, then we can overcome the difficulty. For the sake of comparison we show in Figure 1 the Hilbert transform of a Gaussian, its effective support is about four times the effective support of the Gaussian (defined by making zero those points below 1% of maximum value), which corresponds exactly to the belief that the Radon transform could not be localized, even in the reasonable sense we proposed here.

The key to explain the success of the wavelet method of localization is the following proposition [BW], which in spirit is similar to the general principles about Calderon-Zygmund operators stated in [BCR].

Proposition 3. Supposed that n is an even integer and the compactly supported function $h \in L^2(\mathbf{R})$ is such that for some integer $m \geq 0$ we have that the function $\hat{h}$ is $n + m - 1$ times differentiable and satisfies the two conditions
(a) $\gamma_j \hat{h}^{(k)}(\gamma) \in L^1(\mathbf{R}) \cap L^2(\mathbf{R})$ for $0 \leq j \leq m, 0 \leq m + n - 1$
(b) $\int_{-\infty}^{\infty} t^j h(t) dt = 0$ for $0 \leq j \leq m$
Then
$$I^{1-n} h(t) = o(|t|^{-n-m+1}) \text{ as } |t| \to \infty$$
and
$$t^{n+m-1} I^{1-n} h \in L^2(\mathbf{R}).$$

The proof is rather elementary, it depends on the fact that if h is a function of compact support with $m+1$ vanishing moments then $|\gamma|^{n-1}\hat{h}(\gamma)$ has $n + m - 1$ continuous derivatives.

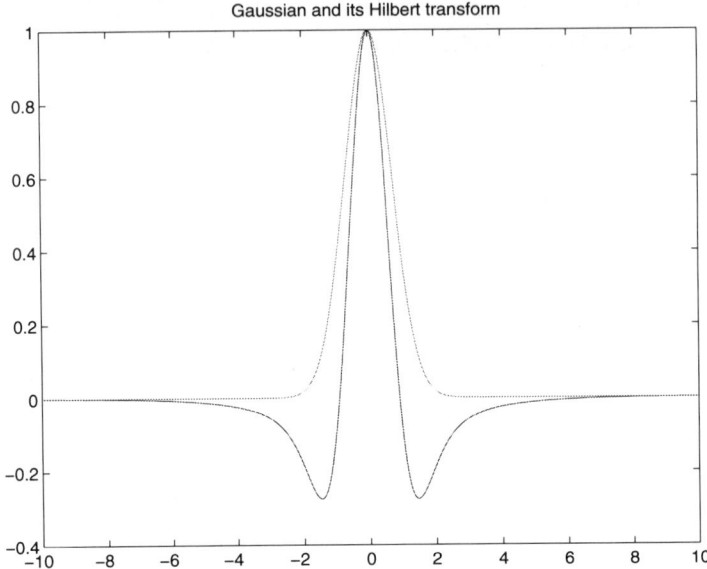

FIGURE 1. Gaussian and its Hilbert transform.

For ease of application it is better to work with the *discrete wavelet transform* (DWT). This is basically obtained by discretizing the CWT or appealing to the multiresolution analysis of Mallat and Meyer [D], [M]. We have done this in detail in [FLBW] using coiflets (a particular kind of wavelets) [D] in order to be able to implement the inversion process using filter banks. One can show that to obtain a relative error of 0.5% one only needs a margin of security of 12 pixels around the region of interest (ROI). For instance, to recover within this error bound an image occupying a disk of radius 20 pixels in a 256×256 image, one only needs about 25% of exposure.

The following figures, obtained using our wavelet reconstruction algorithm, are the reconstruction of a full cross section of the body including the heart and lungs from real CAT scanner data, and the reconstruction of the a subregion near the spine from local data. (Note that Figure 2 includes the support below the patient's body.)

What could be the typical application of this idea? In the case of medical CT it means that one obtains the same quality of image of a region of interest, say one of the lungs, while subjecting the patient to a smaller total radiation dosage. Note that outside the region of interest the patient also receives radiation but it decays exponentially with the distance to the region of interest. In fact, one could probably increase somewhat the strength of the x-rays without violating the medical standards. Why would this have any interest? The local image will then be sharper and allow for application of more accurate image processing techniques. The hope would be to detect tumors earlier than it is done currently. We note that the representation of the image in terms of wavelets is already specially adapted for this purpose. Using that the Radon transform is an elliptic operator and thus edges and ther singularities are preserved in the correspondence between the image plane

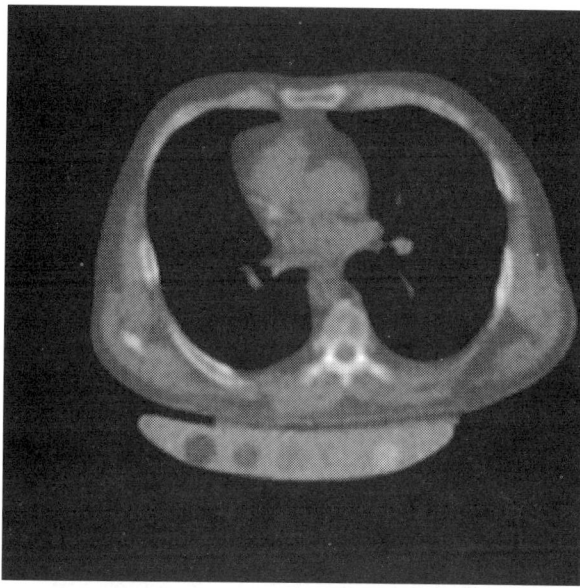

FIGURE 2. Reconstruction of whole cross section.

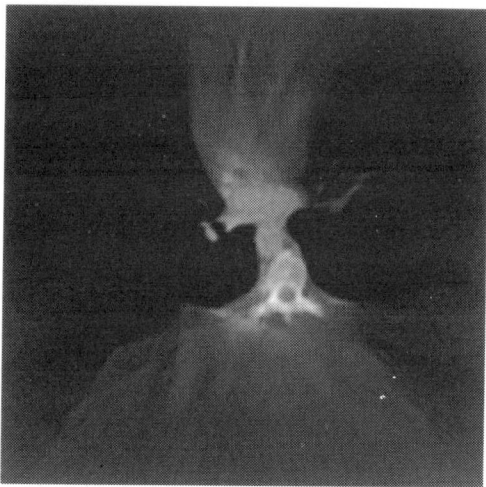

FIGURE 3. The local reconstruction of region just above the vertebra.

and the Radon plane. Since, in principle, the noise in the data increases after every transformation of the data, one can obtain further SNR gains, thus, hopefully, more accurate image processing by using edge detection techniques on the Radon plane, as Walnut and I are currently investigating. (This idea was influenced by the work in [Q].)

In perspective, this approach to image processing has a lot in common with the recent work on ridgelets by Candés [Ca], Donoho and Johnstone. Their work started with a statistical point of view, as a way to improve further image processing done using wavelets. For instance, it incorporates the matching pursuit ideas of

Mallat [MZ]. On the other hand, at least as presented in [Ca] it still suffers with a dependency on "special" directions (e.g., radial). We are currently considering how to consolidate these two approaches. I would like to point out that there is work in signal processing, for instance [CHT], linking wavelets and ridges, which seems to be in the same spirit as that of Candés, Donoho, and Johnstone.

Wavelets are the link between the Euclidean Radon transform ideas I have just presented and a problem associated to the non-Euclidean Radon transform: the Calderon problem [U]. The question is to determine the electrical conductivity profile of a disk (or more generally, a planar conducting plate of arbitrary shape) from measurements on the boundary of the voltage induced by arbitrary currents also applied to the boundary. Briefly recall that if we represent the plate by the unit disk D in the complex plane, the Calderon problem can be posed as follows: a current ψ on ∂D is a function with average zero, the electrical potential u induced by ψ on D solves the boundary value problem

$$(*) \begin{cases} \operatorname{div}(\beta \operatorname{grad} u) = 0 & \text{in } D \\ \beta \frac{\partial u}{\partial n} = \psi & \text{on } \partial D \end{cases}$$

where the positive function β corresponds to the conductivity in the plate D. The potential u is unique up to an additive constant, so that its tangential derivative $\frac{\partial u}{\partial s}$ on ∂D is well defined. The Calderon map is the input-output map (depending on β)

$$\wedge_\beta : \quad \psi \to \frac{\partial u}{\partial s}$$

By using "all possible" inputs ψ we can "determine" the map $\wedge_\beta$. Thus, we have a map (the Calderon map)

$$C : \quad \beta \longmapsto \wedge_\beta$$

and Calderon's problem is to recover β from the knowledge of $\wedge_\beta$. Nachman proved that this is *a priori* possible (i.e., injectivity of the map C), see [N1], [N2] and [U] for background references. There are several methods to invert numerically C, they usually assume that β is either approximately constant or it has some other special properties. What we considered in [BC1] was the first case, that is, we assumed that $\beta = 1 + \gamma$ for some small function γ, satisfying $\gamma = 0$ on ∂D. Then the solution U of $(*)$ can be written as $u = U + v$, U corresponds to setting $\beta \equiv 1$ in $(*)$. Thus, for any input ψ we have

$$(**) \begin{cases} \Delta U = 0 & \text{in } D \\ \frac{\partial U}{\partial n} = \psi & \text{on } \partial D \end{cases}$$

where Δ is the Euclidean Laplacian. The perturbation v satisfies

$$(***) \begin{cases} \Delta u = -<\operatorname{grad} U, \operatorname{grad} \gamma> & \text{in } D \\ \frac{\partial v}{\partial n} = -\gamma \psi & \text{on } \partial D \end{cases}$$

In order to try to determine γ we can use many different inputs ψ. The only restriction is that $\int_{\partial D} \psi ds = 0$. A standard choice is to consider ψ as a linear combination of dipoles. For any point $w \in \partial D$ a dipole supported at w is $-\pi \frac{\partial}{\partial s} \delta_w$, the tangential derivative of the delta function at the point w. In this case the solution $U = U_w$ of $(**)$ satisfies

$$\begin{cases} \Delta U_w = 0 & \text{in } D \\ \frac{\partial U_w}{\partial n} = -\pi \frac{\partial}{\partial s} \delta_w & \text{on } \partial D \end{cases}$$

One can show that the level curves of U_w are precisely the geodesics of the hyperbolic metric on the disk D which end at the point $w \in \partial D$. A bit of reflection shows that the function $\frac{\partial v}{\partial s}|\partial D$ can be considered as a function in the space of geodesics. Thus, it is natural to try to formulate this problem in terms of hyperbolic geometry in the disk and the corresponding hyperbolic Radon transform R_H. For that purpose we endow D with the metric given by

$$ds^2 = \frac{4|dz|^2}{(1-|z|^2)^2},$$

where $|dz|$ denotes the Euclidean arc-length element. This metric is clearly conformal to the Euclidean metric but has constant curvature -1. As it is well known, the geodesics of this metric are the diameters of D and the segments lying in D of the Euclidean circles intersecting the unit circle ∂D perpendicularly. The Laplace-Beltrami operator Δ_H on D can be written in terms of the Euclidean Laplacian Δ as $\Delta_H = \frac{(1-|z|^2)^2}{4} \Delta$. Now one can define the hyperbolic Radon transform R_H by

$$Rf(\gamma) = R_H f(\gamma) = \int_\gamma f(z) ds(z), \gamma \text{ geodesic in } D$$

which is well defined for, say, continuous functions of compact support, or functions decaying sufficiently fast. Denote by Γ the space of all geodesics in D, then the dual transform $R^\# = R_H^\#$ (or backprojection operator) is given by

$$R^\# \phi(z) = \int_{\Gamma_z} \phi(\gamma) d\mu_z(\gamma),$$

where Γ_z is the collection of geodesics through the point z and $d\mu_z$ is the normalized measure of Γ_z. Since a geodesic through z is determined by its starting direction $w \in S^1$, then $\Gamma_z \approx S^1$ and $d\mu_z$ is naturally associated to $\frac{1}{2\pi} dw$ when we use this particular parameterization of Γ_z.

In order to invert R_H one can proceed in the spirit of Radon's inversion formula (19). This was done by Helgason, see [He1], [He2]. Or one can try to find a filtered backprojection type formula like $(*)$. For that purpose we need to define convolution operators with respect to a radial kernel k. For $k \in L^1_{\text{loc}}([0,\infty))$ and $f \in C_0(D)$ we define

$$k * f(z) = k *_H f(z) := \int_D f(w) k(d(z,w)) dm(w)$$

where $dm(w)$ stands for the hyperbolic area measure, given in polar coordinates $dm = \sinh r \, dr \, dw$. One can show [BC1] that

$$R_H^\# R_H f = k * f, \text{ where } k(t) = \frac{1}{\pi \sinh t}$$

so that if $S(t) = \coth t - 1$, we have the exact analogue of $(*)$, namely

$$\frac{1}{2\pi}\Delta_H S *_H R_H^\# R_H = I.$$

In [BC2] we have found the exact relation between the linerization of the Calderon problem mentioned above, and the hyperbolic Radon transform, thus justifying the usual name of EIT (Electrical Impedance Tomography) for the Calderon problem. It turns out that if we introduce the auxiliary function

$$\kappa(t) = \frac{(\cosh t)^{-2} - 3(\cosh t)^{-4}}{8\pi}$$

then the linearization γ of the conductivity satisfies the equation

$$R_H(\kappa *_H \gamma) = \frac{\partial v}{\partial s}$$

where both sides are functions on the space of geodesics (please recall $u = U + v$). Since we know to invert R_H, the problem of finding γ reduces to the inversion of the radial convolution operator with kernel κ. Given the experience obtained in the Euclidean case, it is then natural to study this problem using the wavelet transform in hyperbolic space. In joint work with Rubin [BR] we have studied the *continuous* wavelet transform, but for numerical evaluation one needs the equivalent to the discrete wavelet transform. The difficulty lies in the problem of sampling in hyperbolic space. "Regular" sampling cannot exist having arbitrarily small "steps", so one needs to study irregular sampling. One approach, specific to EIT has been considered by Kuchment et al. [F] and, currently, Rubin, Pesenson and I are working on this problem. Again, we could question whether a "ridgelet" approach may not be helpful to study EIT.

To conclude, it is natural to ask why bother with EIT? Why did Calderon think about this problem? For a lucid presentation we suggest [U] and for a quick answer, we mention the original motivation of Calderon, the determination of cracks in materials. See also the work of Cheney, Isaccson et al. [CIN] on medical applications of EIT like determination of blood circulation, lung volume, and similar. In the same spirit, I'd like to add studies of brain seizures as presented in [EGI]. Final question, can one localize EIT in a manner similar to what we did for the Euclidean Radon transform?

References

[BC1]: C. A. Berenstein and E. Casadio Tarabusi, The inverse conductivity problem and the hyperbolic Radon transform, "75 years of Radon transform", S. Gindikin and P. Michor, editors. International Press, 1994, 39-44.

[BC2]: C. A. Berenstein and E. Casadio Tarabusi, Integral geometry in hyperbolic spaces and electrical impedance tomography, SIAM J. Appl. Math. 56 (1996), 755-764.

[BR]: C. A. Berenstein and B. Rubin, Radon transform of L^p functions on Lobachevsky space and hyperbolic wavelet tranforms, Forum Math. II (1999), 567-590.

[BW1]: C. A. Berenstein and D. Walnut, Local inversion of the Radon tranform in even dimensions using wavelets, "75 years of Radon transform", S. Gindikin and P. Michor, editors, International Press, 1994, 45-69.

[BW2]: C. A. Berenstein and D. Walnut, Wavelets and local tomography, "Wavelets in Medicine and Biology", A. Aldroubi and M. Unser, editors, CRC Press, 1996, 231-261.

[BCR]: G. Beylkin, R. Coifman, and V. Rokhlin, Fast wavelet transforms and numerical algorithms I, Comm. Pure Appl. Math. 44 (1991), 141-183. S

[Ca]: E. J. Candés, Ridgelets: theory and applications, Stanford University Ph.D. thesis, 1998.

[CD]: E. J. Candés and D. L. Donoho, Wavelets: a surprisingly effective nonadaptive representation of objects with edges, in "Curve and surface fitting", St. Malo , A. Cohen et al. eds, Vanderbilt University Press, 2000.

[CHT]: R. Carmona, W. L. Hwang, and B. Torresani, Practical time-frequency analysis", Academic Press, 1998.

[CIN]: M. Cheney, D. Isaacson, and J. C. Newell, Electrical impedance tomography, SIAM Review 41 (1999), 85-101.

[D]: I. Daubechies, "Ten lectures on wavelets", SIAM, 1992.

[DS]: D. C. Dobson and F. Santosa, An image enhancement technique for electrical impedance tomography, Inverse Problems 10 (1994), 317-334.

[EGI]: Lecture by K. Karnofsky, Electrical Geodesics Inc., at Inverse Problem Seminar of Pacific Northwest, Corvallis, OR, June 3-4, 2000.

[FRK]: A Faridani, E. Ritman and K. T. Smith, Local tomography, SIAM J. Applied Math. 52 (1992), 1193-1198.

[FLBW]: F. Rashid-Farrokhi, K. J. R. Liu, C. A. Berenstein and D. Walnut, Wavelet-based multiresolution local tomography, IEEE Trans. Image Proc. 6 (1997), 1412-1430, see also ICIP 95, Washington, DC.

[FLB]: F. Rashid-Farrokhi, K. J. R. Liu and C. A. Berenstein, Local tomography in fan-beam geometry using wavelets, ICIP-96, Lausanne.

[FMP]: B. Fridman, D. Ma, and V. G. Papanicolau, Solution of the linearized inverse conductivity problem in the half space, preprint Wichita St. U., 1995.

[F]: B. Fridman, D. Ma, S. Lissianoi, P. Kuchment, M. Mogilevsky, K. Lancaster, V. Papanicolaou, and I. Ponomaryov, Numeric implementation of harmonic analysis on the hyperbolic disk, to appear in Complex Analysis.

[FV]: A Friedman and M. Vogelius, Determining cracks by boundary measurements, Indiana U. Math. J. 38 (1989), 527-556.

[He1]: S. Helgason, "The Radon transform", Birkhäuser, 1980.

[He2]: S. Helgason, "The Radon transform", 2nd revised edition, Birkhäuser, 1999.

[Ho]: M. Holschneider, Inverse Radon transform through inverse wavelet transforms, Inverse Problems 7 (1991), 853-861.

[J]: F. John, "Plan waves and spherical means", Interscience, 1955.

[Ka]: G. Kaiser, A friendly guide to wavelets, Birkhäuser, 1994.

[KS]: A. C. Kak and M. Slaney, "Principles of computerized tomographic imaging", IEEE Press, 1988.

[M]: Y. Meyer, "Ondelettes et opérateurs", 3 vols, Herman, 1990.

[MZ]: S. Mallat and S. Zhong, Characterization of signals from multiscale edges, IEEE Trans. Patt. Anal. Machine Intell. 14 (1992), 710-732.

[**N1**]: A. I. Nachman, Reconstruction from boundary measurements, Annals Math. 128 (1988), 531-576.

[**N2**]: A. I. Nachman, Global uniquemenss for a two-dimensional inverse boundary value problem, Annals Math. 143 (1996), 71-96.

[**Na**]: F. Natterer, "The mathematics of computerized tomography", Wiley, 1986.

[**P**]: S. K. Patch, ASL Technote 99-24, G. E. Medical Systems.

[**Q**]: E. T. Quinto, Singularities of the x-ray transform and limited data tomography in $\mathbb{R}^2$ and $\mathbb{R}^3$, SIAM J. Math. Anal. 24 (1993), 1215-1225.

[**QCK**]: E. T. Quinto, M. Cheney, and P. Kuchment, eds., "Tomography, impedance imaging, and integral geometry", Lect. Appl. Math. 30, Amer. Math. Soc., 1994.

[**R**]: B. Rubin, Inversion and characterization of Radon transforms via continuous wavelet transforms, Hebrew Univ. TR 13, 1995/96.

[**S**]: F. Santosa, Inverse problem holds key to safe, continuous imaging, SIAM News, July 1994, 1 and 16-18.

[**Si**]: S. Siltanen, Electrical impedance tomography and Fadelo Giren functions, Ann. Acad. Sci. Fennicae, Dissertationes 121, 1999.

[**U**]: G. Uhlmann, Developments in inverse problems since Calderon's foundational paper, Harmonic analysis and partial differential equations, U. Chicago Press, 1999, 295-345.

[**W**]: D. Walnut, Applications of Gabor and wavelet expansions to the Radon transform, in "Probabilistic and stochastic methods in analysis", J. Byrnes et al., ed., Kluwer, 1992,187-205.

[**ZCMB**]: Y. Zhang, M. A. Coplan, J. H. Moore and C. A. Berenstein, Computerized tomographic imaging for space plasma physics, J. Appl. Phys. 68 (1990), 5883-5889.

Department of Mathematics and the Institute for Systems Research, University of Maryland.

Current address: INSTITUTE FOR SYSTEMS RESEARCH, 2221 A. V. WILLIAMS BUILDING, UNIVERSITY OF MARYLAND, COLLEGE PARK, MARYLAND 20742

E-mail address: `carlos@isr.umd.edu`

Tomography problems arising in Synthetic Aperture Radar

Margaret Cheney

ABSTRACT. This paper gives a mathematical tutorial on Synthetic Aperture Radar (SAR). We see that with the usual mathematical model, the SAR reconstruction problem reduces to a problem in integral geometry. A number of mathematical problems are posed; the paper concludes with a short description of the basic idea underlying the algorithms used in most present systems. The challenge to the mathematical community is to find algorithms that might be better.

1. Introduction and Mathematical Model

In the last forty years the engineering community has developed very successful microwave systems for making high-resolution images of the earth from airplanes and satellites. Such systems, which go under the general name Synthetic Aperture Radar (SAR), have received little attention in the mathematical community. The purpose of this paper is to give a tutorial on SAR and to point out some of the associated interesting and challenging mathematical questions.

In strip-mode Synthetic Aperture Radar (SAR) imaging, an antenna (on a plane or satellite) flies along a nominally straight track, which we will assume is along the x_2 axis. The antenna emits pulses of electromagnetic radiation in a more-or-less directed beam perpendicular to the flight track (i.e., in the x_1 direction). These waves scatter off the terrain, and the scattered waves are detected with the same antenna. The received signals are then used to produce an image of the terrain. (See Figure 1.)

The data depend on two variables, namely time and position along the x_2 axis, so we expect to be able to reconstruct a function of two variables.

1.1. The (simplified) partial differential equation. The correct model for radar is of course Maxwell's equations, but the simpler scalar wave equation is commonly used:

$$(1.1) \qquad \left(\nabla^2 - \frac{1}{c^2(x)}\partial_t^2\right) U(t,x) = 0.$$

2000 *Mathematics Subject Classification* 78A45, 94A08.

This work was partially supported by the Office of Naval Research, by Rensselaer Polytechnic Institute, by Lund University through the Lise Meitner Visiting Professorship, and by the Engineering Research Centers Program of the National Science Foundation under award number EEC-9986821.

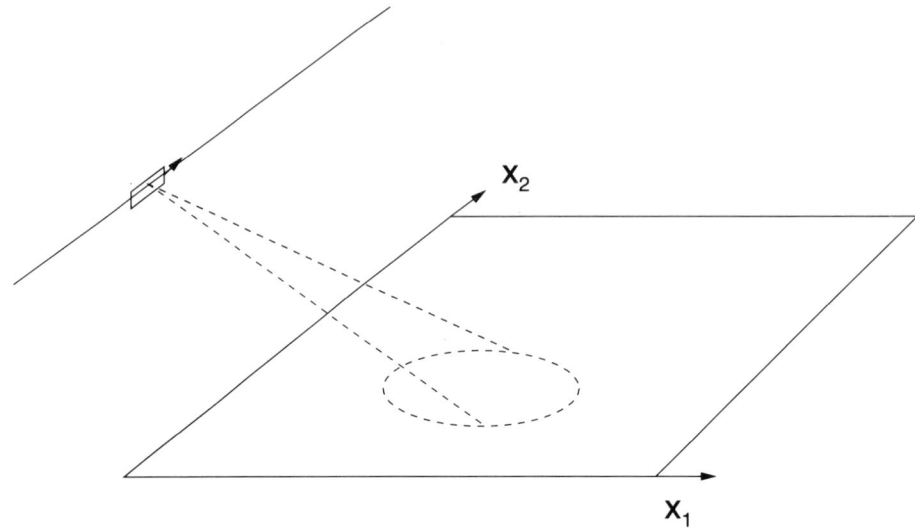

Figure 1: This shows the geometry of a conventional strip-mode SAR system.

This is the equation satisfied by each component of the electric and magnetic fields in free space, and is thus a good model for the wave propagation in dry air. When the electromagnetic waves interact with the ground, their polarization is certainly affected, but if the SAR system does not measure this polarization, then (1.1) is an adequate model.

We assume that the earth is roughly situated at the plane $x_3 = 0$, and that for $x_3 > 0$, the wave speed is $c(x) = c_0$, the speed of light in vacuum (a good approximation for dry air).

The fundamental solution of the free-space wave equation [**19**] is $G_0(t-\tau, x-y)$, given by

$$(1.2) \qquad G_0(t - \tau, x - y) = \frac{\delta(t - \tau - |x - y|/c_0)}{4\pi |x - y|}.$$

It has the physical interpretation of the field at (x, t) due to a delta function point source at position y and time τ. This field satisfies the equation

$$(1.3) \qquad \left(\nabla^2 - \frac{1}{c_0^2}\partial_t^2\right) G_0(t - \tau, x - y) = \delta(t - \tau)\delta(x - y).$$

1.2. The incident wave. The signal sent to the antenna is of the form

$$(1.4) \qquad P(t) = A(t)e^{i\omega_0 t},$$

where ω_0 is the (angular) *carrier frequency* and A is a slowly varying amplitude that is allowed to be complex.

If the source at y has the time history (1.4), then the resulting field $U_y(t, z - y)$ satisfies the equation

$$(1.5) \qquad \left(\nabla^2 - \frac{1}{c_0^2}\partial_t^2\right) U_y(t, z - y) = P(t)\delta(z - y)$$

and is thus given by

$$
\begin{aligned}
U_y(t,z) = (G_0 * P)(t, z-y) &= \int \frac{\delta(t - \tau - |z-y|/c_0)}{4\pi|z-y|} P(\tau) d\tau \\
&= \frac{P(t - |z-y|/c_0)}{4\pi|z-y|} \\
&= \frac{A(t - |z-y|/c_0)}{4\pi|z-y|} e^{i\omega_0(t - |z-y|/c_0)}.
\end{aligned}
\tag{1.6}
$$

The antenna, however, is not a point source. Most conventional SAR antennas are either slotted waveguides [8, 25] or microstrip antennas [18], and in either case, a good mathematical model is a rectangular distribution of point sources. We denote the length and width of the antenna by L and D, respectively. We denote the center of the antenna by x; thus a point on the antenna can be written $y = x + q$, where q is a vector from the center of the antenna to a point on the antenna. We also introduce coordinates on the antenna: $q = s_1 \hat{e}_1 + s_2 \hat{e}_2$, where $\hat{e}_1$ and $\hat{e}_2$ are unit vectors along the width and length of the antenna, respectively. The vector $\hat{e}_2$ points along direction of flight; for the straight flight track shown in Figure 1, this would be the x_2 axis. For side-looking systems as shown in Figure 1, $\hat{e}_1$ is tilted with respect to the x_1 axis so that a vector perpendicular to the antenna points to the side of the flight track.

We consider points z that are far from the antenna; for such points, for which $|q| << |z - x|$, we have the approximation

$$
|z - y| = |z - x| - \widehat{(z-x)} \cdot q + O(L^2/|z-x|),
\tag{1.7}
$$

where the hat denotes a unit vector. We use this expansion in (1.6):

$$
U_y(t,z) \sim \frac{A(t - |z-x|/c_0 + \widehat{z-x} \cdot q/c_0 + \cdots)}{4\pi|z-x|} e^{i\omega_0(t - |z-x|/c_0)} e^{ik\widehat{z-x} \cdot q}
\tag{1.8}
$$

where we have written $k = \omega_0/c_0$. This expansion is valid because we also have $kL^2 << |z-x|$. We now make use of the fact that $|z-x| >> \widehat{z-x} \cdot q$ and that A is assumed to be slowly varying to write

$$
U_y(t,z) \sim \frac{P(t - |z-x|/c_0)}{4\pi|z-x|} e^{ik\widehat{z-x} \cdot q}.
\tag{1.9}
$$

Far from the antenna, the field from the antenna is

$$
\begin{aligned}
U_x^{in}(t,z) &= \int_{-L/2}^{L/2} \int_{-D/2}^{D/2} U_{x+s_1\hat{e}_1+s_2\hat{e}_2}(t,z) ds_1 ds_2 \\
&\sim \int_{-L/2}^{L/2} \int_{-D/2}^{D/2} \frac{P(t - |z-x|/c_0)}{4\pi|z-x|} e^{ik\widehat{z-x} \cdot (s_1\hat{e}_1 + s_2\hat{e}_2)} ds_1 ds_2 \\
&\sim \frac{P(t - |z-x|/c_0)}{4\pi|z-x|} \int_{-L/2}^{L/2} e^{iks_2 \widehat{z-x} \cdot \hat{e}_2} ds_2 \int_{-D/2}^{D/2} e^{iks_1 \widehat{z-x} \cdot \hat{e}_1} ds_1 \\
&\sim \frac{P(t - |z-x|/c_0)}{4\pi|z-x|} w(\widehat{z-x}),
\end{aligned}
\tag{1.10}
$$

where

$$
w(\widehat{z-x}) = 2D\,\text{sinc}(k\widehat{z-x} \cdot e_1 D/2)\, 2L\,\text{sinc}(k\widehat{z-x} \cdot e_2 L/2)
\tag{1.11}
$$

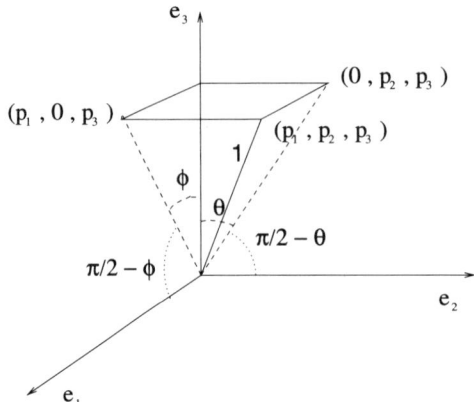

Figure 2: This is a diagram that shows that if $p = \widehat{z-x}$ is a unit vector, then $\hat{p} \cdot \hat{e}_2 = p_2 \approx \sin\theta$ and $\hat{p} \cdot \hat{e}_1 = p_1 \approx \sin\phi$. The antenna lies on the x_1-x_2 plane.

is the antenna beam pattern and where sinc $\beta = (\sin\beta)/\beta$. The sinc function has its main peak at $\beta = 0$ and its first zero at $\beta = \pi$; this value of β gives half the width of the main peak. Thus the main beam of the antenna is directed in the direction perpendicular to the antenna.

We determine the width of the beam by noting that the first zero of sinc($k\widehat{z-x} \cdot \hat{e}_2 L/2$) occurs when $k\widehat{z-x} \cdot \hat{e}_2 L/2 = \pi$. Using the fact that $2\pi/k$ is precisely the wavelength λ, we can write this as $\widehat{z-x} \cdot \hat{e}_2 = \lambda/L$. To understand this condition, we write $\widehat{z-x} \cdot \hat{e}_2 \approx \cos(\pi/2 - \theta) = \sin\theta \approx \theta$, an approximation that is valid for small angles θ. (See Figure 2.) Here θ is the angle between the vector normal to the antenna and the projection of $\widehat{z-x}$ on the plane spanned by $\hat{e}_2$ and $\hat{e}_3$. Thus when $\lambda \ll L$ and thus θ is small, the condition $\widehat{z-x} \cdot \hat{e}_2 = \lambda/L$ reduces to $\theta \approx \lambda/L$. In this case, the main lobe of the antenna beam pattern has angular width $2\lambda/L$ in the $\hat{e}_2$ direction. Similarly the angular width in the $\hat{e}_1$ direction is $2\lambda/D$. We note that smaller wavelengths and larger antennas correspond to more tightly focused beams.

The antenna beam pattern is not always precisely a product of sinc functions: the signal emanating from different parts of the antenna can be weighted so that the integrals appearing in (1.10) are Fourier transforms of functions smoother than characteristic functions [**17**]. Such weighting suppresses the sidelobes at the expense of broadening the main beam slightly.

The Swedish CARABAS system [**13, 22, 21**] uses two parallel wire antennas of length L that are oriented along the flight track. Each antenna can be considered a linear distribution of point sources, so the beam pattern of each antenna is

$$(1.12) \qquad w_C(\widehat{z-x}) = 2L\text{sinc}(k\widehat{z-x} \cdot e_2 L/2)$$

The length L is chosen to be half the wavelength of the carrier wave (i.e., $kL/2 = \pi$), so that the antenna produces only one single main lobe.

1.3. A linearized scattering model. From classical scattering theory we know that a scattering solution of (1.1) can be written

$$(1.13) \qquad \Psi(t,x) = \Psi^{in}(t,x) + \Psi^{sc}(t,x),$$

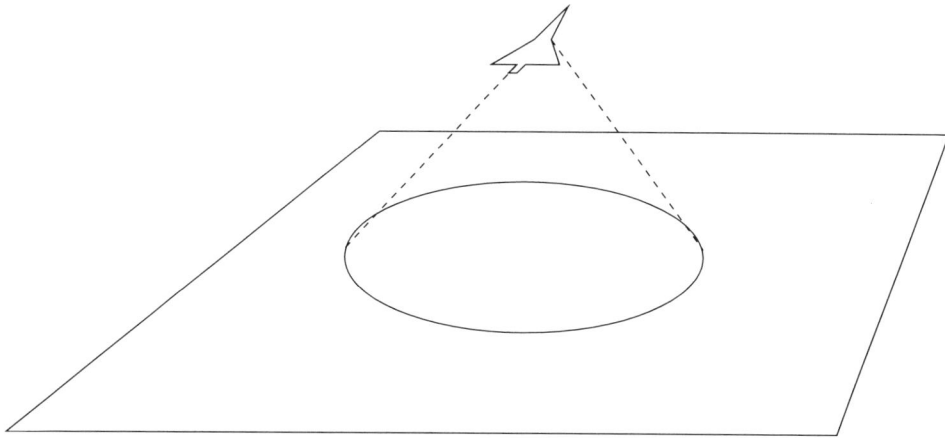

Figure 3: The geometry of the CARABAS SAR system.

where Ψ^{in} satisfies (1.1) with $c(x) = c_0$ and where (see Appendix)

(1.14) $$\Psi^{sc}(t,x) = \int\int G_0(t-\tau, x-z)V(z)\partial_\tau^2 \Psi(\tau,z)d\tau dz$$

and

(1.15) $$V(z) = \frac{1}{c^2(z)} - \frac{1}{c_0^2}.$$

For commonly used carrier frequencies ω_0, the waves decay rapidly as they penetrate into the earth. Thus the support of V can be taken to be a thin layer at the earth's surface. This is discussed in more detail in section 3.

A commonly used approximation, often called the *Born approximation* or the *single scattering approximation*, is

$$\Psi^{sc}(t,x) \approx \Psi^B(t,x) = \int\int G_0(t-\tau, x-z)V(z)\partial_\tau^2 \Psi^{in}(t,x)d\tau dz$$
(1.16)
$$= \int \frac{V(z)}{4\pi|x-z|}\partial_t^2 \Psi^{in}(t-|x-z|/c_0, z)dz.$$

The value of this approximation is that it removes the nonlinearity in the inverse problem: it replaces the product of two unknowns (V and Ψ) by a single unknown (V) multiplied by the known incident field.

The Born approximation makes the problem simpler, but it is not necessarily a good approximation. This issue is discussed briefly in section 3.2. Another linearizing approximation that can be used at this point is the *Kirchhoff approximation*, in which the scattered field is replaced by its geometrical optics approximation [14]. Here, however, we consider only the Born approximation.

In the case of SAR, the antenna emits a series of fields of the form (1.10) as it moves along the flight track. In particular, we assume that the antenna is located at position x^n at time nT, and there emits a field of the form (1.10). In other words, the incident field is

(1.17) $$\Psi^{in}(\tau,z) = \sum_n \Psi_n^{in}(\tau,z),$$

where

(1.18) $$\Psi_n^{in}(\tau, z) \sim \frac{P(\tau - nT - |z - x^n|/c_0)}{4\pi|z - x^n|} w(\widehat{z - x^n})$$

is the nth emission. We use this expression in (1.16) to find an approximation to the scattered field due to the nth emission. The resulting expression involves two time derivatives of $P(t, x)$. In calculating these time derivatives, we use the fact that A is assumed to be slowly varying to obtain

(1.19) $$\partial_t^2 P(t, x) \approx -\omega_0^2 P(t, x).$$

Thus the Born approximation to the scattered field due to the nth emission, measured at the center of the antenna, is

$$S_n(t) \approx \Psi_n^B(t - nT, x^n)$$

(1.20) $$\approx -\int \frac{\omega_0^2 P(t - nT - 2|z - x^n|/c_0)}{4\pi|z - x^n|} \frac{V(z)}{4\pi|z - x^n|} w(\widehat{z - x^n}) dz.$$

In (1.20), we note that $2|z - x^n|/c_0$ is the two-way travel time from the center of the antenna to the point z. The factors $4\pi|z - x^n|$ in the denominator correspond to the geometrical spreading of the spherical wave emanating from the antenna and from the point z.

In practice, the received signal is not measured at a single point in the center of the antenna; rather, the signal is received on the entire antenna. This means that the received signal is subject to the same weighting as the transmitted signal. Thus w in (1.20) should be replaced by w^2. We continue to write simply w.

2. Reduction to a delta function impulse

The SAR reconstruction problem would be a problem in integral geometry if the transmitted signal P were a delta function. Unfortunately, a delta function cannot be produced in practice. Nor can an approximate delta function be used: any short-time, limited-amplitude wave will contain little energy, and the reflected wave contains even less energy. A very low-energy wave will get drowned out by noise.

To circumvent this difficulty, SAR systems use *pulse modulation*, in which the system transmits a complex waveform and then *compresses* the received signal mathematically, to synthesize the response from a short pulse. This processing is explained in this section. The final result is that to a good approximation, P can indeed be replaced by a delta function.

2.1. Matched filter processing. The mathematical processing is done by applying a *matched filter* [20]. Applying a matched filter to the received signal means integrating it against a shifted copy of the complex conjugate of the transmitted signal:

$$O(t', x^n) = \int \overline{P(t - t')} S_n(t) dt$$

$$\approx -\int \overline{P(t - t')} \int \frac{\omega_0^2 P(t - nT - 2|z - x^n|/c_0)}{(4\pi|z - x^n|)^2} V(z) w(\widehat{z - x^n}) dz dt$$

(2.1) $$\approx \int \frac{\omega_0^2 \zeta(nT + 2|z - x^n|/c_0 - t')}{(4\pi|z - x^n|)^2} V(z) w(\widehat{z - x^n}) dz.$$

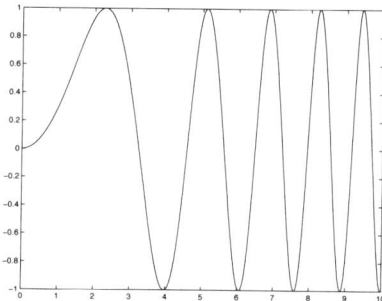

Figure 4: The chirp $\sin(.3t^2)$.

where

$$(2.2) \quad \zeta(s-t') = \int \overline{P(t-t')} P(t-s) dt = \int \overline{P(s')} P(s' - (s-t')) ds'$$

is the *range ambiguity function* [10]. We see that matched-filter processing has the effect of replacing the waveform P in (1.20) by the *compressed* waveform ζ.

To determine ζ, we use (1.4):

$$(2.3) \quad \zeta(s) = \int \overline{A(t) e^{i\omega_0 t}} A(t-s) e^{i\omega_0 (t-s)} dt;$$

we see that ζ is a complex number of modulus one multiplied by the autocorrelation function of A.

The modulation A should be chosen so that ζ is close to a delta function.

The most commonly used modulated pulse is a *chirp*, which involves linear frequency modulation. The idea is to label different parts of the wave by their frequency, and then superimpose these different parts in the compression process.

2.2. Instantaneous frequency. The notion of *instantaneous frequency* of a wave $F(t) = e^{i\phi(t)}$ derives from a stationary phase analysis of the usual Fourier transform integral: we think of the integrand of the Fourier integral

$$(2.4) \quad f(\omega) = \int F(t) e^{-i\omega t} dt = \int e^{i(\phi(t) - \omega t)} dt$$

as being written in the form $\exp(i\lambda(\phi(t) - \omega t))$, where $\lambda = 1$. The usual large-λ stationary phase calculation shows that the leading order contribution comes from the values of t at which the phase is not changing rapidly with respect to t. This occurs when $0 = (d/dt)(\phi(t) - \omega t)$, or in other words, when $\omega = d\phi/dt$. Thus we call $d\phi/dt$ the *instantaneous frequency* of F.

2.3. Chirps. A *chirp* is a finite wavetrain $P(t) = \chi_{[-\tau/2, \tau/2]}(t) \exp(i\phi(t))$ in which the instantaneous frequency changes linearly with time. Here $\chi_{[-\tau/2, \tau/2]}$ denotes the characteristic function of the time interval $[-\tau/2, \tau/2]$, which is one in this time interval and zero outside. In an *upchirp*, the instantaneous frequency increases linearly with time as $d\phi/dt = \omega_0 + Bt/\tau$, where ω_0 is the (angular) carrier frequency and B is called the (angular) *bandwidth*. To determine ϕ, we simply integrate to obtain $\phi(t) = \omega_0 t + Bt^2/(2\tau)$. Thus an upchirp is a wavetrain of the

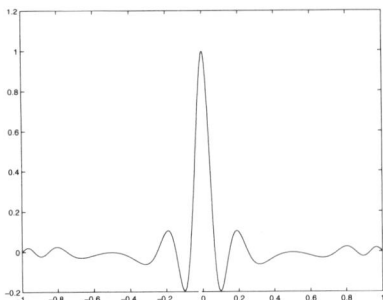

Figure 5: The amplitude of the function ζ for $\tau = 1$ and $B = 100$.

form

(2.5) $$P(t) = \chi_{[-\tau/2,\tau/2]}(t)e^{i\alpha t^2}e^{i\omega_0 t},$$

where $\alpha = B/(2\tau)$. We note that such a pulse is of the form (1.4), where

(2.6) $$A(t) = \chi_{[-\tau/2,\tau/2]}(t)e^{i\alpha t^2}.$$

2.4. The compressed waveform ζ for a chirp. For a chirp, ζ is

(2.7) $$\zeta(s) = e^{-i\omega_0 s}\int \overline{A(t)}A(t-s)dt$$

(2.8) $$= e^{-i\omega_0 s}\int \overline{\chi_{[-\tau/2,\tau/2]}(t)e^{i\alpha t^2}}\chi_{[-\tau/2,\tau/2]}(t-s)e^{i\alpha(t-s)^2}dt.$$

After calculating the integral on the right side of (2.8), we obtain

(2.9) $$\zeta(s) = e^{-i\omega_0 s}e^{iBs^2/(2\tau)}\frac{2\tau\sin(Bs(1-|s|/\tau)/2)}{Bs}\chi_{[-\tau,\tau]}(s)$$

By taking B sufficiently large, we can make ζ peak arbitrarily sharply at zero.

Other modulated waveforms besides chirps can also be used to obtain a range ambiguity function ζ that approximates a delta function.

3. Mathematical Problems

If we replace ζ in (2.1) by a delta function, then (2.1) becomes

(3.1) $$O_c(t', x^n) = \frac{\omega_0^2}{(2\pi^2 c_0(t'-nT))^2}\int \delta((nT-t')-2|z-x^n|/c_0)V(z)w(\widehat{z-x^n})dz$$

We see that the radar reconstruction problem becomes a problem in integral geometry: reconstruct $V(z)$ from its weighted integrals over the spheres $|z - x^n| = c_0(nT - t')/2$. Here $x^n = x(nT)$, where $x(s)$ is a known path (flight track) in space. This problem requires some discussion.

In (3.1), for a straight flight track, O depends on two variables, namely time and position along the line, whereas V appears to depend on three variables. However, as mentioned earlier, V can be considered to have support in a thin layer at the surface. In the case in which the terrain is planar (a good approximation when the flight track is at satellite height), we can replace $V(z)$ by $V(z_1, z_2)\delta(z_3 - 0)$, where $V(z_1, z_2)$ is referred to as the *ground reflectivity function*. In this case (3.1) reduces to a two-dimensional integral.

For lower-altitude flight tracks, the ground topography becomes important. In this case, we replace $V(z)$ by $V(z_1, z_2)\delta(z_3 - h(z_1, z_2))$, where h is the ground altitude. In this case, the problem is to reconstruct two functions of two variables; generally one uses data from two parallel flight tracks.

Some key aspects of the reconstruction problem depend on the weighting function w: the CARABAS system, for example, has a symmetrical beam pattern (1.12), so that from a single antenna moving along a straight line, it is not possible to determine whether a given reflection originated from a point to the left of the antenna or to the right. It is for this reason that the CARABAS system uses two antennas. Conventional side-looking SAR avoids this problem by using a focused beam (1.11) that is directed to one side of the flight track (see Figure 1).

To be of most practical use, reconstruction algorithms should be fast, accurate, and should use as little memory as possible. Ideally they should allow an image to be constructed in real time as the aircraft or satellite flies along the flight track. The amount of data collected by SAR systems can be enormous.

3.1. Integral geometry problems. We simplify the problem by making the change of variables $t = c_0(nT - t')/2$. Then for planar topography, the idealized SAR reconstruction problem is to reconstruct V from

$$(3.2) \qquad \text{data}(t, x(s)) = c \int_{t=|z-x|} V(z_1, z_2)\delta(z_3)w(\widehat{z-x})dz$$

where c is a known constant, $x(s)$ is a known path above the plane $z_3 = 0$ and w is the known antenna beam pattern that depends only on the direction $\widehat{z-x}$. Because the spheres $t = |z - x|$ intersect the plane $z_3 = 0$ in circles, this problem is to reconstruct $V(z_1, z_2)$ from its weighted integrals over circles or circular arcs. Relevant references here are [2, 13, 12, 15, 21].

For non-planar topography, the idealized SAR reconstruction problem is to reconstruct both V and h from the integrals

$$(3.3) \qquad \text{data}(t, x) = c \int_{t=|z-x|} V(z)\delta(z_3 - h(z_1, z_2))w(\widehat{z-x})dz,$$

where x lies on one or more flight tracks above the plane $z_3 = 0$. A relevant reference is [22].

Before making an attack on the full non-planar topography problem, a reasonable warm-up problem is to begin with the case $V = 1$, $w = 1$. This problem is to recover the surface height h from

$$(3.4) \qquad \text{data}(t, x) = c \int_{t=|z-x|} \delta(z_3 - h(z_1, z_2))dz.$$

Probably at least two paths $x(s)$ are needed; a relevant reference is again [22].

In addition to the challenge of developing and analyzing reconstruction algorithms, some relevant questions are the following.

3.1.1. *Problems related to the flight track.* Generally the flight tracks are approximately straight lines. Are straight lines best? Here [2] is relevant. A robust reconstruction algorithm must be able to deal with flight tracks $x(t)$ that deviate from a straight line by known perturbations. Airplanes, in addition, are subject to turbulence in the atmosphere, which means that they are subject to pitch, roll, and yaw. In terms of (1.11), this means that the orientation of the vectors e_1 and e_2 varies with time or with n. In terms of the idealized problem (3.2), this means

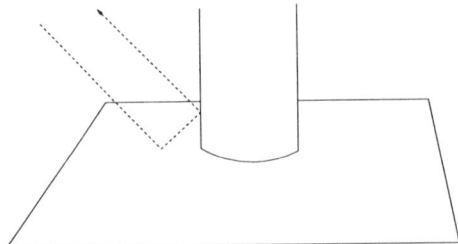

Figure 6: Double scattering from a tree trunk and ground.

that w depends on x as well as on the unit vector $\widehat{z-x}$. A book on SAR with a section about motion correction is [**12**].

Issues related to a non-ideal flight track become even more important in military applications, when enemy fire can induce the aircraft to take evasive action. Can we find a reconstruction algorithm that can use data from a wild flight track? In the study of this issue, [**2**] is relevant.

3.1.2. *Breaking the left-right symmetry*. How can a reconstruction algorithm best take advantage of the two antennas of the CARABAS system? In this case one can assume that the weighting function w is simply one, but that there are two parallel flight tracks. Presumably, if the antennas were moved closer together, it would become more difficult to determine whether a given reflection comes from the left or the right of the flight track. This should be quantified.

3.1.3. *Resolution*. For a given carrier frequency ω_0, antenna pattern w, and bandwidth B, what is the best resolution that can be achieved? Is the resolution improved by incorporating information from many flight tracks? Relevant references are [**16**] and [**21**].

3.2. Other mathematical problems.

3.2.1. *Scatterers that move*. For moving objects such as trains on a railway, the Doppler effect combined with pulse compression techniques results in incorrect estimates for the location of the moving object. How can this be corrected?

For systems that use the correlation between many "looks" to form an image, smaller moving objects tend to disappear. How can these moving objects be imaged? Can their velocity be determined? A relevant reference is [**23**].

3.2.2. *Improvement of the model*. All the current processing is based on a linearizing assumption such as the Born approximation (1.16), which amounts to assuming that the wave scatters only once before returning to the antenna. But in many cases, this model is believed to be too simple. For example, in scanning a forest at low frequencies, the most important scattering mechanism is believed to be double scattering (see Figure 6) in which the wave reflects first from the ground and then from the tree trunk (or vice versa).

The mathematical difficulty is that including multiple scattering makes the inverse problem nonlinear. Can we develop reconstruction algorithms for the nonlinear case?

The entire theory should be extended to the case of Maxwell's equations, for a model of *polarimetric SAR*.

3.2.3. *Theoretical issues.* Little is known about the nature of the inverse scattering problem for the wave equation in the case when only backscattering data is measured. Does backscattering data uniquely determine a scatterer?

4. How the Processing is Done Currently

Present SAR image reconstruction algorithms are based on the notion of matched filters: at each "look" (i.e., at each n), a matched filter is applied to the received signal $S_n(t)$:

$$(4.1) \qquad I_n(y) = \int \overline{P(t - nT - 2|y - x^n|/c_0)} S_n(t) dt.$$

This correlates the received signal with a signal proportional to that due to a "point scatterer" at position y (i.e., take $V(z) = \delta(z - y)$ in the Born approximation).

If we use expression (1.20) (with $|z - x^n|$ replaced by R_0) in (4.1) and interchange the order of integration, we find

$$(4.2) \qquad I_n(y) \approx \int W_n(y, z) \frac{-\omega_0^2 V(z)}{(4\pi R_0)^2} dz.$$

Here

$$(4.3) \qquad W_n(y, z) = w(\widehat{z - x^n}) \int \overline{P(t - nT - 2|y - x^n|/c_0)} P(t - nT - 2|z - x^n|/c_0) dt$$

represents the *point spread function* of this single-look imaging system: if $V(z) = \delta(z - z_0)$, then $I_n(y) = W_n(y, z_0)$ would be proportional to the resulting image of V.

The key idea of SAR is that this point spread function can be made closer to a delta function by summing over n, i.e., by combining information from multiple looks. Thus the final image is formed as

$$(4.4) \qquad I(y) = \sum_n I_n(y) \approx \int W(y, z) \frac{-\omega_0^2 V(z)}{(4\pi R_0)^2} dz,$$

with the point spread function

$$(4.5) \qquad W(y, x) = \sum_n W_n(y, z)$$

This point spread function is called the *generalized ambiguity function* of the SAR system. By taking P to be a chirp or other modulated waveform, W becomes an approximate delta function. An analysis of W, which gives the resolution of the SAR system, can be found in [**4, 6**], and many other sources.

The matched filter reconstruction algorithm results in a complex-valued image. The phase of this image contains information about the distance of the scatterer from the antenna. To recover the ground topography, one uses a technique called *interferometric SAR* [**3, 12**], in which one uses the interference pattern from two complex images from parallel flight tracks.

Standard matched filter reconstruction algorithms cannot be used for the CARABAS system; instead a filtered backprojection scheme [**15**] is currently used.

5. Acknowledgments

I would like to thank Lars Ulander and Hans Hellsten for teaching me about the CARABAS system; many of the open problems listed here were suggested by them. I am also grateful to Ehud Heyman and Anders Derneryd for helpful discussions, and to Frank Natterer, Brett Borden, and Todd Quinto for reading the manuscript and making helpful suggestions regarding the exposition.

Appendix A. Classical scattering theory

To obtain (1.14), we first write $U = \Psi = \Psi^{in} + \Psi^{sc}$ in (1.1); this gives us

$$(\text{A.1}) \qquad \left(\nabla^2 - c_0^{-2}\partial_t^2\right)\Psi^{sc} - V\partial_t^2\Psi = 0$$

We then multiply (A.1) by G_0, multiply (1.3) by Ψ^{sc}, and subtract the resulting two equations to obtain

$$(\text{A.2}) \qquad G_0\nabla^2\Psi^{sc} - \Psi^{sc}\nabla^2 G_0 - G_0 V\partial_t^2\Psi = G_0\Psi^{sc}\delta$$

We then integrate over all time and over a large ball, use Green's identity, and let the radius of the ball go to infinity. The integral involving the spatial derivatives vanishes, and we are left with (1.14).

We note that (1.14) shows that the notion of a a point scatterer is problematic. If we take $V(y) = \delta(y - y^0)$ in (1.14), we obtain

$$(\text{A.3}) \qquad \Psi^{sc}(t,z) = \int G_0(t-\tau, z, y^0)\partial_\tau^2\Psi(\tau, y^0)d\tau = \frac{\partial_t^2\Psi(t - |z - y^0|/c_0, y^0)}{4\pi|z - y^0|},$$

which shows that the scattered field at the point y^0 is singular (unless $\partial_t^2\Psi(t, y^0)$ is zero for all time). But the product of a delta function with a singular function has no conventional meaning. The issue of point scatterers has been studied in [1].

Note, however, that in the Born approximation, the field scattered from a point scatterer is well-defined and nonzero.

References

[1] S. Albeverio, F. Gesztesy, R. Høegh-Krohn, and H. Holden, Solvable Models in Quantum Mechanics, Texts and Monographs in Physics, Springer-Verlag, New York, 1988.

[2] M. Agranovsky and E.T. Quinto, Injectivity sets for the Radon transform over circles and complete systems of radial functions, J. Functional Analysis 139 (1996) 383–414.

[3] R. Bamler and P. Hartl, "Synthetic aperture radar interferometry", Inverse Problems 14 (1998) R1–R54.

[4] M. Cheney, "A mathematical tutorial on Synthetic Aperture Radar", to appear, SIAM Review.

[5] J.C. Curlander and R.N. McDonough, Synthetic Aperture Radar, Wiley, New York, 1991.

[6] L.J. Cutrona, "Synthetic Aperture Radar", in Radar Handbook, second edition, ed. M. Skolnik, McGraw-Hill, New York, 1990.

[7] H. T. Cuong, A. W. Troesch, T. G. Birdsall, "The Generation of Digital Random Time Histories", Ocean Engineering, 9 (1982) 581–588.

[8] A. Derneryd and A. Lagerstedt, "Novel slotted waveguide antenna with polarimetric capabilities", Proc. of IGARSS Conference, Firenze, Italy (1995) 2054–2056.

[9] A.G. Derneryd, R.N.O. Petersson, P. Ingvarson, "Slotted waveguide antennas for remote sensing satellites", Proceedings of PIERS conferences, Noordwijk, The Netherlands (July 1994).

[10] B. Edde, Radar: Principles, Technology, Applications, Prentice Hall, New York, 1993.

[11] C. Elachi, Spaceborne Radar Remote Sensing: Applications and Techniques, IEEE Press, New York, 1987.

[12] G. Franceschetti and R. Lanari, Synthetic Aperture Radar Processing, CRC Press, New York, 1999.
[13] H. Hellsten and L.E. Andersson, "An inverse method for the processing of synthetic aperture radar data", Inverse Problems 3 (1987), 111–124.
[14] K.J. Langenberg, M. Brandfass, K. Mayer, T. Kreutter, A. Brüll, P. Felinger, D. Huo, "Principles of microwave imaging and inverse scattering", EARSeL Advances in Remote Sensing, 2 (1993) 163–186.
[15] Stefan Nilsson, "Application of fast backprojection techniques for some inverse problems of integral geometry", Linköping Studies in Science and Technology, Dissertation No. 499 (1997).
[16] F. Natterer, The Mathematics of Computerized Tomography, Wiley, New York, 1986.
[17] A.V. Oppenheim and R.W. Shafer, Digital Signal Processing, Prentice-Hall, Englewood Cliffs, New Jersey, 1975.
[18] R.N.O. Petersson, A.G. Derneryd, and P. Ingvarson, "Microstrip antennas for remote sensing satellites", Proceedings of PIERS conferences, Noordwijk, The Netherlands (July 1994).
[19] F. Treves, Basic Linear Partial Differential Equations, Academic Press, New York, 1975.
[20] C.W. Therrien, Discrete Random Signals and Statistical Signal Processing, Prentice Hall, Englewood Cliffs, New Jersey, 1992.
[21] L. M. H. Ulander and H. Hellsten, "A new formula for SAR spatial resolution", AEÜ Int. J. Electron. Commun. 50 (1996) no. 2, 117–121.
[22] L.M.H. Ulander and P.-O. Frölund, "Ultra-wideband SAR interferometry", IEEE Trans. on Geoscience and Remote Sensing, vol. 36 no. 5, September 1998, 1540–1550.
[23] L.M.H. Ulander and H. Hellsten, "Low-frequency ultra-wideband array-antenna SAR for stationary and moving target imaging", conference proceedings for the SPIE 13th Annual International Symposium on Aerosense, Orlando, Florida, April 1999.
[24] L.J. Ziomek, Underwater Acoustics: A Linear Systems Theory Approach, Academic Press, Orlando, 1985.
[25] R. Zahn and M. Schlott, "Active antenna for X-band space SAR", Proc. Int. Conf. on Radar, Paris, France, (1994) 36–41.

DEPARTMENT OF MATHEMATICAL SCIENCES, RENSSELAER POLYTECHNIC INSTITUTE, TROY, NY 12180

Current address: (through the end of 2000) Department of Electromagnetic Theory, Lund University, S-221 00 Lund, Sweden

E-mail address: `chenem@rpi.edu`

Introduction to Local Tomography

Adel Faridani, Kory A. Buglione, Pallop Huabsomboon, Ovidiu D. Iancu, and Jeanette McGrath

ABSTRACT. Computed x-ray tomography entails the reconstruction of a density function f from line integrals of f. Ordinary tomography is global since reconstruction at a point x requires integrals over lines far from x. Local tomography uses only integrals over lines close to x. This introduction reviews a number of local tomographic methods developed over the past decade, such as Lambda tomography, pseudolocal tomography, wavelet based methods, and three-dimensional local cone-beam tomography.

1. Introduction

Computed tomography (CT) entails the reconstruction of a generalized density function f from line integrals of f. This reconstruction is not local in the sense that reconstruction of f at a point x requires integrals over lines far from x. In a number of applications only part of an object needs to be imaged. Thus it would be desirable to only use integrals over lines intersecting this region-of-interest (ROI). This entails the loss of uniqueness, but it turns out that the null functions are nearly constant inside the ROI and that the singularities of f inside the ROI can be stably recovered from such data; see, e.g., [**31**, §VI.4], [**36**].

Over the past decade a number of methods to 'localize' the reconstruction have been proposed. These range from methods for 'region-of-interest tomography' which use integrals over all lines passing through a region slightly larger than the ROI, to strictly local methods where reconstruction at a point x only requires integrals over lines very close to x. In this introduction we will review methods of both types. The wavelet-based multiresolution local tomography of [**38**], discussed in section §5, uses all lines passing through a region slightly larger than the ROI, while Lambda tomography, reviewed in §3, is strictly local. Pseudolocal tomography, described in §4, can be used in both modes. Some other techniques use a limited amount of data outside the ROI [**33**], or extrapolate the missing data [**31**, §VI.4]. Extensions of local tomographic methods to more general settings have been presented in [**22, 26, 29**].

2000 *Mathematics Subject Classification.* 45L05, 45Q05, 45-02, 65R10, 65R30.
Adel Faridani was supported by NSF grant DMS-9803352.

© 2001 American Mathematical Society

This article is organized as follows: In the next section we review some background material on the x-ray transform and its inversion.

Section 3 is devoted to Lambda tomography. Here not the function f itself but the related function $Lf = \Lambda f + \mu \Lambda^{-1} f$ is reconstructed, where $\Lambda = \sqrt{-\Delta}$, and Δ denotes the Laplacian. This reconstruction is strictly local, and our discussion centers on what features of f can be found from reconstructions of Lf. Particular attention will be given to the computation of density jumps.

In section 4 we discuss the pseudolocal tomography of [25, 37] and relate this method for computing density jumps to the methods based on Lambda tomography.

Section 5 briefly reviews wavelet based multiresolution tomography [38], one of the wavelet based methods for region-of-interest tomography. The goal of this method is to reconstruct the function f itself up to an almost constant error. For further applications of wavelets in local tomography see, e.g., [1, 2, 6, 33, 34, 52, 53].

In section 6 we turn to the three dimensional case and examine local conebeam tomography with sources on a curve. Here additional problems arise since it is usually impractical to collect sufficiently many data to ensure stable recovery of all singularities of f inside the region of interest. We discuss which singularities are stably determined and compare the two leading reconstruction algorithms.

2. The x-ray transform

We begin by introducing some notation and background material. $\mathbb{R}^n$ consists of n-tuples of real numbers, usually designated by single letters, $x = (x_1, \ldots, x_n)$, $y = (y_1, \ldots, y_n)$, etc. The inner product and absolute value are defined by $\langle x, y \rangle = \sum_1^n x_i y_i$ and $|x| = \sqrt{\langle x, x \rangle}$. The unit sphere S^{n-1} consists of the points with absolute value 1. $C_0^\infty(\mathbb{R}^n)$ denotes the set of infinitely differentiable functions on $\mathbb{R}^n$ with compact support. A continuous linear functional on C_0^∞ is called a distribution. If X is a set, X° denotes its interior, $\overline{X}$ its closure, and X^c its complement. χ_X and χ_n denote the characteristic functions (indicator functions) of X, and of the unit ball in $\mathbb{R}^n$, respectively. I.e., $\chi_X(x) = 1$ if $x \in X$, and $\chi_X(x) = 0$ if $x \notin X$. $|X|$ denotes the n-dimensional Lebesgue measure of $X \subset \mathbb{R}^n$. However, when it is clear that X should be treated as a set of dimension $m < n$, $|X|$ is the m-dimensional area measure. Thus

$$|S^{k-1}| = 2\pi^{k/2}/\Gamma(k/2)$$

is the $(k-1)$-dimensional area of the $(k-1)$-dimensional sphere.

The convolution of two functions is given by

$$f * g(x) = \int_{\mathbb{R}^n} f(x-y)g(y) dy.$$

The Fourier transform is defined by

$$\hat{f}(\xi) = (2\pi)^{-n/2} \int_{\mathbb{R}^n} f(x) e^{-i\langle x, \xi \rangle} dx$$

for integrable functions f, and is extended to larger classes of functions or distributions by continuity or duality.

The integral transform most relevant for local tomography is the x-ray transform.

DEFINITION 2.1. Let $\theta \in S^{n-1}$ and $\Theta^\perp$ the hyperplane through the origin orthogonal to θ. We parametrize a line $l(\theta, y)$ in $\mathbb{R}^n$ by specifying its direction $\theta \in S^{n-1}$ and the point y where the line intersects the hyperplane $\Theta^\perp$.

The x-ray transform of a function $f \in L_1(\mathbb{R}^n)$ is given by

$$Pf(\theta, y) = P_\theta f(y) = \int_{\mathbb{R}} f(y + t\theta)dt, \quad y \in \Theta^\perp. \tag{2.1}$$

We see that $Pf(\theta, x)$ is the integral of f over the line $l(\theta, y)$ parallel to θ which passes through $y \in \Theta^\perp$.

The inversion formula for the x-ray transform reads as follows:

$$f(x) = \left(2\pi |S^{n-2}|\right)^{-1} \int_{S^{n-1}} \Lambda P_\theta f(E_{\Theta^\perp} x) d\theta, \tag{2.2}$$

where $E_{\Theta^\perp} x$ denotes the orthogonal projection of x onto the subspace $\Theta^\perp$, and Calderón's operator Λ is defined in terms of Fourier transforms by

$$\widehat{\Lambda g}(\xi) = |\xi| \hat{g}(\xi), \quad g \in C_0^\infty(\mathbb{R}^k).$$

It is extended by duality to the class of functions g for which $(1 + |x|)^{-1-k} g$ is integrable [8]. In (2.2) the operator Λ acts on the function $g(y) = P_\theta f(y)$ defined on the subspace $\Theta^\perp$ of dimension $k = n - 1$. Note that

$$\Lambda^2 = -\Delta, \quad \Delta = \text{Laplacian}. \tag{2.3}$$

For a derivation of (2.2) and its numerical implementation, as well as for other inversion formulas see [**31**, §II.2 and Ch. V].

In two dimensions we parametrize $\theta \in S^1$ by its polar angle φ and define a vector $\theta^\perp$ orthogonal to θ such that

$$\theta = (\cos\varphi, \sin\varphi), \quad \theta^\perp = (-\sin\varphi, \cos\varphi). \tag{2.4}$$

Then the points in the subspace $\Theta^\perp$ are given by $\Theta^\perp = \{s\theta^\perp, s \in \mathbb{R}\}$. When working in two dimensions, we will often use the simplified notation $Pf(\theta, s)$ or $P_\theta f(s)$ instead of $Pf(\theta, s\theta^\perp)$. Occasionally we will also replace θ by the polar angle φ according to (2.4) and write $Pf(\varphi, s)$.

For g a function of one variable we have $\Lambda g = \mathcal{H}\partial g$, where ∂g denotes the derivative of g and $\mathcal{H}$ denotes the Hilbert transform

$$\mathcal{H}g(s) = \frac{1}{\pi} \int_{\mathbb{R}} \frac{g(t)}{s - t} dt, \tag{2.5}$$

where the integral is understood as a principal value.

In dimension $n = 2$, i.e., when f is a function of 2 variables, $P_\theta f$ is a function of one variable and the inversion formula (2.2) becomes

$$f(x) = \frac{1}{4\pi^2} \int_0^{2\pi} \int_{\mathbb{R}} \frac{\partial_s P_\theta f(s)}{\langle x, \theta^\perp \rangle - s} ds d\varphi. \tag{2.6}$$

From equation (2.6) we see that computation of $f(x)$ requires integrals over lines far from x, because the Hilbert transform kernel has unbounded support. Note that $P_\theta f(\langle x, \theta^\perp \rangle)$ is the integral over the line with direction θ which passes through x. Hence the inversion formula is not "local". A local inversion formula would utilize only integrals over lines passing close to x, i.e., values $P_\theta f(s)$ with s close to $\langle x, \theta^\perp \rangle$. For dimension $n > 2$ the inversion formula (2.2) is also not local.

The operator Λ is not continuous in an L_2 setting. Hence, in order to use the inversion formula in practice we have to stabilize it. This involves a well-known trade-off between stability and accuracy of the reconstruction. Here we give up the goal of recovering the function f itself, and aim instead at reconstructing an approximation $e * f$, where e is an approximate delta function whose Fourier transform $\hat{e}(\xi)$ decays sufficiently fast for large $|\xi|$. The price to pay for the stabilization is limited resolution, so e must be chosen carefully, depending on the amount and accuracy of the available measurements.

In order to allow for local reconstruction formulas we reconstruct $\Lambda^m f$ instead of f, with $m > -1$ an integer. This yields the approximate inversion formula

$$e * \Lambda^m f(x) = \int_{S^{n-1}} (k * P_\theta f)(E_{\Theta^\perp} x) d\theta, \quad m \geq -1, \qquad (2.7)$$

with the convolution kernel

$$k(y) = (2\pi |S^{n-2}|)^{-1} \Lambda^{m+1} P_\theta e(y), \quad y \in \Theta^\perp. \qquad (2.8)$$

If e is a radial function, then $P_\theta e$ and the convolution kernel k are independent of θ. Of greatest interest are the case $m = 0$, which gives the formulas for reconstructing the function f itself, and the cases $m = \pm 1$ which give local reconstruction formulas. The approximate inversion formula (2.7) is the basis for the popular filtered backprojection reconstruction algorithm (in dimension $n = 2$); see [10] for an error analysis, and [31] for a general discussion and references.

Since the parameters θ and $y \in \Theta^\perp$ of a line passing through a point x must satisfy the equation $E_{\Theta^\perp} x = y$, reconstruction according to (2.7) will be local if the kernel k is supported in a small neighborhood of the origin. However, for m even and $\int_{\mathbb{R}^n} e(x) dx \neq 0$, $\hat{k}$ is not analytic, so k cannot have compact support. This again reflects the fact that ordinary tomography is global, not local. On the other hand, it follows from (2.8) and (2.3) that k does have compact support if $m \geq -1$ is odd and e has compact support. This explains the interest in the cases $m = \pm 1$. Computing $\Lambda^{-1} f(x)$ consists of taking the average of all integrals over lines passing through x. This was done in early imaging techniques preceding CT. However, since Λ^{-1}, the inverse of Λ, is given by convolution with the Riesz kernel R_1,

$$\Lambda^{-1} f = R_1 * f, \quad R_1(x) = (\pi |S^{n-2}|)^{-1} |x|^{1-n}, \qquad (2.9)$$

the result is a very blurry image of f which by itself is of limited usefulness; see the bottom left image in Fig. 1. Current Lambda tomography avoids this disadvantage by computing a linear combination of Λf and $\Lambda^{-1} f$.

3. Lambda tomography

Lambda tomography was introduced independently in [49] and [46], further developed in work including [8, 9, 10, 24, 37, 50], and generalized in [22, 26]. It does not attempt to reconstruct the function f itself but instead produces the related function $Lf = \Lambda f + \mu \Lambda^{-1} f$. This has the advantage that the reconstruction is strictly local in the sense that computation of $Lf(x)$ requires only integrals over lines passing arbitrarily close to x. Lambda tomography has found applications in medical imaging [47], nondestructive testing [42, 50], and microtomography [9, 10, 41, 43]. (The term microtomography refers to the use of x-ray tomography to produce very high resolution images of small objects [13, 19]. While the spatial resolution in medical tomography is about 1 mm, the spatial resolution of microtomographic images is a few micrometers.) Local reconstructions from efficiently sampled data are analyzed in [**10**]. The choice of suitable convolution kernels for the filtered backprojection algorithm has been investigated in [**39, 40**].

Intelligent use of Lambda tomography requires knowledge of what kind of useful information about f is retained in Lf. Let us consider an example. The upper left of Fig. 1 shows an ordinary, global reconstruction of the density function f of a calibration object used by the Siemens company. The data come from an old generation Siemens hospital scanner. Units are such that the radius of the global reconstruction circle is one. The figure displays the reconstruction inside the rectangle $[-.5,.5]^2$. The scanning geometry is a fan-beam geometry with source radius $R = 2.868$, $p = 720$ source positions, and $2q = 512$ rays per source; cf. [**31**, p. 75]. The upper right of Fig. 1 shows a reconstruction of Λf. Reconstructions of $\Lambda^{-1}f$ and $Lf = \Lambda f + 46\Lambda^{-1}f$ are shown in the lower left and lower right, respectively. The similarity between the images of f and Λf is at first glance surprising. We expect that a good local reconstruction method should detect the singularities of f, since these are stably determined by the data. Indeed, since Λ is an invertible elliptic pseudo-differential operator, f and Λf have precisely the same singular set. However, we see that Λf is cupped where f is constant, and that the singularities are amplified in Λf. The image of $\Lambda^{-1}f$ by itself seems less useful, but it provides a countercup for the cup in Λf. Thus, the image of Lf shows less cupping and looks even more similar to f than the image of Λf. For example, the image of Lf indicates that the density just inside the boundary of the object is larger than the density outside the object, while this can not be clearly seen from the image of Λf. To achieve this effect, a good selection of μ is necessary. Here $\mu = 46$ was chosen by trial and error. The following prescription for selecting at least a good starting value for μ can be found in [**9**, §4]. The idea is to choose μ such that the reconstruction of the characteristic function of a disk with radius r_0 is as flat as possible in the interior of the disk. The radius r_0 should be chosen to lie between the radius r_i of the region of interest under consideration, and the radius r_w of a ball circumscribing the whole object, i.e., $r_i \leq r_0 \leq r_w$. Then μ scales as $\mu = cr_0^{-2}$, and experiments showed that $c = 6$ is a good choice. For the calibration object in Figure 1 we have $r_0 \simeq 0.36$, which gives $\mu = 45$, in good agreement with the experimental value $\mu = 46$.

A more detailed understanding of images of Λf or Lf is obtained from studying quantitative relations between Λf, $\Lambda^{-1}f$ and f [**8, 9**]. Some of the results for Λf are stated in Theorem 3.1 and discussed in Remark 3.2 below. For corresponding results on Λ^{-1} see [**8**].

THEOREM 3.1. *([**8**]) Let X and Y be measurable subsets of $\mathbb{R}^n$, $n \geq 2$, and let $(1+|x|)^{-1-n}f$ be integrable. Let X° and X^c denote the interior and the complement of X, respectively, and X^{co} the interior of X^c.*

(a) If $f_r(x) = f(x/r)$, then $\Lambda f_r(x) = r^{-1}\Lambda f(x/r)$.

(b) $\Lambda \chi_X(x) > 0$ on X°, and < 0 on X^{co}; $\Lambda \chi_{X^c} = -\Lambda \chi_X$.

(c) $\Lambda \chi_X$ is subharmonic (Laplacian ≥ 0) on X°, and superharmonic on X^{co}. This implies that $\Lambda \chi_X$ cannot have a local maximum in X°, nor a local minimum in X^{co}.

(d) If x is outside the support of f, then

$$\Lambda f(x) = \frac{1-n}{\pi |S^{n-2}|} \int_{\mathbb{R}^n} |x-y|^{-1-n} f(y) dy.$$

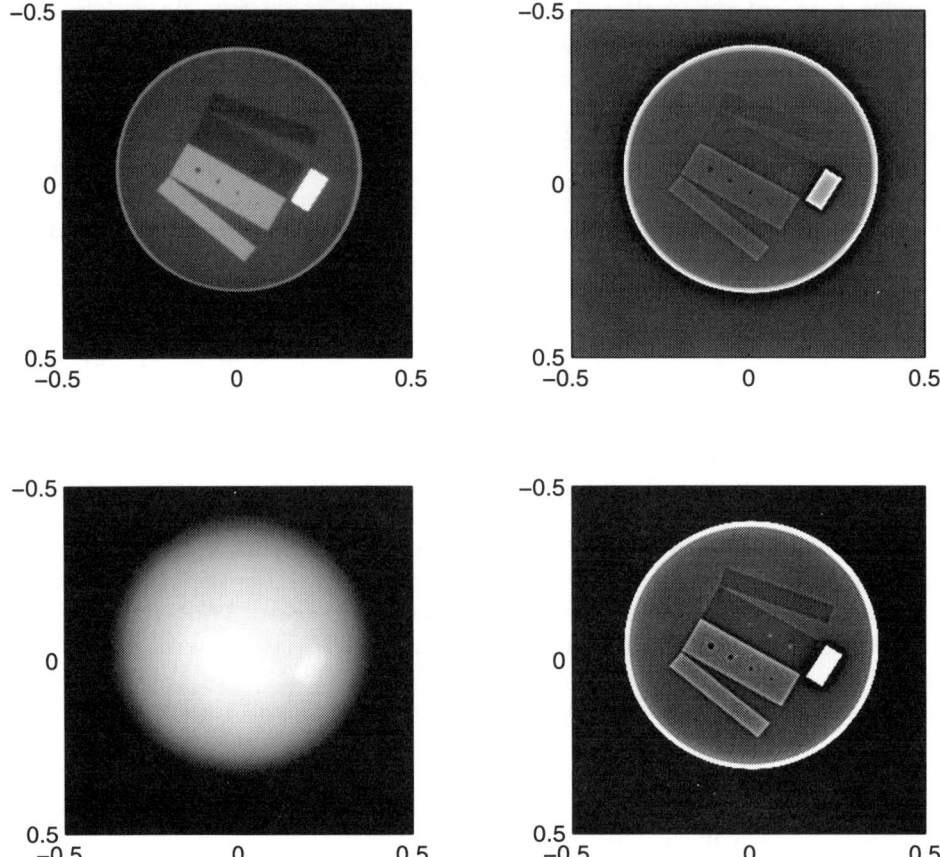

FIGURE 1. Top left: Global reconstruction of density $f(x)$ of calibration object. Top right: Reconstruction of Λf. Bottom left: Reconstruction of $\Lambda^{-1} f$. Bottom right: Reconstruction of $Lf = \Lambda f + \mu \Lambda^{-1} f$, $\mu = 46$.

(e) Near ∂X, $|\Lambda \chi_X(y)| \sim \frac{1}{d(y, \partial X)}$, where $d(x, \partial X)$ denotes the distance of x to ∂X.

REMARK 3.2. The results for $\Lambda \chi_X$ are of practical interest, since in many applications the function f can be modeled as a linear combination of characteristic functions.

- As a consequence of (a), small features are amplified in images of Λf. This is beneficial for the detection of small, low contrast details. For example, in Fig. 1 the small holes in the rectangular pieces are more clearly visible in the image of Λf than in the image of f.
- Part (b) indicates that the jumps of Λf at discontinuities of f have the same direction as those of f.
- Part (c) explains why there are no oscillations which could be mistaken for actual details in images of Λf.

- Part (d) shows that if f has compact support, then Λf cannot. This means that there are global effects in images of Λf in the sense that the value of $\Lambda f(x_0)$ depends on the values of f everywhere. However, Part d) implies that $\Lambda f(x)$ will decay at least as $O(|x|^{-1-n})$ for $|x| \to \infty$. More refined estimates are derived in [9] and used to develop a procedure to reduce the global effects.
- Part (e) shows that a finite jump in f causes an infinite jump in Λf. In a neighborhood of ∂X, Λf is not a function but a principal value distribution [8].

While Lf retains the signs of jumps in density, it does not give direct information about the size of these jumps. However, such information about density differences may be extracted in certain cases. In this and the following sections we will describe several methods. We assume that f is a linear combination of a smooth function and of characteristic functions of sets:

$$f = f_0 + \sum c_i \chi_{X_i}, \quad f_0 \in C_0^\infty, \quad |\partial X_i| = 0, \quad X_i = \overline{X_i^\circ}, \quad X_i^\circ \cap X_j^\circ = \emptyset \text{ if } i \neq j. \tag{3.1}$$

We are interested in estimating $c_j - c_i$ when X_j, X_i have a common nontrivial boundary Γ,

$$\Gamma = \partial X_i \cap \partial X_j \cap W \neq \emptyset, \quad W = (X_i \cup X_j)^\circ. \tag{3.2}$$

We first discuss the method developed in [9]. It is based on Theorem 3.3 below. The theorem expresses the fact that for x sufficiently close to Γ,

$$c_j - c_i = \frac{\Lambda f(x)}{\Lambda \chi_{X_j}(x)} + O(d), \quad \text{and}$$

$$|c_j - c_i| = \frac{|\nabla \Lambda f(x)|}{|\nabla \Lambda \chi_{X_j}(x)|} + O(d^2),$$

where d is the distance from x to Γ.

We say that a set Y has curvature $\leq 1/r$ along a subset Y_0 of ∂Y if for each point $\bar{y} \in Y_0$ there are open balls $B \subset Y$ and $B' \subset Y^c$ of radius r with $\bar{y} \in \bar{B} \cap \overline{B'}$. The distance of a point x to a set Y is denoted by $d(x, Y)$.

THEOREM 3.3. ([9]) Let f be as in (3.1). Fix i, j, let $W = (X_i \cup X_j)^\circ$ and assume that

$$\Gamma = \partial X_i \cap \partial X_j \cap W \neq \emptyset.$$

Let X_j have curvature $\leq 1/r$, $r > 0$, along a closed subset Γ_0 of Γ. Let $x \in W \backslash \Gamma$ be such that $d(x, \partial X_j) = d(x, \Gamma_0) = d$. Then

$$\left| \frac{\Lambda f(x)}{\Lambda \chi_{X_j}(x)} - (c_j - c_i) \right|$$

$$\leq F_1(d/r) \left(\max |\Lambda f_0| + C_1 \frac{(\max_{k \neq j} |c_k|)}{d(x, \partial W)} \right) d \tag{3.3}$$

$$\left| \frac{|\nabla \Lambda f(x)|}{|\nabla \Lambda \chi_{X_j}(x)|} - |c_j - c_i| \right|$$

$$\leq F_2(d/r) \left(\max |\nabla \Lambda f_0| + C_2 \frac{(\max_{k \neq j} |c_k|)}{d(x, \partial W)^2} \right) d^2 \tag{3.4}$$

The constants C_1 and C_2 and the functions F_1, F_2 can be given explicitly. E.g., for $n = 2$, $C_1 = 2$, and $C_2 = 3$. Furthermore,

$$\lim_{t \to 0^+} F_1(t) = \lim_{t \to 0^+} F_2(t) = \pi.$$

The error terms on the right-hand sides of (3.3) and (3.4) indicate that in general the estimate (3.4) should be more accurate than (3.3) when d is small. The terms involving $d(x, \partial W)$ come from the influence of other boundaries than Γ and reflect the global effects mentioned above.

Numerical implementation of (3.3) or (3.4) requires computation of reconstructions of Λf and $\Lambda \chi_{X_j}$ inside a region of interest R. In the following let $\bar{\Lambda} f$ and $\bar{\Lambda} \chi_{X_j}$ denote these reconstructions, rather than the functions Λf and $\Lambda \chi_{X_j}$ themselves. It is also assumed that f has the form (3.1) with sets X_i such that $X_i \subset R$ or $X_i \cap R = \emptyset$. This entails no loss of generality since any set X_i violating this condition can be replaced by the two sets $X_i \cap R$ and $X_i \cap R^c$. $\bar{\Lambda}\chi_{X_j}$ is computed using simulated x-ray data, after ∂X_j has been found from $\bar{\Lambda} f$. In principle, either (3.3) or (3.4) can be used, but as mentioned above the method based on (3.4) is likely to be more accurate. This gives only $|c_j - c_i|$, but since the sign of $c_j - c_i$ is preserved in Λf, this is all that is needed.

The method consists of the following steps:

(1) Compute $\bar{\Lambda} f$ from local data inside a region of interest R.
(2) Determine X_j by finding ∂X_j from $\bar{\Lambda} f$.
(3) Compute $\bar{\Lambda}\chi_{X_j}$ inside the region of interest from simulated x-ray data, using the same sampling geometry as for the original data.
(4) If $x \in \partial X_j$, take the ratio $|\nabla \bar{\Lambda} f(x)|/|\nabla \bar{\Lambda}\chi_{X_j}(x)|$ as an estimate for the magnitude of the density jump. It is advisable to use suitable averages of the gradients over points near the boundary of X_j instead of the gradient at a single point x. This reduces effects due to measurement noise.

A detailed discussion of the implementation of this method and numerical tests using real-world data have been reported in [9]; see also [7, 41].

The method described above can be simplified by making a priori assumptions about the unknown boundary ∂X_j. This can be used to simplify the edge detection in step 2 and to avoid the reconstruction from simulated data in step 3. For example, X_j could be assumed to be a halfspace H. If the filtered backprojection algorithm, i.e., a discretization of the approximate inversion formula (2.7) is used, then the reconstruction $\bar{\Lambda} f$ will, apart from discretization errors, be equal to $e * \Lambda f$. Hence Λf and $\Lambda \chi_{X_j}$ in (3.3) and (3.4) can be replaced by $e * \Lambda f$ and $e * \Lambda \chi_H$, respectively. We can compute $e * \Lambda \chi_H$ analytically in the following way: For $x \notin \partial H$ one has ([8, Theorem 4.5])

$$\Lambda \chi_H(x) = (\pi \tilde{d}(x))^{-1},$$

where $\tilde{d}(x)$ is the signed distance of x from ∂H, i.e., $\tilde{d}(x) = d(x, \partial H)$ for $x \in H$, and $\tilde{d}(x) = -d(x, \partial H)$ for $x \notin H$. Computing $e * \Lambda \chi_H$ involves the Radon transform of e. It is given by

$$R_\theta e(s) = \int_{\Theta^\perp} e(s\theta + y) dy, \quad \theta \in S^{n-1}, \quad s \in \mathbb{R}.$$

We assume that e is radial, so that $R_\theta e$ does not depend on θ. Therefore the subscript θ will be suppressed and $Re(s)$ viewed as a function of the one variable

s. It now follows that
$$e * \Lambda \chi_H(x) = \mathcal{H} Re(\tilde{d}(x)), \tag{3.5}$$
where $\mathcal{H}$ denotes the Hilbert transform as defined in (2.5). Recalling that for functions g of one variable $\Lambda g(t) = \frac{d}{dt} \mathcal{H} g(t)$ gives
$$|\nabla(e * \Lambda \chi_H(x))| = |\Lambda Re(\tilde{d}(x))|. \tag{3.6}$$

Replacing Λf and $\Lambda \chi_{X_j}$ in (3.3) and (3.4) by $e * \Lambda f$ and $e * \Lambda \chi_H$, and using (3.5) and (3.6) gives the approximate formulas
$$c_j - c_i \simeq \frac{e * \Lambda f(x)}{\mathcal{H} Re(\tilde{d}(x))}, \tag{3.7}$$
$$|c_j - c_i| \simeq \frac{|\nabla(e * \Lambda f(x))|}{|\Lambda Re(\tilde{d}(x))|}. \tag{3.8}$$

These two formulas are the basis of two of the algorithms proposed in [24, 37] for dimension $n = 2$, cf. formulas (2.17) and (2.21) in [24]. The derivation in [24, 37] is different and employs an asymptotic expansion for Λf, where f is smooth except for jumps across smooth boundaries. An algorithm based on (3.8) given in [37] uses the fact that $|\nabla(e * \Lambda f(x))|$ will be maximal for $x \in \Gamma$, and that $\tilde{d}(x) = 0$ for $x \in \Gamma$. Hence one can find the points $x \in \Gamma$ by looking for the local maxima of $|\nabla \bar{\Lambda} f|$ and then estimate the jump by
$$|c_j - c_i| \simeq \frac{|\nabla \bar{\Lambda} f(x))|}{|\Lambda Re(0)|}.$$

In our numerical experiments this algorithm tended to be somewhat less accurate than the more elaborate method of [9].

4. Pseudolocal tomography

Another method to compute jumps of a function from essentially local data is *pseudolocal tomography*. It was introduced in [25] and further developed in [37]. Here we follow the presentation given in [4] which allows us to understand the numerical implementation of this method in the framework of (3.7) and (3.8).

The starting point for pseudolocal tomography is the two-dimensional inversion formula (2.6) which we repeat here:
$$f(x) = \frac{1}{4\pi} \int_{S^1} \mathcal{H} \partial P_\theta f(\langle x, \theta^\perp \rangle) d\theta$$
$$= \frac{1}{4\pi^2} \int_0^{2\pi} \int_{\mathbb{R}} \frac{\frac{d}{ds} P_\theta f(s)}{\langle x, \theta^\perp \rangle - s} ds \, d\varphi.$$

Now truncate the Hilbert transform integral and define
$$f_d(x) = \frac{1}{4\pi^2} \int_0^{2\pi} \int_{\langle x, \theta^\perp \rangle - d}^{\langle x, \theta^\perp \rangle + d} \frac{\frac{d}{ds} P_\theta f(s)}{\langle x, \theta^\perp \rangle - s} ds \, d\varphi. \tag{4.1}$$

It was shown in [25] that $f - f_d$ is continuous, hence f_d has the same jumps as f. Recalling that $P_\theta f(\langle x, \theta^\perp \rangle)$ is the integral over the line in direction θ which passes through x, we see that computation of $f_d(x)$ requires only integrals over lines with distance at most d from x ("pseudo-local" reconstruction.)

In practice one has to use an approximate inversion formula and computes

$$f_{d,r}(x) = e_r * f_d(x) = \int_0^{2\pi} \int_{\mathbb{R}} \tilde{k}_{d,r}(\langle x, \theta^\perp \rangle - s) P_\theta f(s) \, ds \, d\varphi, \quad (4.2)$$

$$\tilde{k}_{d,r}(t) = \frac{1}{4\pi^2} \int_{t-d}^{t+d} \frac{\frac{d}{ds} P_\theta e_r(s)}{t-s} \, ds,$$

where e_r is a *radial* function satisfying

$$e_r(x) = r^{-2} e_1(x/r), \quad e_1(x) = 0 \text{ for } |x| > 1, \quad \int_{\mathbb{R}^2} e_1 \, dx = 1.$$

Note that $\tilde{k}_{d,r}(t) = 0$ for $|t| > d+r$, i.e., computation of $f_{d,r}(x)$ requires integrals over lines with distance at most $d+r$ from x. Furthermore, $\lim_{d\to\infty} \tilde{k}_{d,r}(t) = (4\pi)^{-1} \mathcal{H} \partial P_\theta e_r(t)$. Hence (2.7) gives that $\lim_{d\to\infty} f_{d,r}(x) = e_r * f(x)$. Indeed, the convolution kernel $\tilde{k}_{d,r}$ can be obtained from the kernel k in (2.8) by letting $m = 0$ and truncating the Hilbert transform integral. The relation $f_{d,r} = e_r * f_d$ was shown in [25].

It turns out that for small d (i.e., local data), f_d is significantly different from zero only in a narrow region near a boundary (cf. [25, Fig. 3]), and that the convolution with the point spread function e_r alters these values so much that the jumps cannot just be simply read off the reconstructed image $f_{d,r}$. We need an algorithm to obtain information about the jumps of f. The methods developed by Katsevich and Ramm [25, 37] can be understood in the framework developed for Lambda tomography. According to (3.7) and (3.8) we have for x close to Γ

$$c_j - c_i \simeq \frac{E * \Lambda f(x)}{\mathcal{H} R E(\tilde{d}(x))} \quad (4.3)$$

$$|c_j - c_i| \simeq \frac{|\nabla E * \Lambda f(x)|}{|\Lambda R E(\tilde{d}(x))|} \quad (4.4)$$

The task now is to find $E_{d,r}$ such that $E_{d,r} * \Lambda f = f_{d,r} = e_r * f_d$.

PROPOSITION 4.1. ([37, 4]) Define $E_{d,r}$ by

$$P_\theta E_{d,r} = (P_\theta e_r) * M_d$$

with

$$M_d(s) = -\frac{1}{\pi} \ln(|s/d|) \chi_{[-d,d]}(s).$$

Then

$$f_{d,r}(x) = E_{d,r} * \Lambda f(x).$$

With this result (4.3) and (4.4) give

$$c_j - c_i \simeq \frac{f_{d,r}(x)}{\mathcal{H} R E_{d,r}(\tilde{d}(x))} \quad (4.5)$$

$$|c_j - c_i| \simeq \frac{|\nabla f_{d,r}(x)|}{|\Lambda R E_{d,r}(\tilde{d}(x))|} \quad (4.6)$$

and we can apply the same algorithms for recovering the jumps as in Lambda tomography.

Some remarks are in order.

(1) Note that because $E_{d,r}$ is radial, $\mathcal{H}RE_{d,r}(0) = 0$, so $f_{d,r}(x) \simeq 0$ for $x \in \Gamma$. This makes it difficult to use the relation (4.5) in practice, since finding $\tilde{d}(x)$ is not easy, cf. the algorithm given [25] and further discussed in [4]. However, since $|\nabla f_{d,r}|$ is maximal for $x \in \Gamma$ one can find the points $x \in \Gamma$ by looking for the local maxima of $|\nabla f_{d,r}|$ and then estimate the jump by

$$|c_j - c_i| \simeq \frac{|\nabla f_{d,r}(x)|}{|\Lambda R E_{d,r}(0)|}, \quad x \in \Gamma.$$

This approach has essentially been used in [37] for pseudolocal tomography and in [24] for Lambda tomography.

(2) The property that f_d has the same jumps as f is not used in the algorithm.

(3) $E_{d,r}(x) = 0$ for $|x| > d + r$. Hence our derivation of the algorithm is only justified for $d+r$ sufficiently small. In practice the method seems to work also for much larger values of $d + r$.

5. Wavelet-based multiresolution local tomography

Wavelet-based multiresolution local tomography is a method for region of interest tomography developed in [38]. The goal here is to reconstruct the function f itself within the region of interest up to an almost constant error. The method illustrates the possible uses of wavelets to 'localize' the x-ray transform, or, more precisely, to separate the features which are well determined by local data from those who are not. The following discussion assumes some background on wavelets which can be found in [51] or other texts on this subject.

Consider a (two-dimensional) multiresolution analysis of nested subspaces V_j, $j \in \mathbb{Z}$ of $L_2(\mathbb{R}^2)$. We use the notation

$$f_{j,k}(x) = 2^j f(2^j x - k), \quad j \in \mathbb{Z}, \ k \in \mathbb{Z}^2, \ x \in \mathbb{R}^2.$$

Let Φ be the scaling function and Ψ^μ, $\mu = 1, 2, 3$ the associated wavelets. Since the $\Phi_{j+1,k}$, $k \in \mathbb{Z}^2$ are a Riesz basis of the subspace V_{j+1}, a function $f \in V_{j+1}$ can be written as

$$f(x) = \sum_{k \in \mathbb{Z}^2} \tilde{A}_{j+1,k} \Phi_{j+1,k}(x).$$

The so-called *approximation coefficients* $\tilde{A}_{j,k}$ are given by

$$\tilde{A}_{j,k} = \langle f, \tilde{\Phi}_{j,k} \rangle$$

where $\langle , \rangle$ denotes the inner product in L_2 and $\tilde{\Phi}$ is the biorthogonal scaling function. Alternatively we can use the relation $V_{j+1} = V_j + W_j$ and obtain the expansion

$$f(x) = \sum_{k \in \mathbb{Z}^2} \tilde{A}_{j,k} \Phi_{j,k}(x) + \sum_{\mu=1}^{3} \sum_{k \in \mathbb{Z}^2} \tilde{D}^\mu_{j,k} \Psi^\mu_{j,k}(x).$$

We can interpret the first sum as an approximation to f in $V_j \subset V_{j+1}$, i.e., at a lower resolution. The second sum supplies the missing detail information. Therefore the coefficients

$$\tilde{D}^\mu_{j,k} = \langle f, \tilde{\Psi}^\mu_{j,k} \rangle$$

are called *detail coefficients*. The Fast Wavelet Transform and its inverse allow efficient computation of the $\tilde{A}_{j,k}$ and $\tilde{D}^\mu_{j,k}$, $k \in \mathbb{Z}^2$ from the $\tilde{A}_{j+1,k}$, $k \in \mathbb{Z}^2$, and vice versa.

We now observe that the approximation and detail coefficients can be computed directly from the x-ray data. Let $f^\vee(x) = f(-x)$. Then

$$\tilde{A}_{j,k} = \langle f, \tilde{\Phi}_{j,k} \rangle = \left(f * \overline{\tilde{\Phi}_{j,0}^\vee} \right)(2^{-j}k) \tag{5.1}$$

Similarly,

$$\tilde{D}_{j,k}^\mu = \langle f, \tilde{\psi}_{j,k}^\mu \rangle = \left(f * \overline{\left(\tilde{\Psi}_{j,0}^\mu\right)^\vee} \right)(2^{-j}k) \tag{5.2}$$

Hence we can use the approximate inversion formula (2.7) with $e(x) = \overline{\tilde{\phi}_{j,0}^\vee}(x)$ and reconstruction on the grid $x = 2^{-j}k$, $k \in \mathbb{Z}^2$, to obtain the approximation coefficients directly from the x-ray data. For the detail coefficients we let $e = \overline{\left(\tilde{\Psi}_{j,0}^\mu\right)^\vee}$. Alternatively one could first compute the approximation coefficients $\tilde{A}_{j+1,k}$ by letting $e(x) = \overline{\tilde{\phi}_{j+1,0}^\vee}(x)$ and choosing the finer grid $x = 2^{-j-1}k$, $k \in \mathbb{Z}^2$, and then use the Fast Wavelet Transform to obtain the approximation and detail coefficients at level j. Since the additional computational burden of applying the Fast Wavelet Transform is negligible compared to the effort required for the reconstruction from the x-ray data, this alternative method seems preferable, since only one point-spread function and corresponding convolution kernel need to be used. However, if not all coefficients on level j are needed, the first method will be more efficient.

The next question is how this approach allows to 'localize' the x-ray transform, i.e., to separate features which are determined by local data from those which are not. It was observed in [33] that the detail coefficients for sufficiently large j should be well determined by local data, if the wavelets Ψ^μ have vanishing moments. Let us see why.

DEFINITION 5.1. *A function f of n variables has vanishing moments of order up to N, if*

$$\int_{\mathbb{R}^n} x^\alpha f(x) dx = 0$$

for all multiindices $\alpha = (\alpha_1, \ldots, \alpha_n)$ with $|\alpha| = \sum \alpha_i \leq N$. Recall that the α_i are non-negative integers and that $x^\alpha = x_1^{\alpha_1} x_2^{\alpha_2} \ldots x_n^{\alpha_n}$.

The nonlocality in the approximate inversion formula comes from the convolution kernel k in (2.8) in case of $m = 0$. In two dimensions this is caused by the presence of the Hilbert transform in the formula $k = (4\pi)^{-1} \Lambda P_\theta e = (4\pi)^{-1} \mathcal{H} \partial P_\theta e$. The key observation now is that the Hilbert transform of a function with vanishing moments decays fast.

LEMMA 5.2. *([38, p. 1418]) Let $f(t) \in L_2(\mathbb{R})$ vanish for $|t| > A$ and have vanishing moments of order up to N. Then, for $|s| > A$,*

$$|\mathcal{H}f(s)| \leq \frac{1}{\pi |s - A|^{N+2}} \int_{-A}^{A} |f(t) t^{N+1}| dt$$

It is well known how to construct wavelets with vanishing moments, and it turns out that the functions $\partial P_\theta \left(\tilde{\Psi}_{j,0}^\mu\right)^\vee$ inherit the vanishing moments from the $\tilde{\Psi}^\mu$. Therefore the convolution kernels $k = (4\pi)^{-1} \mathcal{H} \partial P_\theta \left(\tilde{\Psi}_{j,0}^\mu\right)^\vee$ will decay rapidly outside the support of $P_\theta \left(\tilde{\Psi}_{j,0}^\mu\right)^\vee$.

So we see that the detail coefficients for large j, when $\widetilde{\Psi}_{j,0}^{\mu}$ has small support, are well determined by local data. This is intuitively plausible since these coefficients contain high-frequency information, and we know already from Lambda tomography that high-frequency information is well-determined. So the nonlocality shows its greatest impact in the approximation coefficients.

If the scaling function $\widetilde{\Phi}(x)$ is sufficiently smooth and has compact support, then the zero order moment of $\partial_s P_\theta \widetilde{\Phi}(s)$ will vanish. However, since the scaling function satisfies $\int \widetilde{\Phi}(x) dx = 1$, the first order moment of $\partial_s P_\theta \widetilde{\Phi}$ is always non-zero. Hence the corresponding convolution kernel $k(s)$ will decrease no faster than $O(s^{-2})$ for large $|s|$. One could still choose the scaling $\widetilde{\Phi}$ so that its moments of order 1 through N vanish. It is shown in [**38**, p. 1419] that in such a case the resulting convolution kernel k satisfies

$$|k(s)| = O(s^{-2}) + O(s^{-N-3}).$$

It seems that this does not achieve much, since we cannot remove the leading $O(s^{-2})$ term. Nevertheless, the authors of [**38**] found that some scaling functions having vanishing moments lead to convolution kernels with sufficiently rapid decay for practical purposes. In their reconstructions the authors of [**38**] also extrapolated the missing data by constant values, thus reducing cupping artifacts. While it is suggested in [**38**] to first compute the approximation and detail coefficients at level j and then use an Inverse Fast Wavelet Transform to obtain the approximation coefficients at level $j+1$, our numerical tests in [**44**] indicated that the simpler approach of directly computing the approximation coefficients at level $j+1$ yields equivalent results. We observe that this can be accomplished without using wavelets in the algorithm, namely just by specifying the particular point spread function $e = \widetilde{\Phi}_{j+1,0}^{\vee}$ in the standard reconstruction formula (2.7).

6. Cone-beam local tomography with sources on a curve

A problem of great practical interest which still poses many open problems is three-dimensional cone-beam reconstruction with sources on a curve. See, e.g., [**48**] for an inversion formula, [**12**] for a general stability result, [**36**] for conditions to detect singularities, and [**5, 11, 14, 28, 32, 53**] for reconstruction algorithms and other developments.

To describe data collection with an x-ray source moving on a curve, the parameterization of lines by $\theta \in S^{n-1}$ and $y \in \Theta^\perp$ is less convenient. It is more suitable to introduce the *divergent beam x-ray transform*

$$Df(a,\theta) = D_a f(\theta) = \int_0^\infty f(a+t\theta)dt, \quad \theta \in S^{n-1}, \tag{6.1}$$

which gives the integral of f over the ray with direction θ emanating from the source point a. If f is supported in the unit ball, and the source points a lie on a sphere A with center in the origin and radius $R > 1$, then the approximate inversion formula for the divergent beam x-ray transform reads [**46**]

$$e * \Lambda^m f(x) = R^{-1} \int_A \int_{S^{n-1}} D_a f(\theta) \, |\langle a,\theta\rangle| \, k(E_{\Theta^\perp}(x-a)) \, d\theta \, da, \tag{6.2}$$

with $m \geq -1$ and k as in (2.8). This formula is very useful in two dimensions, but not so in three dimensions. It needs integrals over all lines, but in three dimensions the lines form a four parameter family, so (6.2) requires far more data than should be needed to determine a function of three variables. In practical 3D tomography an x-ray source moves on a curve, so only integrals over lines intersecting the curve are measured. The conditions on the source curve Γ for stable inversion are restrictive, so that in most practical situations one has an incomplete data problem.

Microlocal analysis has proved to be a useful tool in determining which singularities of f are stably determined by the available data. Based on the exposition in [36] we now state the relevant microlocal concepts and apply them to this situation. The reader interested in a deeper treatment may wish to first read [36] and [18], and then proceed to articles such as [3, 15, 16, 17, 35].

The following concept of a wavefront set uses the fact that the Fourier transform of a C_0^∞ function decays rapidly. A local version of this fact can be obtained by first multiplying f with a C_0^∞ cut-off function Φ with small support, and seeing if the Fourier transform of the product Φf decays rapidly. The wavefront set gives even more specific, so-called microlocal information, inasmuch as it identifies the directions in which the Fourier transform of Φf does not decrease rapidly.

DEFINITION 6.1. Let f be a distribution and let $x_0, \xi_0 \in \mathbb{R}^n$, $\xi_0 \neq 0$. Then (x_0, ξ_0) is in the wavefront set of f if and only if for each cut-off function Φ in C_0^∞ with $\Phi(x_0) \neq 0$, the Fourier transform of Φf does not decrease rapidly in any conic neighborhood of the ray $\{t\xi_0, t > 0\}$.

Loosely speaking, we say that a singularity of f can be stably detected from available x-ray data, if there exists a corresponding singularity of comparable strength in the data. The strength of a singularity can be quantified microlocally using Sobolev space concepts:

DEFINITION 6.2. A distribution f is in the Sobolev space H^s microlocally near (x_0, ξ_0) if and only if there is a cut-off function $\Phi \in C_0^\infty(\mathbb{R}^n)$ with $\Phi(x_0) \neq 0$ and function $u(\xi)$ homogeneous of degree zero and smooth on $\mathbb{R}^n \setminus \{0\}$ and with $u(\xi_0) \neq 0$ such that $u(\xi)\widehat{(\Phi f)}(\xi) \in L^2(\mathbb{R}^n, (1+|\xi|^2)^s)$.

First, one localizes near x_0 by multiplying f by Φ, then one microlocalizes near ξ_0 by forming $u\widehat{\Phi f}$ and sees how rapidly $\widehat{\Phi f}$ decays at infinity.

For 3D tomography with sources on a curve we have the following result:

THEOREM 6.3. *(cf.* [36, Theorem 4.1], *and* [3, 15]*) Let Γ be a smooth curve in $\mathbb{R}^3$ and f a distribution whose support is compact and disjoint from Γ. Then any wavefront set of f at (x_0, ξ_0) is stably detected from divergent beam x-ray data Df with sources on Γ if and only if*

the plane $\mathcal{P}$ through x_0 and orthogonal to ξ_0, intersects Γ transversally.

If data are taken over an open set of rays with sources on Γ, then a ray in $\mathcal{P}$ from Γ to x_0 must be in the data set for stable detection to apply. In these cases f is in H^s microlocally near (x_0, ξ_0) if and only if the corresponding singularity of Df is in $H^{s+1/2}$.

We see that the corresponding singularities of Df are weaker by 1/2 Sobolev order, but this is still strong enough to allow stable detection in practice.

It is now interesting to ask if the available numerical algorithms can actually reconstruct all the stable singularities. The results for a general class of restricted x-ray transforms obtained in [15, 16, 17] show that microlocal analysis is also a powerful tool to answer such a question. For an introduction to these results see [18]. Explicit calculations analysing an algorithm for contour reconstruction proposed by Louis and Maass in [28] and some closely related methods have recently been given in [23, 27].

The algorithm of [28] aims to reconstruct the function

$$f_R = -\Delta D^* D f, \qquad (6.3)$$

with

$$D^* g(x) = \int_\Gamma \|x - a\|^{-1} g\left(a, \frac{x-a}{\|x-a\|}\right) da.$$

An advantage of the formula (6.3) is that reconstruction of f_R is local. In [28] it is shown that f_R approximates Λf in certain cases.

The results in [15, 23, 27] show that the wavefront set of f_R consists of two parts. The first part contains those wavefronts (x, ξ) of f for which the plane through x and normal to ξ intersects Γ. The second part may introduce new singularities, namely on the line from a source point $a \in \Gamma$ to x, the location of the original singularity in f. This will happen if the plane through x and normal to ξ contains a and the tangent vector to Γ at a is orthogonal to ξ, i.e., the plane touches Γ but does not intersect Γ transversally. In addition, the acceleration vector of the curve at a should not be orthogonal to ξ. The Sobolev strength of these additional singularities is the same as the reconstructed part of the original wavefront set [16, 17, 23], and they appear as artifacts in numerical simulations [21, 23].

Another, and apparently the historically first method for 3D local tomography is an adaptation of the algorithm by Feldkamp, Davis and Kress [11] (FDK algorithm) which was developed by P.J. Thomas at the Mayo Clinic. While the details of this local FDK algorithm have not been published, it has been used in various papers, e.g., [47, 8]. A recent implementation of a local FDK algorithm has been reported in [20]. The modification from the original FDK algorithm consists in replacing the global convolution kernel corresponding to $m = 0$ in (2.8) with a local kernel corresponding to $m = 1$. A different adaptation using wavelet based kernels has been given in [53].

Figures 2 and 3 provide a comparison of the two algorithms, using the implementations in [21] and [20], respectively. The experiments use a mathematical phantom consisting of a superposition of four balls with the following parameters:

Center	Radius	Density
(0, 0, 0)	0.5	1
(0, 0, 0.125)	0.1	-1
(-0.3, 0, -0.125)	0.02	-1
(0.3, 0, 0.2)	0.01	-1

The source is assumed to move on a circle in the x-y plane with radius $R = 3$ and center in the origin. We used 400 equidistant source positions and a 240×240 detector array. The local FDK algorithm [20] used a planar detector array, while our code for the Louis-Maass algorithm assumed a spherical array; see [21]. The images consist of 131×131 pixels.

Figure 2 shows reconstructions in the vertical plane $y = 0$, with the Louis-Masss method in the upper left, and the local FDK algorithm in the upper right. The additional singularities predicted for the Louis-Mass algorithm by the references given above are clearly visible in the upper left image as lines tangential to each ball and intersecting the source curve. The reconstruction with the local FDK algorithm does not show these lines, but does appear to have greater distortions with regard

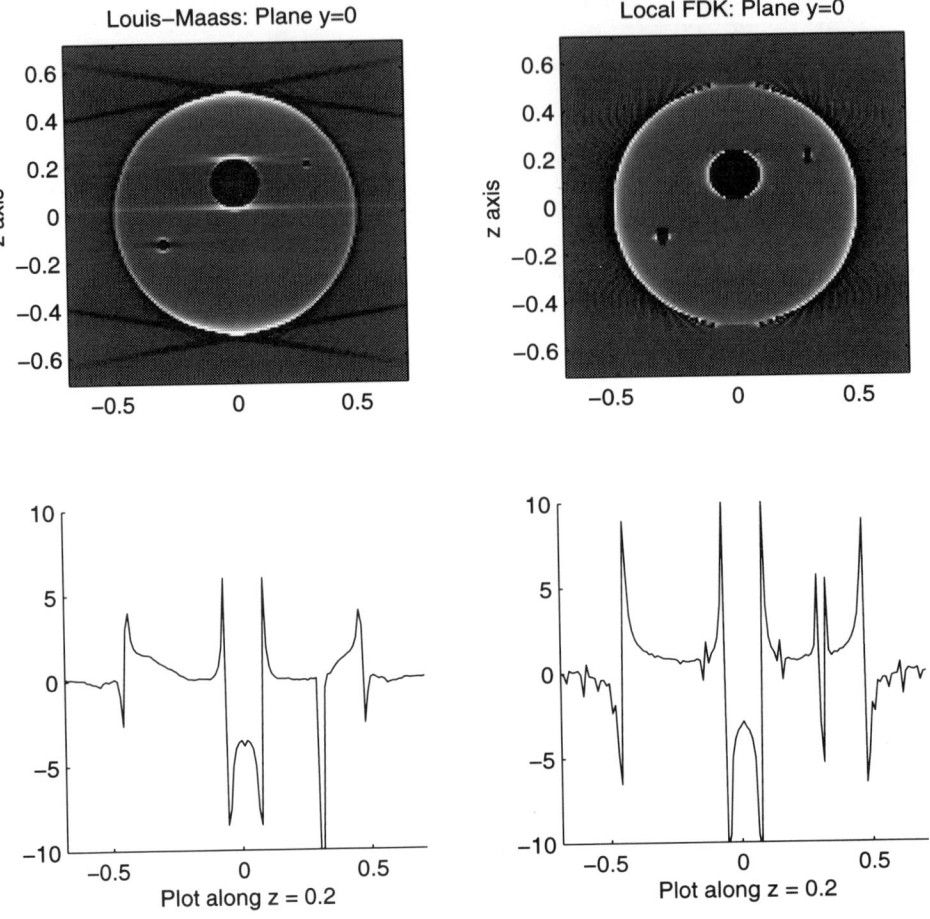

FIGURE 2.

to the small objects. This is confirmed in Figure 3 which shows reconstructions of the horizontal plane $z = 0.2$.

References

[1] C. Berenstein and D. Walnut, *Local inversion of the Radon transform in even dimensions using wavelets*, in: 75 Years of Radon Transform, S. Gindikin and P. Michor (eds.), Conference Proceedings and Lecture Notes in Mathematical Physics, Vol. 4, International Press, Boston, 1994, pp. 45-69.
[2] C. Berenstein and D. Walnut, *Wavelets and local tomography*, in: Wavelets in Medicine and Biology, A. Aldroubi and M. Unser (eds.), CRC Press, Boca Raton, 1996.
[3] J. Boman and E.T. Quinto, *Support theorems for real-analytic Radon transforms on line complexes in three-space*, Trans. Amer. Math. Soc., 335(1993), pp. 877-890.
[4] K. Buglione, *Pseudolocal tomography*, M.S. paper, Dept. of Mathematics, Oregon State University, Corvallis, OR 97331, U.S.A., (1998).
[5] M. Defrise and R. Clack, *A cone-beam reconstruction algorithm using shift-variant filtering and cone-beam backprojection*, IEEE Trans. Med. Imag., MI-13 (1994), pp. 186-195.
[6] A. Delaney and Y. Bresler, *Multiresolution tomographic reconstruction using wavelets*, IEEE Trans. Image Proc., 4 (1995), 799-813.

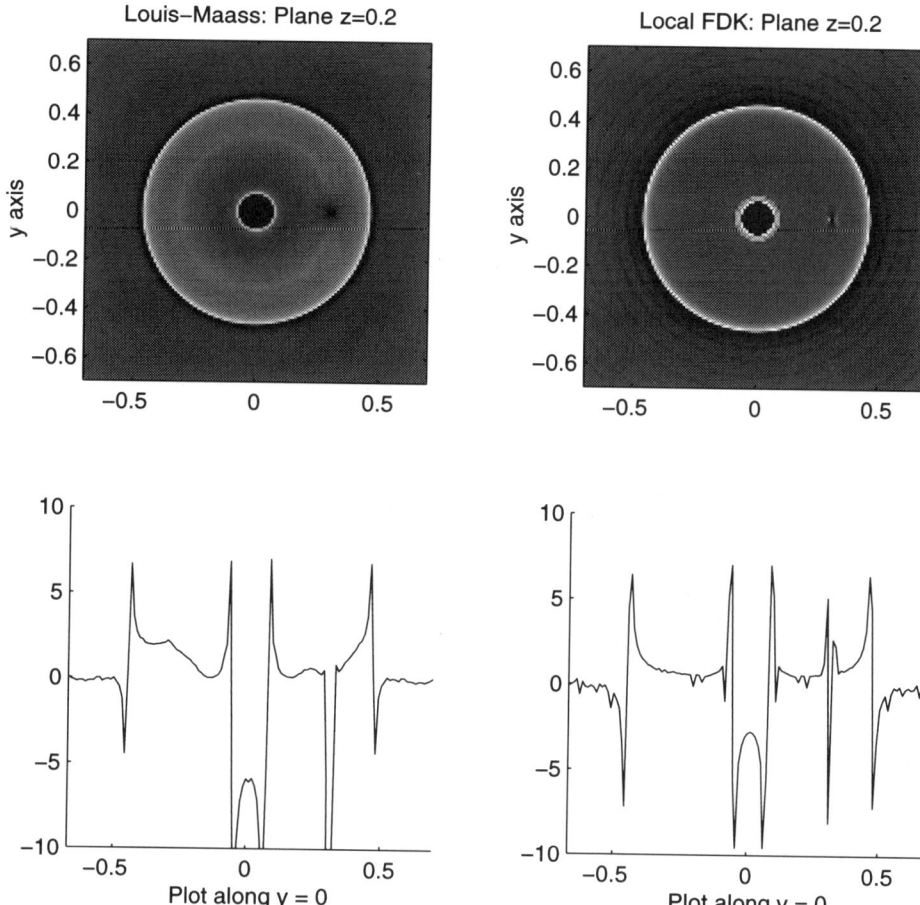

FIGURE 3.

[7] A. Faridani, *Results, old and new, in computed tomography*, in: Inverse Problems in Wave Propagation, G. Chavent et al. (editors), The IMA Volumes in Mathematics and its Applications, Vol. 90, Springer Verlag, New York, 1997, pp. 167-193.
[8] A. Faridani, E. L. Ritman, and K. T. Smith, *Local tomography*, SIAM J. Appl. Math., 52 (1992), pp. 459-484. *Examples of local tomography*, SIAM J. Appl. Math., 52 (1992), pp. 1193-1198.
[9] A. Faridani, D.V. Finch, E. L. Ritman, and K. T. Smith, *Local tomography II*, SIAM J. Appl. Math., 57 (1997), pp. 1095-1127.
[10] A. Faridani and E. L. Ritman, *High-resolution computed tomography from efficient sampling*, Inverse Problems, 16(2000), pp. 635-650.
[11] L. A. Feldkamp, L. C. Davis, and J. W. Kress, *Practical cone-beam algorithm*, J. Opt. Soc. Am. A, 1 (1984), pp. 612-619.
[12] D. V. Finch, *Cone beam reconstruction with sources on a curve*, SIAM J. Appl. Math., 45(1985), pp. 665-673.
[13] B.P. Flannery, H.W. Deckman, W.G. Roberge, and K.L. D'Amico, *Three-dimensional x-ray microtomography*, Science, 237 (1987), pp. 1439-1444.
[14] P. Grangeat, *Mathematical framework of cone beam 3D reconstruction via the first derivative of the Radon transform*, in: Mathematical Methods in Tomography, G.T. Herman, A.K. Louis, and F. Natterer (eds.), Lecture Notes in Mathematics, Vol. 1497, Springer, 1991, pp. 66-97.

[15] A. Greenleaf and G. Uhlmann, *Nonlocal inversion formulas for the X-ray transform*, Duke Math. J., 58(1989), pp. 205-240.
[16] A. Greenleaf and G. Uhlmann, *Estimates for singular Radon transforms and pseudodifferential operators with singular symbols*, J. Funct. Anal., 89(1990), pp. 202-232.
[17] A. Greenleaf and G. Uhlmann, *Composition of some singular Fourier integral operators and estimates for restricted X-ray transforms.*, Ann. Inst. Fourier, 40(1990), pp. 443-466.
[18] A. Greenleaf and G. Uhlmann, *Microlocal techniques in integral geometry.*, in: Integral Geometry and Tomography, E. Grinberg and E.T. Quinto (eds.), Contemporary Mathematics, Vol. 113, Amer. Math. Soc., Providence, R.I., 1990, pp.121-135.
[19] L. Grodzins, *Optimum energies for x-ray transmission tomography of small samples*, Nuclear Instruments and Methods, 206 (1983), pp. S41-S45.
[20] P. Huabsomboon, *3D Filtered Backprojection Algorithm for Local Tomography*, M.S. paper, Dept. of Mathematics, Oregon State University, Corvallis, OR 97331, U.S.A., (2000).
[21] O. D. Iancu, *Contour reconstruction in 3D x-ray computed tomography*. M.S. paper, Dept. of Mathematics, Oregon State University, Corvallis, OR 97331, U.S.A., (1999).
[22] A. I. Katsevich, *Local Tomography for the generalized Radon transform*, SIAM J. Appl. Math. 57(1997), pp. 1128-1162.
[23] A. Katsevich, *Cone beam local tomography*, SIAM J. Appl. Math, 59(1999), pp. 2224-2246.
[24] A.I. Katsevich and A. G. Ramm, *New methods for finding jumps of a function from its local tomographic data*, Inverse Problems, 11 (1995), pp. 1005-1023.
[25] A.I. Katsevich and A. G. Ramm, *Pseudolocal tomography*, SIAM J. Appl. Math., 56, (1996), pp. 167-191.
[26] P. Kuchment, K. Lancaster and L. Mogilevskaya, *On local tomography*, Inverse Problems, 11 (1995), pp. 571-589.
[27] I. Lan, *On an operator associated to a restricted x-ray transform*, Ph.D. thesis, Dept. of Mathematics, Oregon State University, Corvallis, OR 97331, U.S.A., (1999).
[28] A. K. Louis, and P. Maass, *Contour reconstruction in 3-D x-ray CT*, IEEE Trans. Med. Imag., MI-12 (1993), pp. 764–769.
[29] A. K. Louis and E. T. Quinto, *Local tomographic methods in SONAR*, in: Surveys on Solution Methods for Inverse Problems, D. Colton et al. (eds.), Springer, 2000.
[30] W. R. Madych, *Tomography, approximate reconstruction, and continuous wavelet transforms*, Appl. Comp. Harm. Anal., 7 (1999), 54-100.
[31] F. Natterer, *The Mathematics of Computerized Tomography*, Wiley, 1986.
[32] F. Natterer, *Recent developments in x-ray tomography*, in: Tomography, Impedance Imaging, and Integral Geometry, E.T. Quinto, M. Cheney, and P. Kuchment (eds.), Lectures in Applied Mathematics, Vol. 30, Amer. Math. Soc., 1994, pp. 177-198.
[33] T. Olson and J. de Stefano, *Wavelet localization of the Radon transform*, IEEE Trans. Sig. Proc., 42 (1994), pp. 2055-2067 .
[34] T. Olson, *Optimal time-frequency projections for localized tomography*, in: Wavelets in Medicine and Biology, A. Aldroubi and M. Unser (eds.), CRC Press, Boca Raton, 1996, pp. 263-296.
[35] E.T. Quinto, *The dependence of the generalized Radon transform on defining measures*, Trans. Amer. Math. Soc. 257 (1980), pp. 331-346.
[36] E. T. Quinto, *Singularities of the x-ray transform and limited data tomography in $\mathbb{R}^2$ and $\mathbb{R}^3$*, SIAM J. Math. Anal., 24 (1993), pp. 1215-1225.
[37] A. G. Ramm and A. I. Katsevich, *The Radon Transform and Local Tomography*, CRC Press, Boca Raton, 1996.
[38] F. Rashid-Farrokhi, K. J. R. Liu, C. A. Berenstein, and D. Walnut, *Wavelet-based multiresolution local tomography*, IEEE Transactions on Image Processing, 6 (1997), pp. 1412-1430.
[39] A. Rieder, R. Dietz, and T. Schuster, *Approximate inverse meets local tomography*, Math. Meth. Appl. Sci., 23 (2000), pp. 1373-1387.
[40] A. Rieder and T. Schuster, *The approximate inverse in action with an application to computerized tomography*, SIAM J. Numer. Anal., 37 (2000), pp. 1909-1929.
[41] E. L. Ritman, J. H. Dunsmuir, A. Faridani, D. V. Finch, K. T. Smith, and P. J. Thomas, *Local reconstruction applied to microtomography*, in: Inverse Problems in Wave Propagation, G. Chavent et al. (editors), The IMA Volumes in Mathematics and its Applications, Vol. 90, Springer Verlag, New York, 1997, pp. 443-452.

[42] E. A. Sivers, D. L. Halloway, W. A. Ellingson, and J. Ling, *Development and application of local 3-D CT reconstruction software for imaging critical regions in large ceramic turbine rotors*, in Rev. Prog. Quant. Nondest. Eval.:, D.O. Thompson and D.E. Chimenti (eds.), Plenum, New York, 1993, pp. 357-364.

[43] E. A. Sivers, D. L. Halloway, W. A. Ellingson, *Obtaining high-resolution images of ceramics from 3-D x-ray microtomography by region-of-interest reconstruction*, Ceramic Eng. Sci. Proc., 14, no. 7-8, (1993), pp. 463-472.

[44] J. Skaggs, *Region of interest tomography using biorthogonal wavelets*, M.S. paper, Dept. of Mathematics, Oregon State University, Corvallis, OR 97331, U.S.A., (1997).

[45] K. T. Smith, D.C. Solmon, and S. L. Wagner, *Practical and mathematical aspects of the problem of reconstructing objects from radiographs*, Bull. Amer. Math. Soc., 83(1977), pp. 1227-1270. Addendum in Bull. Amer. Math. Soc., 84(1978), p. 691.

[46] K. T. Smith and F. Keinert, *Mathematical foundations of computed tomography*, Appl. Optics 24 (1985), pp. 3950-3957.

[47] W. J. T. Spyra, A. Faridani, E. L. Ritman, and K. T. Smith, *Computed tomographic imaging of the coronary arterial tree - use of local tomography*, IEEE Trans. Med. Imag., 9 (1990), pp. 1-4.

[48] H. K. Tuy, *An inversion formula for cone beam reconstruction*. SIAM J. Appl. Math. 43(1983), pp. 546-552.

[49] É. I. Vainberg, I. A. Kazak, and V. P. Kurozaev, *Reconstruction of the internal three-dimensional structure of objects based on real-time internal projections*, Soviet J. Nondestructive Testing, 17 (1981), pp. 415-423.

[50] É. I. Vainberg, I. A. Kazak, and M. L. Faingoiz, *X-ray computerized back projection tomography with filtration by double differentiation. Procedure and information features*, Soviet J. Nondestructive Testing, 21 (1985), pp. 106-113.

[51] M. Vetterli and J. Kovacevic, *Wavelets and Subband coding*, Prentice Hall, 1995.

[52] D. Walnut, *Applications of Gabor and wavelet expansions to the Radon transform*, in: Probabilistic and Stochastic Methods in Analysis, J. Byrnes et al. (eds.), Kluwer, Boston, 1992, pp. 187-205.

[53] S. Zhao and G. Wang, *Feldkamp-type type cone-beam tomography in the wavelet framework*. IEEE Trans. Med. Imag., 19 (2000), pp. 922-929.

DEPT. OF MATHEMATICS, OREGON STATE UNIVERSITY, CORVALLIS, OR 97331
E-mail address: faridani@math.orst.edu
URL: http://ucs.orst.edu/~faridana

DEPT. OF MATHEMATICS, OREGON STATE UNIVERSITY, CORVALLIS, OR 97331

DEPT. OF MATHEMATICS, OREGON STATE UNIVERSITY, CORVALLIS, OR 97331

DEPT. OF MATHEMATICS, OREGON STATE UNIVERSITY, CORVALLIS, OR 97331

DEPT. OF MATHEMATICS, OREGON STATE UNIVERSITY, CORVALLIS, OR 97331

Algorithms in Ultrasound Tomography

Frank Natterer

Abstract *In ultrasound tomography one has to solve the inverse problem of the Helmholtz equation at fixed frequency. Linear approximations, such as the Born and Rytov approximation, lead to diffraction tomography that numerically can be done with the nonequispaced fast Fourier transform. The fully nonlinear problem can be solved by iterative methods. We give a survey of recently suggested iterative methods and their mutual interdependencies. Finally we describe a combination of Kaczmarz's method with an initial value technique for solving the Helmholtz equation.*

1. Ultrasound Tomography

In ultrasound tomography we want to identify an object which is given by its complex valued refractive index $1 + f$ where $f = f(x)$. Typically,

$$f(x) = \frac{c_0^2}{c^2} - 1 - i\frac{2\alpha c_0}{kc}$$

where $c(x)$ is the speed of sound in the object that is contained in the sphere Ω of radius ρ, c_0 is the (constant) speed of sound in the surrounding medium, α is the attenuation of the tissue, and k is the wave number of the irradiating plane wave. For other models see Nachman, Smith and Waag (1990). The pressure field u satisfies the Helmholtz equation

(1)
$$\Delta u_\theta + k^2(1+f)u_\theta = 0 \quad \text{in } \mathbf{R}^n$$

$$u_\theta = u_{i,\theta} + v_\theta, \quad u_{i,\theta}(x) = e^{ikx\cdot\theta}$$

where $\theta \in S^{n-1}$ is the direction of the irradiating wave and v is the scattered field satisfying the Sommerfeld radiation condition at infinity. The scattered field is measured on the detector $\rho\theta + \theta^\perp$ where $\theta^\perp$ is the subspace perpendicular to θ. Thus the available data is

(2) $$(R_\theta(f))(y) = v_\theta(\rho\theta + y), \quad \theta \in S^{n-1}, \quad y \in \theta^\perp.$$

The problem is to recover f from $R_\theta(f)$, $\theta \in S^{n-1}$. In practice, $R_\theta(f)$ is given for a finite set of directions $\theta_1, \ldots, \theta_p$. This paper concentrates on numerical methods. For theoretical aspects see Nachman (1996).

1991 *Mathematics Subject Classification.* Primary 81U40, 92C55; Secondary 65N99.
Key words and phrases. Inverse acoustics, ultrasound tomography.

2. Linear Approximations

In most of the practical work the scattered field v is assumed to be small:

(3) $$|v_\theta| \ll 1 .$$

Then we can neglect v on the right hand side of the differential equation for v_θ, namely
$$\Delta v_\theta + k^2 v_\theta = -k^2 f(e^{ikx\cdot\theta} + v_\theta) ,$$
obtaining

(4) $$\Delta v_\theta + k^2 v_\theta = -k^2 f e^{ikx\cdot\theta} .$$

The conditions for (3) to hold are not fully explored. By heuristic reasoning Kak and Slaney (1987) arrive at $\frac{k}{\pi}\rho|f| \ll 1$. This condition ensures that the phases of the incoming and the total wave are close. This condition is widely accepted. In many practical applications, e.g. in breast screening, this condition is far from being satisfied.

(4) is called the Born approximation to the (direct) scattering problem. It gives rise to the propagation operator
$$(U_r f)(\theta, y) = \frac{1}{k} \frac{v_\theta(r\theta + y)}{e^{ikr}}$$
where v_θ is the solution to (4) satisfying the radiation condition. U_r leads to a simple solution of the inverse scattering problem (in the Born approximation) by the following generalization of the projection theorem of X-ray tomography.

THEOREM 2.1. *Let $n = 2, 3$. Let $|r| > \rho$ and $\eta \in \theta^\perp$. Then,*

$$(U_r f)^\wedge(\theta, \eta) = ik\sqrt{\frac{\pi}{2}} e^{ir(\varepsilon a(\eta) - k)} \frac{\hat{f}((\varepsilon a(\eta) - k)\theta + \eta)}{a(\eta)} ,$$

$$a(\eta) = \begin{cases} \sqrt{k^2 - |\eta|^2} , & |\eta| \leq k , \\ i\sqrt{|\eta|^2 - k^2} , & |\eta| \geq k , \end{cases} \quad \varepsilon = \begin{cases} 1 , & r > \rho , \\ -1 , & r < -\rho . \end{cases}$$

Here, $(U_r f)^\wedge$ is the Fourier transform in $\theta^\perp$, while $\hat{f}$ is the nD Fourier transform of f.

By Theorem 2.1, $\hat{f}$ is determined by the data on the half spheres
$$(\varepsilon a(\eta) - k)\theta + \eta , \quad \eta \in \theta^\perp$$
centered at $k\theta$ with radius k. As θ runs through S^{n-1} these half spheres fill the ball of radius $2k$, permitting the stable determination of $\hat{f}(\xi)$ for $|\xi| \leq 2k$. According to the Nyquist criterion of communication theory this corresponds to a spatial resolution of π/k. Recently, Palamodov (1999) proved stability for this resolution also in the fully nonlinear case.

Computationally, the reconstruction of f by means of Theorem 2.1 requires a nonequispaced fast Fourier transform (FFT). In the 1D case such an FFT is

(5) $$\hat{y}_k = \sum_{\ell=-q}^{q-1} e^{\pi i x_\ell k/q} y_\ell , \quad k = -k, \ldots, q-1 .$$

For $x_\ell = \ell$ we have the usual equispaced FFT which can be done in $O(q \log q)$ operations. This is true for the general case, too, as follows from the following Theorem which is due to Fourmont (2000).

THEOREM 2.2. *Let $0 < \beta < \alpha$, $\alpha + \beta \leq 2\pi$. Let ϕ be a continuous function in $[-\alpha, \alpha]$, $\phi \neq 0$ in $[-\beta, \beta]$, and $\hat{\phi} = 0$ outside $[-\alpha, \alpha]$. Then,*

$$\hat{y}_k = \frac{(2\pi)^{-1/2}}{\phi(\beta k/q)} z_k,$$

$$z_k = \sum_m e^{-i\beta mk/q} w_m,$$

$$w_m = \sum_{\ell=-q}^{q-1} \hat{\phi}\left(\frac{\pi}{\beta} x_\ell - m\right) y_\ell.$$

The theorem leads immediately to an $O(q \log q)$ algorithm for (5). We choose ϕ in such a way that $\hat{\phi}$ is small outside $[-K, K]$. A good choice for ϕ is the Kaiser-Bessel-window

$$\phi(\xi) = I_0\left(K\sqrt{\alpha^2 - \xi^2}\right)/I_0(K\alpha)$$

with I_0 the modified Bessel function of order 0. Then, the evaluation of w_m requires only $O(K)$ operations. The computation of z_k from w_m is an equispaced FFT of length $2q$ which can be done with $O(q \log q)$ operations. The algorithm ends with a scaling operation, requiring $O(q)$ operations. Thus the total operation count is $O(q \log q)$, as claimed.

We note that this algorithm has many predecessors, most notably the gridding method of O'Sullivan (1985), Dutt and Rokhlin (1993), Beylkin (1995), Steidl (1998). Reconstruction algorithms based on these novel Fourier techniques are now available and can easily compete with more established methods, such as filtered backpropagation (Devaney (1982)).

Instead of the Born approximation the Rytov approximation can be used; see Kak and Slaney (1987). The algorithm is exactly the same. Only the physical interpretation is different.

3. Iterative methods for the fully nonlinear problem

Most of these methods work on the integral equation formulation of the problem. Using the operator

$$(Gu)(x) = -k^2 \int_{\mathbf{R}^3} \frac{e^{ik|x-y|}}{4\pi |x-y|} u(y) dy \quad \text{for } n = 3$$

$$= -k^2 \frac{i}{4} \int_{\mathbf{R}^2} H_0(k|x-y|) u(y) dy \quad \text{for } n = 2$$

with H_0 the zero order Hankel function of the first kind we can rewrite (1) as

(6) $$u_\theta = u_{i,\theta} + Gfu_\theta.$$

We have to determine f from knowing

(7) $$g_\theta = u_\theta \text{ on } \partial\Omega, \quad \theta \in S^{n-1}.$$

All the iterative methods produce a sequence u^ℓ, f^ℓ which (hopefully) converge to u, f, the solution of (6), (7).

3.1. The iterated Born method (IB). An iterative method which suggests itself is

$$u_\theta^\ell = u_{i,\theta} + Gf^{\ell-1}u_\theta^\ell \text{ in } \Omega \tag{8}$$

$$g_\theta = u_{i,\theta} + Gf^\ell u_\theta^\ell \quad \text{on } \partial\Omega, \quad \theta \in S^{n-1}. \tag{9}$$

For $f^0 = 0$, f^1 is just the Born approximation. Solving (8) for u_θ^ℓ is just the solution of an integral equation, the Lippman-Schwinger equation. This is possible but time-consuming, in particular for k large. Determining f^ℓ from (9) is more of a problem except in the case $f^0 = 0$, of course. See Tijhuis (1989) for details.

3.2. Source type integral equation method (STIE). On introducing the "equivalent source" $w_\theta = fu_\theta$ and multiplying (6) inside Ω with f we get

$$w_\theta = f(u_{i,\theta} + Gw_\theta) \text{ in } \Omega \tag{10}$$

$$g_\theta = u_{i,\theta} + Gw_\theta \quad \text{on } \partial\Omega \tag{11}$$

which hold for each θ. With w_θ^+ the Moore-Penrose generalized solution of (11) we put

$$w_\theta = w_\theta^+ + w_\theta^{NR}$$

where w_θ^{NR} (NR stands for non-radiating) is such that $Gw_\theta = 0$ on $\partial\Omega$. Each such w_θ satisfies (11), so we are left with (10). We iterate according to

$$w_\theta^\ell = f^{\ell-1}(u_{i,\theta} + Gw_\theta^\ell) \text{ in } \Omega, \tag{12}$$

$$u_\theta^\ell = u_{i,\theta} + Gw_\theta^\ell \quad \text{in } \Omega, \tag{13}$$

$$f^\ell = w_\theta^\ell/u_\theta^\ell. \tag{14}$$

These equations are used in consecutive order. The last equation needs explanations. As it stands it doesn't make sense since f^ℓ has to be independent of θ. Thus (14) has to be understood in the least squares sense. It is hoped that in the course of the iteration the dependence of $w_\theta^\ell/u_\theta^\ell$ on θ disappears.

The computations are done entirely on the non-radiating fields for which a finite series expansion is used. STIE can be understood as IB with the critical step of solving (9) for f^ℓ replaced by the simple (but questionable) rule (14). For details see Habashy, Oristaglio and de Hoop (1989).

3.3. The conjugate gradient method (CG). Again we start out from the STIE formulation (10), (11). Let $f^{\ell-1}$, $w^{\ell-1}$ be given. Rather than solving (10), (11) exactly for w_θ we define w_θ^ℓ by one step of the usual CG method for the linear system (10), (11) with $f = f^{\ell-1}$. This reads

$$w_\theta^\ell = w_\theta^{\ell-1} + \alpha_{\ell,\theta} v_{\ell,\theta} \text{ in } \Omega$$

where $v_{\ell,\theta}$ is the CG direction and $\alpha_{\ell,\theta}$ a scalar. From here on the algorithm is identical to STIE, i.e. we compute f^ℓ from (14). Of course the same procedure could be done for each iterative method, not only for CG. As for the usual CG algorithm, convergence is very slow for large values of k. On the other hand each step of the iteration can be done quickly since G is a convolution operator. For details see van den Berg and Kleinman (1997).

3.4. The modified Newton method.

First we extend g_θ from $\partial\Omega$ to the exterior of Ω by solving the exterior Helmholtz problem with $f = 0$ and boundary values g_θ on $\partial\Omega$. Then $u_\theta = g_\theta$ is known in $\mathbf{R}^2 \setminus \Omega$ and we can compute $\partial u_\theta / \partial \nu = \partial g_\theta / \partial \nu$ on $\partial\Omega$. Our problem can now be recast as: Find f in Ω satisfying

$$\Delta u_\theta + k^2(1+f)u_\theta = 0 \quad \text{in } \Omega, \tag{15}$$

$$u_\theta = g_\theta, \quad \frac{\partial u_\theta}{\partial \nu} = \frac{\partial g_\theta}{\partial \nu} \quad \text{on } \partial\Omega \tag{16}$$

for $\theta \in S^{n-1}$.

Introducing the operator

$$A \begin{pmatrix} u_\theta \\ f \end{pmatrix} = \Delta u_\theta + k^2(1+f)u_\theta$$

and restricting u_θ to those functions which satisfy (16) our problem simply reads

$$A \begin{pmatrix} u_\theta \\ f \end{pmatrix} = 0, \quad \theta \in S^{n-1}. \tag{17}$$

The Fréchet derivative of A is

$$A' \begin{pmatrix} u_\theta \\ f \end{pmatrix} \begin{pmatrix} w_\theta \\ h \end{pmatrix} = \Delta w_\theta + k^2(1+f)w_\theta + k^2 h u_\theta.$$

The Newton method for (17) reads

$$A' \begin{pmatrix} u_\theta^\ell \\ f^\ell \end{pmatrix} \begin{pmatrix} w_\theta \\ h \end{pmatrix} = -A \begin{pmatrix} u_\theta^\ell \\ f^\ell \end{pmatrix},$$

$$w_\theta = u_\theta^{\ell+1} - u_\theta^\ell, \quad h = f^{\ell+1} - f^\ell \tag{18}$$

i.e.

$$\Delta w_\theta + k^2(1+f^\ell)w_\theta + k^2 h u_\theta^\ell = -\Delta u_\theta^\ell - k^2(1+f^\ell)u_\theta^\ell,$$

$$w_\theta = 0, \quad \frac{\partial w_\theta}{\partial \nu} = 0 \quad \text{on } \partial\Omega. \tag{19}$$

This is a non-standard boundary value problem. It is not clear how to solve it for w_θ and h. However if we simplify (18) by replacing

$$A' \begin{pmatrix} u_\theta^\ell \\ f^\ell \end{pmatrix} \quad \text{by} \quad A' \begin{pmatrix} u_{i,\theta} \\ 0 \end{pmatrix}$$

(modified Newton method) (19) becomes

$$\Delta w_\theta + k^2 w_\theta + k^2 h u_{i,\theta} = -\Delta u_\theta^\ell - k^2(1+f^\ell)u_\theta^\ell \quad \text{in } \Omega,$$

$$w = 0, \quad \frac{\partial w}{\partial \nu} = 0 \quad \text{on } \partial\Omega. \tag{20}$$

This equation can be solved for h by a simple trick: Multiply the differential equation by $e^{-ik\omega \cdot x}$, $\omega \in S^{n-1}$ and integrate over Ω to obtain

$$\int_\Omega (\Delta w_\theta + k^2 w_\theta)e^{-ik\omega \cdot x} dx + k^2 \int_\Omega h e^{ik(\theta - \omega) \cdot x} dx$$

$$= -\int_\Omega (\Delta u_\theta^\ell + k^2(1+f^\ell)u_\theta^\ell)e^{-ik\omega \cdot x} dx.$$

Due to the boundary conditions of w_θ the first integral on the left hand side vanishes, as can be seen by an integration by parts. Hence

$$\hat{h}(k(\theta - \omega)) = -\frac{1}{2\pi k^2} \int_\Omega (\Delta u_\theta^\ell + k^2(1 + f^\ell)u_\theta^\ell)e^{-ik\omega \cdot x}dx .$$

This yields $\hat{h}$ in the ball of radius $2k$. Note that $\{k(\theta - \omega) : \omega \in S^{n-1}\}$ is just the Ewald sphere which is of paramount interest in diffraction tomography. Thus we can determine a low-pass filtered version of h, and the bandwidth of the filter is $2k$. This corresponds exactly to the resolution limit in diffraction tomography.

Once h has been determined, w_θ can be computed from (20) by any Helmholtz solver, ignoring one of the boundary conditions. The other one will be satisfied automatically (at least approximately) due to our choice of h.

The problem with this elegant approach is that the modified Newton method can't possibly converge for objects f which are far from satisfying the assumptions underlying the Born approximation: Neglecting f^ℓ on the left hand side of (19) corresponds to using the Born approximation. This means that the phase of the solution w_θ of (20) is off by more than π, making it useless as update for w_θ^ℓ. For details see Gutman and Klibanov (1994).

4. Kaczmarz's Method

Kaczmarz's method is extensively used in X-ray tomography. It's the basis of the ART algorithm; see Herman (1980). In ultrasound tomography we extend the Kaczmarz's method to nonlinear problems of the form

(21) $$R_j(f) = g_j , \quad j = 1, \ldots, p$$

where $R_j : H \to H_j$ is a nonlinear operator between the Hilbert spaces H, H_j. The Kaczmarz update reads

(22) $$f^j = f^{j-1} - \omega R'_j(f_{j-1})^* C_j^{-1} (R_j(f_{j-1}) - g_j) , \quad j = 1, 2, \ldots$$

where $R'_j(f_{j-1})$ the derivative of R_j at f_{j-1} and $R'_j(f_{j-1})^*$ its adjoint. C_j is a positive definite operator that is chosen as an approximation to $R'_j(f_{j-1})^* R_j(f_{j-1})$. The subscripts of R_j, g_j, C_j are to be understood modulo p.

In order to carry out (22) one needs a routine returning $R_j(f)$ for arbitrary f (forward problem) and another routine that returns $R'_j(f)^* r$ for arbitrary r (adjoint problem). For ultrasound tomography a natural choice for R_j would be $R_j = R_{\theta_j}$ with R_θ from (2). However, this choice would necessitate the solution of the Helmholtz boundary value problem (1) as the forward problem. The numerical solution of (1) is a challenging problem of numerical analysis, in particular at large frequencies k. Therefore we prefer a different choice of R_j.

Our choice of R_j is based on the observation that the Cauchy initial value problem for the Helmholtz equation is stable for spatial frequencies $< k$. To see this we consider the Cauchy problem

$$\Delta u + k^2(1 + f)u = 0 \text{ in } \mathbf{R}^2 ,$$

$$u = g , \quad \frac{\partial u}{\partial x_2} = h \quad \text{for } x_2 = 0 .$$

Doing a 1D Fourier transform with respect to x_1 we get

$$\frac{\partial^2 \hat{u}}{\partial x_2^2}(\xi_1, x_2) + (k^2 - \xi_1^2)\hat{u}(\xi_1, x_2) = -k^2(fu)^\wedge(\xi_1, x_2) ,$$

$$\hat{u}(\xi_1, 0) = \hat{g}(\xi_1) , \quad \frac{\partial \hat{u}}{\partial x_2}(\xi_1, 0) = \hat{h}(\xi_1) .$$

It follows immediately that

$$\begin{aligned}\hat{u}(\xi_1, x_2) &= \hat{g}(\xi_1)\cos(\kappa x_2) + \frac{\hat{h}(\xi_1)}{\kappa}\sin(\kappa x_2) \\ &\quad - \frac{k^2}{\kappa}\int_0^{x_2}(fu)^\wedge(\xi_1, x_2')\sin(\kappa(x_2 - x_2'))dx_2'\end{aligned}$$

where $\kappa = \sqrt{k^2 - \xi_1^2}$. This is an integral equation of Volterra type that can be solved stably provided that $|\xi_1| < k$. For details see Natterer (1997).

Based on this stability result we can turn the boundary value problem (1) into an initial value problem in the following way. First we compute u in the exterior of Ω by solving the exterior boundary value problem

$$\begin{aligned}\Delta u_\theta + k^2 u_\theta &= 0 , \quad |x| > \rho , \\ u_\theta &= g_\theta , \quad |x| = \rho\end{aligned}$$

where g_θ is the given data function. Then we decompose $\partial\Omega$ into Γ_θ^+ and Γ_θ^- where

$$\Gamma_\theta^\pm = \{x \in \partial\Omega : \pm x \cdot \theta \geq 0\} .$$

Since u_θ is known outside Ω we can compute the normal derivatives of u_θ, denoted by $\partial g_\theta/\partial \nu$, on Γ_j^-. Now we solve the initial value problem

(23)
$$\Delta u_\theta + k^2(1+f)u_\theta = 0 \text{ in } \Omega$$

$$u_\theta = g_\theta , \quad \frac{\partial u_\theta}{\partial \nu} = \frac{\partial g_\theta}{\partial \nu} \text{ on } \Gamma_j^-$$

and put

$$R_j(f) = u_{\theta_j} \text{ on } \Gamma_j^+ .$$

$R_j(f)$ can be computed by a finite difference marching scheme for (23). For details see Natterer and Wübbeling (1995). In this paper it is also shown that the evaluation of $R_j'(f)^*r$ can be done in the following way. Solve the final value problem

(24)
$$\Delta z + k^2(1+\overline{f})z = 0 \text{ in } \Omega$$

$$z = 0 , \quad \frac{\partial z}{\partial \nu} = r \text{ on } \Gamma_j^+$$

and put

$$R'(f)^*r = k^2 \overline{u}_{\theta_j} z .$$

Again, this can be done by a finite difference marching scheme.

The Kaczmarz's method as outlined above is very efficient. One can do reconstructions on a 128×128 grid in a few minutes on a workstation. The method can be shown to achieve the resolution $\lambda/2$ where $\lambda = 2k/\pi$ is the wavelength of the incoming plane waves; see Grinberg, Natterer and Palamodov (1997).

So far our numerical experiments are based on simulated data. We are confident that we will get measured data in the near future.

References

[1] Beylkin, G. (1995): *On the fast Fourier transform of functions with singularities*, Appl. Comp. Harm. Anal. **2**, 363-381.

[2] Devaney, A.J. (1982): *A filtered backprojection algorithm for diffraction tomography*, Ultrasonic Imaging **4**, 336-350.

[3] Dutt, A. and Rokhlin, V. (1993): *Fast Fourier transforms for nonequispaced data*, SIAM J. Sci. Comput. **14**, 1368-1393.

[4] Fourmont, K. (2000): *Schnelle Fourier - Transformation bei nichtäquidistanten Gittern und tomographische Anwendungen*, Dissertation, Fachbereich Mathematik der Universität Münster.

[5] Gutman, S. and Klibanov, M. (1994): *Iterative method for multi-dimensional scattering problems at fixed frequencies*, Inverse Problems **10**, 573-599.

[6] Grinberg, N., Natterer, F., and Palamodov, V.P. (1997): *Spatial resolution of the PBP algorithm for acoustical inverse scattering*, Technical Report, University Münster.

[7] Habashy, T.M, Oristaglio, and de Hoop, A.T. (1989): *Simultaneous nonlinear reconstruction of two-dimensional permittivity and conductivity*, Radio Science **29**, 1101-1118.

[8] Herman, G.T. (1980): Image Reconstruction from Projections. The Fundamentals of Computerized Tomography. Academic Press.

[9] Kak, A.C. and Slaney, M. (1987): Principles of Computerized Tomography Imaging. IEEE Press, New York.

[10] Nachman, A.I. (1996): *Global uniqueness for a two-dimensional inverse boundary value problem*, Annals of Math. (2) **143**, 71-96 (1996).

[11] Nachman, A.I., Smith, J.F., and Waag, R.C. (1990): *An equation for acoustic propagation in inhomogeneous media with relaxation losses*, J. Acoust. Soc. Amer. **88**, 1584-1595.

[12] Natterer, F. and Wübbeling, F. (1995): *A propagation - backpropagation method for ultrasound tomography*, Inverse Problems **11**, 1225-1232.

[13] Natterer, F. (1997): *An initial value approach to the inverse Helmholtz problem at fixed frequency*, in: H. Engl et al. (eds.): Inverse Problems in Medical Imaging and Nondestructive Testing, p. 159-167, Springer.

[14] Palamodov, V.P. (1999): *On stability of reconstruction in acoustic tomography*, Preprint 1999.

[15] O'Sullivan, J.D. (1985): *A fast sinc function gridding algorithm for Fourier inversion in computer tomography*, IEEE Trans. Med. Imag. **4**, 200-207.

[16] Steidl, G. (1998): *A note on fast Fourier transforms for nonequispaced grids*, Advances in Computational Mathematics **9**, 337-352.

[17] Tijhuis, A.G. (1989): *Born-type reconstruction of material parameters of an inhomogeneous lossy dielectric slab from reflected field data*, Wave Motion **11**, 151-173.

[18] van den Berg, P.M. and Kleinman, R.E. (1997): *A contrast source inversion method*, Inverse Problems **13**, 1607-1620.

INSTITUT FÜR NUMERISCHE UND INSTRUMENTELLE MATHEMATIK, WESTF. WILHELMS-UNIVERSITÄT MÜNSTER, EINSTEINSTRASSE 62, D-48149 MÜNSTER, GERMANY

E-mail address: `nattere@math.uni-muenster.de`

Radon Transforms, Differential Equations, and Microlocal Analysis

Eric Todd Quinto

ABSTRACT. This article is an expanded version of my talk at the 2000 AMS/-IMS/SIAM conference on Radon transforms and Tomography. The goal is to show how microlocal analysis can be used in integral geometry and to show how uniqueness theorems for Radon transforms (proven using microlocal analysis) can be applied to partial differential equations. Microlocal analysis is used to prove a support theorem of Boman and Quinto for the Radon transform on lines. A uniqueness theorem for the circle transform of Agranovsky and Quinto is described and applied to the wave equation.

1. Introduction

This is an expanded version of an introductory talk I gave at the AMS/IMS/-SIAM conference at Mt. Holyoke College on June 22, 2000. This reports joint work with Jan Boman [9] and with Mark Agranovsky [2, 3].

As one of the conference organizers, I want to thank all the participants for coming! I was happy to see several participants of this conference who also attended the 1993 conference on Tomography and Integral Geometry when they were students.

Radon transforms are integral transforms that integrate functions over sets. The most popular example is the line transform that integrates a plane function over lines. One can generalize this transform in many ways. For example, one can integrate with respect to different weights on the line as we discuss in §2 (see (2.1)). One can also integrate with respect to different curves, such as circles, which we consider in §3, or surfaces in a manifold.

The basic results to investigate are injectivity (and inversion formulas) range theorems, and support theorems. One uses a support theorem to infer support restrictions on a function f from support restrictions on its Radon transform, Rf. The classical support theorem [22] says that if all line integrals of a compactly

---

2000 *Mathematics Subject Classification.* Primary: 44A12, Secondary: 58J40, 35S30.

Key words and phrases. Radon transform, support theorem, microlocal analysis.

The author thanks the American Mathematical Society for its support and help organizing this conference. Their staff, including Wayne Drady, made our job as organizers easy and pleasurable. The author was partially supported by NSF grant 9877155.

supported continuous function $f \in C_c(\mathbb{R}^2)$ are zero for lines lying outside a convex set, the f is zero outside of the convex set.

Radon transforms can be applied to solve partial differential equations. For example, the classical Radon transforms reduces the wave equation in $\mathbb{R}^n$ to the one-dimensional wave equation (with a parameter) (e.g., [24]). One can use this reduction and properties of the Radon transform to understand solutions of the wave equation.

The goal of this paper is to show some of the ways microlocal analysis can be used in integral geometry and to demonstrate how integral geometry is applied to other areas in mathematics. We use microlocal analysis to prove a support theorem for the line transform [9]. We will apply a uniqueness theorem for the Radon transform on spheres [2] (that was proven using microlocal analysis) to derive properties of solutions of the wave equation.

Many researchers have used microlocal analysis to solve problems in integral geometry and tomography. Victor Guillemin and Shlomo Sternberg started it all when they showed that the Radon transform is a special type of Fourier integral operator (FIO), a so-called conormal distribution (since the Lagrangian manifold for this FIO is a conormal bundle [18, 19]). The geometric properties of this Lagrangian manifold determine the microlocal properties of the Radon transform. For example, the Lagrangian for the hyperplane transform we discuss in §2 has simpler geometric properties than the Lagrangian for the circle transform in §3. So, the proofs for this transform are easier.

Guillemin and Sternberg also used microlocal analysis to prove general range characterizations for Radon transforms [20]

D.H. Phong and Elias Stein [32], Allan Greenleaf and Gunther Uhlmann and others have used microlocal analysis to understand mapping properties of singular and nonsingular Radon transforms. Greenleaf and Uhlmann developed the microlocal theory for the geodesic line transform on admissible complexes [15]. They have recently proven a connection between uniqueness for the Dirichlet-to-Neumann problem and support theorems for the two-plane transform. They have used this to give conditions when a density is determined by limited Dirichlet-to-Neumann data [16].

Other researchers [2, 9, 10, 14, 17, 36] have used microlocal analysis to prove support theorems for Radon transforms with real-analytic weights. Microlocal analysis allows us to infer smoothness of a function from smoothness of its Radon transform. The proper concept of non-smoothness is the analytic wavefront set, Definition 2.3. These authors use the microlocal properties of the specific Radon transform to tell what smoothness of f is reflected in smoothness of its Radon transform, e.g., at points where when $Rf = 0$. Then a powerful theorem of Kawai, Kashiwara, and Hörmander, Theorem 2.6, is used to infer that f is zero if f is smooth in certain directions at points on the boundary of its support (i.e., those directions are not in the analytic wavefront set of f). We'll see how this procedure works for the Radon transform on lines in §2.

Microlocal analysis is used in tomography to understand which singularities of a function are stably reconstructed from limited Radon data. A correspondence is worked out between singularities of f and singularities of Rf where R is the classical Radon transform on lines. Sobolev spaces provide an L^2 grading of smoothness, and Sobolev wavefront sets tell the smoothness of singularities in this grading. In

[35] one sees how this smoothness is reflected in the smoothness of singularities in the data. This was generalized to lambda CT and the attenuated Radon transform in [29].

David Finch [13], Ih-Ren Lan [30], and Alexander Katsevich [27] have taken the results in [15, 35] on three-dimensional X-ray CT farther by using microlocal analysis to analyze the specific artifacts that from backprojection inversion algorithms. Microlocal ideas have been applied to singularity detection in sonar by Louis and Quinto [31].

In Section 2, we will use microlocal ideas to prove a support theorem for the Radon transform on lines, and in Section 3 we will see a more complicated uniqueness theorem for a Radon transform on circles. This proof requires not only microlocal analysis but also the relationship between this Radon transform and harmonic polynomials. This result will be applied to learn qualitative properties of solutions to the wave equation in section 3.3.

2. A Support Theorem for the Radon transforms on lines in the plane

The Radon transform on lines models X-ray computed tomography, and it is one of the most fundamental transforms in integral geometry. Let Ξ be the set of lines in the plane and let μ be a nowhere zero real-analytic function on $\mathbb{R}^2 \times \Xi$. We define the Radon transform on lines

$$(2.1) \qquad R_\mu f(\ell) = \int_{x \in \ell} f(x)\mu(x,\ell) ds$$

for $f \in C_c(\mathbb{R}^2)$. The classical Radon transform on lines is (2.1) with $\mu = 1$. Inversion formulas, range theorems, and support theorems are known for this classical transform (e.g., [22, 23]).

Transforms with general weights, $\mu \neq 1$, come up in emission tomography (the attenuation of the body adds an exponential weight in (2.1)) and in other applications.

One theme of my research has been how properties of a Radon transform depend on the weight. In 1980, using microlocal analysis, I showed the relationship between the weight for a Radon transform (such as the line transform) and whether the transform was invertible by a differential operator. Then, I proved that nonzero rotation invariant transforms on hyperplanes were invertible and satisfied a support theorem [33]. This early result made me believe that Radon transforms on lines would be invertible as long as the weight μ were smooth and positive.

However, in 1984 Jan Boman constructed a very important example of a positive C^∞ weight μ and a function $f \in C_c^\infty(\mathbb{R}^2)$, $f \neq 0$, such that $R_\mu f = 0$ [8]. This demonstrated that the relationship between invertibility and the weight is extremely subtle. He proved invertibility if the weight is real-analytic and nowhere zero [7]. Together, we proved Theorem 2.1 and then support theorems for the line transform integrating over lines through a real-analytic curve (or lines tangent to a surface) in $\mathbb{R}^3$ [10].

THEOREM 2.1 ([9]). *Let $\mu(x,\ell)$ be a nowhere zero real-analytic function on $\mathbb{R}^2 \times \Xi$. Let $f \in \mathcal{E}'(\mathbb{R}^2)$. Let $\mathcal{A} \subset \Xi$ be an open connected set of lines. Assume some $\ell_0 \in \mathcal{A}$ is disjoint from $\operatorname{supp} f$ and assume $R_\mu f(\ell) = 0 \ \forall \ell \in \mathcal{A}$. Then, all lines in $\mathcal{A}$ are disjoint from $\operatorname{supp} f$.*

The point is that, if all line integrals of f are zero and if f is zero near some line in $\mathcal{A}$, then one can "analytically continue" f to be zero near all lines in $\mathcal{A}$. The theorem is true for a large class of Radon transforms including the hyperplane transform [**34**]. These transforms all satisfy the Bolker Assumption, a geometric assumption on the Lagrangian manifold associated to the Radon transform [**18, 19**]. For the classical transform Theorem 2.1 follows from Helgason's support theorem ([**24**], Lemma 2.11).

EXAMPLE 2.2. *What support restrictions do we get on $f \in C_c(\mathbb{R}^2)$ if $R_\mu f(\ell) = 0$ for all lines in the sets below*

(1) $\mathcal{A}$ *is the set of all lines ℓ of angle within one degree of the x-axis?*
(2) $\mathcal{B}$ *is the set of lines not meeting $W = [0,1]^2 \cup ([3,4] \times [0,1])$?*

The set of lines in (1), $\mathcal{A}$, is open and connected, and it fills $\mathbb{R}^2$. Since $f \in C_c(\mathbb{R}^2)$, some line in this set is disjoint from $\mathrm{supp}\, f$. Therefore Theorem 2.1 can be used to show $f = 0$ on all lines in $\mathcal{A}$, that is $f \equiv 0$ on $\mathbb{R}^2$. The set of lines, $\mathcal{B}$, in (2) has two connected components, one consisting of the lines outside the convex hull of W and the other consisting of all lines between the two squares in W. Since f has compact support, Theorem 2.1 implies that f is zero outside of the convex hull of W. However, the theorem says nothing about the part of $\mathrm{supp}\, f$ between the squares. If we know that f is zero near one line between the squares, then $f \equiv 0$ outside of W by Theorem 2.1. Can you construct a counterexample of a continuous function f that is supported in the convex hull of W and for which $R_1 f(\ell) = 0$ for all lines in $\mathcal{B}$? (HINT: Think about thin strips with value $+1$ and -1.)

PROOF OF THEOREM 2.1. There are three key ideas. First, we define microlocal analytic singularity: the analytic wavefront set. Second, we give a theorem that determines what microlocal singularities R_μ detects, and finally, we quote a strong theorem of Kawai, Kashiwara, and Hörmander [**25**] on analytic wavefront at the boundary of the support of a function.

Real-analytic wavefront sets allow us to characterize singularities in the real-analytic category. Let $\mathcal{F}f(\xi) = \int_{x \in \mathbb{R}^2} f(x) e^{-ix\cdot\xi} dx$ be the Fourier transform of $f \in L^1(\mathbb{R}^2)$. If

(2.2) $\quad\quad \forall \xi \in \mathbb{R}^2 \; |\mathcal{F}f(\xi)| \leq Ce^{-c|\xi|}$ for some $C > 0$ and $c > 0$,

then f is real-analytic since f can be extended to be holomorphic in a strip in $\mathbb{C}^2$, $\{x + iw \in \mathbb{C}^2 \mid |w| < c\}$:

(2.3) $\quad\quad f(x+iw) = \dfrac{1}{(2\pi)^2} \displaystyle\int_{\xi \in \mathbb{R}^2} \mathcal{F}f(\xi) e^{i(x+iw)\cdot\xi} d\xi.$

If $|w| < c$, the integral (2.3) converges because $\mathcal{F}f$ decreases more quickly by (2.2) than $e^{i(x+iw)\cdot\xi}$ increases. Since $e^{i(x+iw)\cdot\xi}$ is holomorphic in $z = x + iw$, (2.3) gives f as the restriction of a function that is holomorphic on this strip.

Formula (2.2) relates global exponential decrease of $\mathcal{F}f$ with global real-analyticity of f.

To localize this idea, we need a real-analytic cut off function. Of course, there is a problem because any real-analytic function that is zero on an open set is zero everywhere. So, we use a Gaussian as an almost-cutoff function–we ask whether $\mathcal{F}f$ times a Gaussian (that becomes more localized as a limit is taken) is exponentially decreasing. We don't just want to know where f is real-analytic, but we want to try to find directions in which f is real-analytic, locally. This leads to the Fourier-Bros-Iagolnitzer transform (2.4) ([**25**] Theorem 9.6.3).

DEFINITION 2.3. Let $f \in \mathcal{E}'(\mathbb{R}^2)$ and let $x_1 \in \mathbb{R}^2$ and $\xi_1 \in \mathbb{R}^2 \setminus 0$. Then, $(x_1, \xi_1) \notin \mathrm{WF}_A(f)$ **if and only if** there are neighborhoods U of x_1 and V of ξ_1 and $\exists C > 0$, $\exists c > 0$ such that for all $x \in U$, $\xi \in V$:

$$\left| \int_{y \in \mathbb{R}^2} \left(f(y) e^{-\lambda |x-y|^2} \right) e^{-iy \cdot (\lambda \xi)} dy \right| \leq C e^{-c\lambda}. \tag{2.4}$$

The function f is localized as Gaussian becomes more focused at y as $\lambda \to \infty$. The exponential decrease of the localized Fourier transform is given by the inequality in (2.4). And the direction (microlocalization) is represented by the requirement that the Fourier transform exponentially decreases in directions near the vector ξ_1: $\lambda \xi$ with ξ near ξ_1 as $\lambda \to \infty$.

EXAMPLE 2.4. *Calculate* $\mathrm{WF}_A(f)$, *if f is the characteristic function of the lower half plane in* $\mathbb{R}^2$. If you calculate the FBI transform of f, (2.4) you will see it is exponentially decreasing for all points not on the x-axis. For points on the x-axis, the only directions in which this FBI transform is not exponentially decreasing are the vertical directions. That is, $\mathrm{WF}_A(f)$ is the conormal bundle of the x-axis, the set of all covectors conormal to the x-axis. In a similar way, one can show that if f is the characteristic function of a disk in the plane, then the analytic wavefront set of f is the conormal bundle of the boundary circle, the set of nonzero vectors perpendicular to the boundary.

Note that we will identify covectors (in $T^*\mathbb{R}^n$) with vectors (in $T\mathbb{R}^n$) and conormal covectors to a set with normal vectors to that set.

Now, we look at how the Radon transform (2.1) detects analytic wavefront set. In general, Radon transforms detect singularities perpendicular to the sets being integrated over but not in other directions. This makes sense from elementary considerations, even though the proof requires microlocal analysis, in general. Let's look at a simple example. Let f be the characteristic function of the unit disk. According to Example 2.4, $\mathrm{WF}_A(f)$ consists of $\{(x, cx) \mid |x| = 1, c \in \mathbb{R}, c \neq 0\}$, the normals at the boundary. Furthermore, $R_1 f(p) = 2\sqrt{1-p^2}$ where p is the distance from the line to the origin. This function is real-analytic except when $p = \pm 1$, that is except for all lines tangent to the boundary of the unit disk. These lines have normals that are normal to the boundary, i.e., the normals to the lines at points of intersection with $|x| = 1$ correspond to the singularities of f that are detected by the Radon data. In fact, a more subtle version of this observation and its converse are true in general.

THEOREM 2.5 (Microlocal Regularity Theorem, [**9**]). *Let $f \in \mathcal{E}'(\mathbb{R}^2)$. Let μ be a nowhere zero real-analytic function on $\mathbb{R}^2 \times \Xi$ and let R_μ be the associated Radon transform on lines. Let $\ell_1 \in \Xi$, and let ξ_1 be conormal to ℓ_1. If $R_\mu f$ is zero near ℓ_1, then $(x_1, \xi_1) \notin \mathrm{WF}_A(f)$ $\forall x_1 \in \ell_1$.*

Morally, this is related to the projection slice theorem (the relation between R_1 and $\mathcal{F}$). Theorem 2.5 is true because R_μ is an elliptic Fourier integral operator associated to a specific conormal bundle, and R_μ detects singularities conormal to the line being integrated over but not other singularities. So, if $R_\mu f$ is smooth near a line ℓ_1, then f is smooth at all points on ℓ_1, at least in the directions conormal to ℓ_1.

An important theorem of Kawai, Kashiwara and Hörmander gives precise information about analytic wavefront set at bd supp f.

THEOREM 2.6 (Microlocal Holmgren Theorem, [**25**] Theorem 8.5.6). *Let $f \in \mathcal{D}'(\mathbb{R}^2)$. Let ℓ_1 be a line that meets $\operatorname{supp} f$ at a point $x_1 \in \ell_1 \cap (\operatorname{bd} \operatorname{supp} f)$ and let ξ_1 be conormal to ℓ_1. Assume $\operatorname{supp} f$ lies on one side of ℓ_1 locally near x_1 (that is: there is a connected neighborhood U of x_1 such that ℓ_1 divides U into two disjoint open sets and one of them does not meet $\operatorname{supp} f$). Then, $(x_1, \xi_1) \in \operatorname{WF}_A(f)$.*

We know f cannot be real-analytic at x_1 since x_1 is a boundary point of $\operatorname{supp} f$. This theorem gives the stronger result that the conormal wavefront direction must be in $\operatorname{WF}_A(f)$.

Now, we prove Theorem 2.1 by contradiction. Assume there is a line $\ell_2 \in \mathcal{A}$ that meets $\operatorname{supp} f$. Then, because $\mathcal{A}$ is connected, there is a line $\ell_1 \in \mathcal{A}$ such that ℓ_1 touches $\operatorname{supp} f$ and $\operatorname{supp} f$ lies on one side of ℓ_1 locally near a point $x_1 \in \operatorname{bd} \operatorname{supp} f$. To see this, imagine moving the line ℓ_0 on a path through $\mathcal{A}$ to the line ℓ_2; ℓ_0 does not meet $\operatorname{supp} f$, but ℓ_2 does. So, there must be a first line in the path, ℓ_1, that meets $\operatorname{supp} f$. Let ξ_1 be conormal to ℓ_1.

By Theorem 2.6 and because $\operatorname{supp} f$ lies on one side of ℓ_1 locally near x_1, $(x_1, \xi_1) \in \operatorname{WF}_A(f)$.

However, by Theorem 2.5 and because $R_\mu f = 0$ near ℓ_1, $(x_1, \xi_1) \notin \operatorname{WF}_A(f)$.

This contradiction proves the theorem. A loose summary of the proof is as follows. Theorem 2.6 shows that if a line, ℓ_1 in $\mathcal{A}$ is tangent to the boundary of $\operatorname{supp} f$, then f has wavefront set normal to ℓ_1. But Theorem 2.5 shows that f does not have any wavefront set in directions normal to lines in $\mathcal{A}$. Therefore, no line in $\mathcal{A}$ can be tangent to the boundary of $\operatorname{supp} f$. Since f is zero near ℓ_0 and $\mathcal{A}$ is connected, no line in $\mathcal{A}$ can meet $\operatorname{supp} f$. □

3. The Radon transform on circles in the plane

Now, we investigate a circular transform and use it to prove properties of zero sets of solutions to the wave equation. Let $a \in \mathbb{R}^2$ and let $r > 0$. Let $C(a, r)$ be the circle centered at a and with radius r. We define the circular Radon transform of $f \in C_c(\mathbb{R}^2)$

$$(3.1) \qquad Rf(a,r) = \int_{x \in C(a,r)} f(x) ds$$

This transform has a rich theory. If integrals of $f \in C(\mathbb{R}^2)$ are given over all circles, then an elementary inversion formula is given by $f(a) = \lim_{r \to 0^+} \frac{1}{2\pi r} Rf(a, r)$. Inversion is easy because this transform is geometrically overdetermined. The set of all circles in the plane has dimension 3 (2 dimensions for $a \in \mathbb{R}^2$ plus 1 for $r \in (0, \infty)$), but the plane has dimension two. Injectivity holds even when the radii are bounded away from zero [**39**]. So, it is natural to ask when will injectivity hold with restricted sets of centers and radii?

First, let's explore what happens when we fix the radius r_0 but let the centers vary in an open set in $\mathbb{R}^2$. If $f \in C_c(\mathbb{R}^2)$, then $Rf(a, r_0) = f * \mathcal{M}$ where $\mathcal{M}$ is the measure on the sphere $S(0, r_0)$. The Fourier transform $\mathcal{F}(f * \mathcal{M})$ is the product of $\mathcal{F}f$ and $\mathcal{F}\mathcal{M}$, and the function $\mathcal{F}\mathcal{M}$ is a Bessel function [**40**]. If $Rf(a, r_0) = 0$, $\forall a \in \mathbb{R}^2$, then this argument shows that $\mathcal{F}f$ must be zero almost everywhere because the Bessel function $\mathcal{F}\mathcal{M}$ is nonzero almost everywhere. Therefore, f is zero.

However, there are functions, f, not of compact support such that $Rf(a, r_0) \equiv 0$ [40]. On the other hand, one can show that data $Rf(a, r_0)$ for $a \in \mathbb{R}^2$ determines f, if f is assumed to be zero on a neighborhood of one disk of radius r_0 [26], (see [36] for a similar result on real-analytic manifolds).

In general, integrals over spheres of one radius do not determine f, but integrals over spheres of two "well chosen" radii determine $f \in C(\mathbb{R}^n)$ [12, 39, 40]. Berenstein and Zalcman generalized this to symmetric spaces of real rank one [6]. Properties including mean value theorems, injectivity, and inversion formulas were proven by Asgeirsson and John [26] for $\mathbb{R}^n$, and by Helgason [21] for homogeneous spaces. Local inversion formulas and support theorems are known for transforms with standard weights on disks in symmetric spaces [38], on disks of two radii in $\mathbb{R}^n$ [5]. Schneider uses spherical harmonics to prove injectivity of a large class of spherical transforms on the sphere, [37]. These results are, in general, proven using the harmonic analysis of the specific ambient spaces. The article [40] is a wonderful introduction to the spherical and related transforms, and [41] and [4] provide comprehensive bibliographies and summaries of results. Larry Zalcman's article in these proceedings updates these bibliographies and provides valuable perspective.

Several authors have used microlocal techniques in this area. Microlocal techniques have been used to prove specialized two-radius theorems for spherical transforms on $\mathbb{R}^n$ [42] and real-analytic manifolds [43] for transforms with arbitrary nowhere zero real-analytic weights. In [17], the authors prove a fairly general support theorem on a real-analytic manifold, M.[1]

Now, let's examine what happens in the plane when we restrict the set of centers but allow the radii to be arbitrary for the classical circular transform, (3.1).

DEFINITION 3.1. Let $S \subset \mathbb{R}^2$. S is called a *set of injectivity* for R **if and only if** $Rf(a, r) = 0$, $\forall a \in S$, $\forall r > 0$ implies $f = 0$ for all $f \in C_c(\mathbb{R}^2)$. If S is not a set of injectivity, S is called a set of noninjectivity.

EXAMPLE 3.2. *Are the following sets of injectivity?*

(1) $S = \mathbb{R}^2$
(2) $S = $ *one point?*
(3) $S = $ *a line?*
(4) $S = $ *two parallel lines?*

We have already seen that $S = \mathbb{R}^2$ is a set of injectivity (any continuous function f is determined by integrals over all circles). If S is one point, we would guess S is not a set of injectivity. In fact, radial functions about that point map surjectively onto the range of this one-point circular transform, so one point is not a set of injectivity. $S = $ a finite set is also not a set of injectivity. The case of a line is more interesting. The set S is one-dimensional, so we have data Rf over a two dimensional set of circles. Courant and Hilbert proved the following theorem.

---

[1]Let $S(a, r)$ be the geodesic sphere of radius r centered at $a \in M$. Assume $\mathcal{A} \subset M \times (0, I_M)$ is an open connected set. Here I_M is the injectivity radius of the manifold M (so, any sphere on M of radius less than I_M is diffeomorphic to a Euclidean sphere and its interior is diffeomorphic to an open disk in Euclidean space). See [28] for more information. If $Rf(a, r) = 0$ for all $(a, r) \in \mathcal{A}$ and if for some $(a_0, r_0) \in \mathcal{A}$, the sphere $S(a, r)$ is disjoint from supp f, then $f = 0$ on the union of spheres in $\mathcal{A}$.

THEOREM 3.3 ([11]). *If ℓ is a line, then ℓ is not a set of injectivity even for arbitrary continuous functions. The null space for R is the set of continuous functions that are odd about this line.*

We can use Theorem 3.3 to answer item 4. Let $S = \ell_1 \cup \ell_2$ be two parallel lines. Let f have compact support and assume Rf is zero for all centers on S. Then, f is odd about both parallel lines. By successively reflecting f in both lines, we see f must be zero because f has compact support.

3.1. Relation of R to Harmonic Polynomials. An important way to learn more about sets of injectivity for R is to develop their relationship to harmonic polynomials.

DEFINITION 3.4. For $f \in C_c(\mathbb{R}^2)$ and $k \in \{0, 1, 2, ...\}$ let

$$(3.2) \qquad Q_k[f] = f * |x|^{2k}, \qquad S[f] = \{a \in \mathbb{R}^2 \mid Rf(a,r) = 0, \ \forall r > 0\}.$$

Notice that $Q_k[f]$ is a polynomial of degree at most $2k$ since $|x|^{2k}$ is such a polynomial. The set $S[f]$ is the set of centers at which f has zero circular integrals: the set of noninjectivity for f.

THEOREM 3.5. *For $f \in C_c(\mathbb{R}^2)$,*
(1) $S[f] = \bigcap_{k=0}^{\infty} V(Q_k[f])$ *where $V(Q)$ is the zero set of the polynomial Q.*
(2) *Assume $f \not\equiv 0$. Then, $Q_k[f] \neq 0$ for some k. If k_0 is the smallest integer such that $Q_k[f] \neq 0$, then $Q_{k_0}[f] = f * |x|^{2k_0}$ is a harmonic polynomial.*

If $f \not\equiv 0$, then Theorem 3.5 demonstrates that $S[f]$ is a subset of the zero set of the nontrivial harmonic polynomial, $Q_{k_0}[f]$. This was first proved by Lin and Zobin.

PROOF OF THEOREM 3.5. By definition

$$a \in S[f] \quad \textbf{if and only if} \quad \int_{|x-a|=r} f(x)ds = 0 \ \text{ for all } \ r > 0.$$

If we integrate with respect to r^{2k} from 0 to ∞, using polar coordinates, then we get

$$(3.3) \qquad \int_0^\infty Rf(a,r)r^{2k} dr = \int_0^\infty \left(\int_{|x-a|=r} f(x)ds \right) |x-a|^{2k} dr$$
$$= f * |x|^{2k}(a) = Q_k[f](a)$$

Since f has compact support, $Rf(a,r) = 0$ for all r **if and only if** all these moments $Q_k[f](a) \equiv 0$. Thus, (1) holds.

If $Q_k[f] = f * |x|^{2k} \equiv 0$ for all k, then f is zero because Theorem 3.5, (1) shows $Rf(a,r) = 0$ for all (a,r). So, if $f \not\equiv 0$, then some $Q_k[f]$ is nonzero. Let k_0 be the smallest index such that $Q_k[f] \not\equiv 0$. To prove (2), we just note that

$$\Delta Q_{k_0}[f] = f * (\Delta |x|^{2k_0}) = \text{const} f * |x|^{2k_0 - 2} = 0$$

as k_0 is the smallest integer, k such that $Q_k[f] \neq 0$. Therefore $Q_{k_0-1}[f] = 0$ and $Q_{k_0}[f]$ is harmonic. □

Theorem 3.5 allows to get better information about sets of injectivity.

EXAMPLE 3.6. *Is an equilateral triangle, T, a set of injectivity?* We prove this by contradiction. Assume f is a nonzero function such that $T \subset S[f]$. Let $Q_{k_0}[f]$ be the nontrivial harmonic polynomial in Theorem 3.5 (2). By this theorem, T is a subset of the zero set of $Q_{k_0}[f]$. By the maximum principle, the zero set of a nontrivial harmonic polynomial contains no closed curve. This contradiction shows T is a set of injectivity. In fact, this proof shows that every closed curve a set of injectivity.

3.2. Characterization of sets of injectivity. The following sets will be the basis of our sets of noninjectivity.

DEFINITION 3.7. The *Coxeter Set* Σ_M ($M \in \mathbb{N}$) is the set of lines through the origin and M^{th} roots of unity. We define $\Sigma_0 = \emptyset$.

One can show any Coxeter system is not a set of injectivity. One just chooses a function compactly supported in one of the "Vs" of the system and reflects it oddly about each of the lines in the system. The resulting function is odd about each of the lines. Our next theorem shows that Coxeter systems are essentially the only sets of noninjectivity.

THEOREM 3.8 ([2]). *The condition*

(3.4)   the set S is not contained in any set of the form $k(\Sigma_M) \cup F$
for any $M = 0, 1, 2, \ldots$, where k is a rigid motion and F is a finite set

is necessary and sufficient for S to be a set of injectivity for the Radon transform over circles.

EXAMPLE 3.9. *Are the following sets of injectivity?*
  (1) *A small "curved" curve?*
  (2) *An "L" (right angle)?*
  (3) *A "W"?*

According to Theorem 3.8, any set of noninjectivity must contain only lines (and a finite set). So, a curved curve is a set of injectivity. An "L" is not a set of injectivity since it is a subset of some rigid motion of Σ_2. The segments in a "W" do not meet at a single point, so they do form a set of injectivity.

Theorem 3.8 demonstrates that lines (and finite points) are the basic building blocks of all sets of noninjectivity. The only sets of noninjectivity are the examples we saw, very symmetric unions of lines and finite sets. In fact, we show [2] that for each $f \in C_c(\mathbb{R}^2)$, $f \neq 0$,

(3.5)   if $S[f] \neq \emptyset$, then $S[f] = k(\Sigma_M) \cup F$ for some $M = 0, 1, 2, \ldots$

where k is a rigid motion of $\mathbb{R}^2$ and F is a finite set. This implies that if $S[f]$ is an infinite set, $S[f]$ is the zero set of a homogeneous harmonic polynomial (after translation) union a finite set (because these are the sets in (3.5)).

We outline the proof of Theorem 3.8. We let $f \neq 0$ and assume $S[f]$ is an infinite set. By Theorem 3.5, we know that $S[f]$ is a subset of the zero set of a harmonic polynomial. If this polynomial is not homogeneous, we prove that this subset, $S[f]$, must satisfy a specific geometric condition. Then, we use microlocal analysis to prove that any set satisfying this condition is a set of injectivity. This essentially shows $f = 0$ or (3.5) holds.

The microlocal analysis for the circular transform is more difficult than for the line transform. For example, the microlocal regularity theorem for this transform that corresponds to Theorem 2.5 is more complicated geometrically ([**2**], Lemma 4.3). So, we must use not only microlocal analysis but also the relation of R to harmonic polynomials to prove the uniqueness theorem.

For functions not of compact support, injectivity sets are harder to characterize. For example, concentric circles can be sets of noninjectivity.

Sets of noninjectivity have been characterized in $\mathbb{R}^n$ in special cases [**1, 3**], but the analogue of Theorem 3.8 has only been conjectured [**2**]. The problem is that the geometry and harmonic analysis are much more complicated in $\mathbb{R}^n$ than in the plane.

3.3. Applications to the wave equation.

Consider the wave equation IVP:

(3.6)
$$\Delta u = \frac{\partial^2 u}{\partial t^2}$$
$$u(x,0) = 0, \quad \frac{\partial}{\partial t} u(x,0) = f(x), \ f \in C_c(\mathbb{R}^2)$$

DEFINITION 3.10. The *stationary set* for the IVP for f, (3.6), is the set
$$N[f] = \{a \in \mathbb{R}^2 \mid u(a,t) = 0 \ \forall t > 0\}$$

Our goal is to characterize stationary sets when $f \in C_c(\mathbb{R}^2)$. By the Poisson-Kirchoff formula, the solution to (3.6) satisfies:

$$u(a,t) = \frac{1}{t} \int_{|a-\xi| \leq t} \frac{f(\xi) d\xi}{\sqrt{t^2 - |a-\xi|^2}}$$

(3.7)
$$= \frac{1}{t} \int_0^t \frac{Rf(a,r) dr}{\sqrt{t^2 - r^2}}$$

But, (3.7) is an invertible Abel equation of the first kind. Therefore,

$$Rf(a,r) = 0 \ \forall r > 0 \ \textbf{if and only if} \ u(a,t) = 0 \ \forall t > 0.$$

This shows that

(3.8)
$$S[f] = N[f]$$

and (3.5) and (3.8) gives us the theorem:

THEOREM 3.11 ([**2**]). *Let* $f \in C_c(\mathbb{R}^2)$ *and let* u *be the solution to the IVP (3.6). If* $N[f] \neq \emptyset$, *then for some* $M = 0, 1, 2, \ldots$,
$$N[f] = k\left(\Sigma_M\right) \cup F$$
where k *is a rigid motion and* F *is a finite set.*

The correspondence (3.8) is valid in $\mathbb{R}^n$ and for arbitrary continuous functions. Mark Agranovsky and I have used this to generalize Theorem 3.11 to the IVP (3.6) in $\mathbb{R}^n$ when the initial data, f, is a sum of point distributions [**3**] and we are working on the Dirichlet problem on some nice domains in $\mathbb{R}^2$.

References

[1] M.L. Agranovsky, C.A. Berenstein, and P. Kuchment, Approximation by spherical waves in L^p spaces, *J. Geom. Analysis*, **6**(1996), 365-383.

[2] M.L. Agranovsky and E.T. Quinto, Injectivity sets for the Radon transform over circles and complete systems of radial functions, *J. Functional Anal.* **139**(1996), 383-414.

[3] M.L. Agranovsky and E.T. Quinto, Geometry of Stationary Sets for the Wave Equation in $\mathbb{R}^n$. The Case of Finitely Supported Initial Data, *Duke Math. J.*, to appear.

[4] C.A. Berenstein, D-C. Chang, D. Pasucas, and L. Zalcman, Variations on the theorem of Morera *Contemporary Math.*, **137**(1992), 63-78.

[5] C.A. Berenstein, R. Guy, and A. Yger Inversion of the local Pompeiu transform, *J. Analyse Math.* **54**(1990), 259-287.

[6] C.A. Berenstein and L. Zalcman, Pompeiu's problem on symmetric spaces, *Comment. Math. Helvetici*, **55**(1980), 593-621.

[7] J. Boman, Uniqueness theorems for generalized Radon transforms, in *Proceedings of the conference on the Constructive Theory, of Functions*, Varma, Bulgaria, 1984.

[8] J. Boman, An Example of Non-uniqueness for a generalized Radon transform, *J. d'Analyse Mathématique* (1993), 395-401.

[9] J. Boman and E.T. Quinto, Support theorems for real analytic Radon transforms, *Duke Math. J.* **55**(1987), 943-948.

[10] J. Boman and E.T. Quinto, Support theorems for real analytic Radon transforms on line complexes in $\mathbb{R}^3$, *Trans. Amer. Math. Soc.* **335**(1993), 877-890.

[11] R. Courant and D. Hilbert, *Methods of Mathematical Physics, Volume II Partial Differential Equations*, Interscience, New York, 1962.

[12] J. Delsarte, and J.L. Lions, Moyennes généralisées, *Comment. Math. Helv.* **33**(1959), 59-69.

[13] D. Finch, talk at 2000 AMS/IMS/SIAM Radon transform and Tomography Conference.

[14] J. Globevnik, A support theorem for the X-ray transform, *J. Math. Anal. Appl.* **165**,(1992), 284-287.

[15] A. Greenleaf and G. Uhlmann, Non-local inversion formulas for the X-ray transform, *Duke Math. J.* **58**(1989), 205-240.

[16] A. Greenleaf and G. Uhlmann, Local uniqueness for the Dirichlet-to-Neumann map via the two-plane transform, *Duke Math. J.*, to appear.

[17] E. Grinberg and E.T. Quinto, Morera theorems for complex manifolds, *J. Functional Analysis*, to appear.

[18] V. Guillemin, Some remarks on integral geometry, Technical Report, MIT, 1975.

[19] V. Guillemin and S. Sternberg, *Geometric Asymptotics*, American Mathematical Society, Providence, RI, 1977.

[20] V. Guillemin and S. Sternberg, Some problems in integral geometry and some related problems in micro-local analysis, *Amer. J. Math.* **101**,(1979), 915-955.

[21] S. Helgason, Differential operators on homogeneous spaces *Acta Math.*, **102**(1959), 239-299¿

[22] S. Helgason, The Radon transform on Euclidean spaces, compact two-point homogeneous spaces and Grassman manifolds, *Acta Math.* **113**(1965), 153-180.

[23] S. Helgason, *Groups and Geometric Analysis*, Academic Press, New York, 1984.

[24] S. Helgason, *The Radon Transform, Second Edition*, Birkhäuser Boston, 1999.

[25] L. Hörmander, *The Analysis of Linear Partial Differential Operators I*, Springer Verlag, New York, 1983.

[26] F. John, *Plane Waves and Spherical Means*, Interscience, 1955.

[27] A. Katsevich, Cone Beam Local Tomography, *SIAM J. Appl. Math.* **59**(1999), 2224-2246.

[28] S. Kobayashi and K. Nomizu, *Foundations of Differential Geometry, Vol. I*, Interscience, New York 1963.

[29] P. Kuchment, K. Lancaster, and L. Mogilevskaya, On the structure of local tomography, *Inverse Problems*, **11**(1995), 571-589.

[30] Ih-Ren Lan, *On an Operator Associated to a Restricted X-ray Transform*, Ph.D. Thesis, Oregon State University, 1999.

[31] A.K. Louis and E.T. Quinto, Local Tomographic Methods in SONAR, joint with Alfred Louis, in "Surveys on Solution Methods for Inverse Problems" Eds.: D. Colton, H. Engl, A. Louis, J. McLaughlin, W. Rundell, Springer Vienna/New York 2000

[32] D.H. Phong and E.M. Stein, Singular integrals related to the Radon transform and boundary value problems, *Proc. National Academy Science (USA)* **80**,(1983), 7694-7696.
[33] E.T. Quinto, The invertibility of rotation invariant Radon transforms, *J. Math. Anal. Appl.* **91**(1983), 510-522. Erratum, *J. Math. Anal. Appl.* **94**(1983), 602-603.
[34] E.T. Quinto, Radon transforms satisfying the Bolker assumption, pp. 263-270 in *Proceedings of conference "Seventy-five Years of Radon Transforms"*, International Press Co. Ltd., Hong Kong, 1994
[35] E.T. Quinto, Singularities of the X-ray transform and limited data tomography in $\mathbb{R}^2$ and $\mathbb{R}^3$, *SIAM J. Math. Anal.* **24**(1993), 1215-1225.
[36] E.T. Quinto, Pompeiu transforms on geodesic spheres in real analytic manifolds, *Israel J. Math.* **84**(1993), 353-363.
[37] R. Schneider, Functions on a sphere with vanishing integrals over certain subspheres, *J. Math. Anal. Appl.*, **26**(1969), 381–384.
[38] M. Shahshahani, and A. Sitaram The Pompeiu problem in exterior domains in symmetric spaces *Contemporary Math.*, **63**(1987), 267-277.
[39] L. Zalcman, Analyticity and the Pompeiu problem, *Arch. Rat. Mech. Anal.*, **47**(1972), 237–254.
[40] L. Zalcman, Offbeat Integral Geometry, *Amer. Math. Monthly*, **87**(1980), 161-175.
[41] L. Zalcman, A bibliographic survey of the Pompeiu problem, in *Approximation of Solutions of Partial Differential Equations,* editors B. Fuglede, M. Goldstein, W. Haussmann, W. K. Hayman, and L. Rogge, Vol. 365, Series C: Mathematics and Physical Sciences, NATO ASI Series, Kluwer Academic, Boston, 1992, 185-194.
[42] Y. Zhou, Two Radius Support Theorem for the Sphere Transform, *J. Math. Anal. Appl.*, to appear.
[43] Y. Zhou and E.T. Quinto, Two-radius Support Theorems for Spherical Radon transforms on Manifolds, *Contemporary Math.* **251**(2000), 501-508.

DEPARTMENT OF MATHEMATICS, TUFTS UNIVERSITY, MEDFORD, MA 02155
E-mail address: equinto@math.tufts.edu, http://www.tufts.edu/~equinto

Contemporary Mathematics
Volume **278**, 2001

Supplementary Bibliography to "A Bibliographic Survey of the Pompeiu Problem"

Lawrence Zalcman

This supplementary bibliography updates and completes our previous bibliographic survey of the Pompeiu problem [**Z2**] through the year 2000. For the reader's convenience, we also include a very abbreviated account of the subject to which the bibliography is devoted (further details and references are in [**Z2**]), some remarks on recent developments, and complete entries for those items listed in [**Z2**] which appeared in print only after the publication of that bibliography.

1. The Pompeiu Problem

A compact set $K \subset \mathbb{R}^2$ has the *Pompeiu property* (PP) if, whenever $f \in C(\mathbb{R}^2)$ and

$$(1.1) \qquad \iint_{\sigma(K)} f(x,y) dx dy = 0$$

for all rigid motions σ of $\mathbb{R}^2$, it follows that $f \equiv 0$. The set of regions possessing (PP) includes (but is by no means limited to) all Jordan regions with non-real-analytic Lipschitz boundary, as well as noncircular ellipses and cigars (convex regions at least twice as long as they are wide). On the other hand, discs fail to possess (PP); indeed, when D is a disc, (1.1) holds for an appropriate exponential. The *Pompeiu Problem* asks whether discs are the only Jordan regions which fail to have (PP).

Sets having (PP) can be characterized in terms of the Fourier transforms of their characteristic functions. Specifically, $K \subset \mathbb{R}^2$ has (PP) if and only if $\hat{\chi}_K$ does not vanish identically on any of the analytic varieties

$$C_\alpha = \{(\zeta_1, \zeta_2) \in \mathbb{C}^2 : \zeta_1^2 + \zeta_2^2 = \alpha\}, \quad \alpha \neq 0.$$

Here the Fourier transform

$$\hat{\chi}_K(\zeta_1, \zeta_2) = \iint_K e^{-i(x\zeta_1 + y\zeta_2)} dx dy$$

is an entire function on $\mathbb{C}^2$.

It can be shown that if the smoothly bounded Jordan region D fails to have (PP), so that $\hat{\chi}_D$ vanishes identically on C_α for some $\alpha \neq 0$, then $\alpha > 0$ and the

---

1991 *Mathematics Subject Classification.* Primary 53C65; Secondary 35R35.

© 2001 American Mathematical Society

overdetermined boundary problem

$$\Delta u + \alpha u = 0 \quad \text{on } D$$

(1.2)
$$u = c, \quad \frac{\partial u}{\partial n} = 0 \quad \text{on } D$$

has a nontrivial solution. The assertion that if (1.2) has a nontrivial solution for some $\alpha > 0$, then D is disc is known as the Schiffer Conjecture. It is equivalent to the Pompeiu Problem for domains with Lipschitz boundary.

For a more discursive account of this material and complete references, see [**Z2**].

2. Recent Developments

[**Z2**] was written with the express purpose of bringing the Pompeiu Problem and related questions to the attention of workers in PDE. To judge by the number of papers which have appeared in recent years dealing with this aspect of the subject, it seems to have succeeded in this aim. These include [**A**], [**AS**], [**CafKS**], [**Cd1**], [**Cd2**], [**Cn1**], [**Cn2**], [**CnCH**], [**CnH**], [**CH**], [**Da1**], [**Da2**], [**Da3**], [**E1**], [**E2**], [**E3**], [**J**], [**Ka**], [**N**], [**Rm1**], [**Rm2**], [**Se**], [**WCG**], and [**WG**].

Most of the work on the Pompeiu Problem over the past decade has followed the lines sketched in [**Z2**]; a detailed account would be far beyond the scope of this paper. We do wish, however, to call attention to the work of Garofalo and Segala ([**Go**], [**GoS**]; cf. [32], [33], and [34] in [**Z2**]), Ebenfelt ([**E1**], [**E2**], [**E3**]), and Dalmasso [**Da3**], which has greatly enriched the collection of available examples of Jordan domains with real-analytic boundary having the Pompeiu property, as well as to the extraordinary contribution of Valery Volchkov, whose wide-ranging studies of the Pompeiu Problem and related matters ([**V1**]–[**V34**]) account for almost a third of the papers listed below.

3. Filling the Gaps

The following short list provides full information on items which were incomplete in the original bibliography.

[1] M. Agranovsky, *On the stability of the spectrum in the Pompeiu problem*, J. Math. Anal. Appl. **178** (1993), 269–279.

[3] E. Badertscher, *The Pompeiu problem on locally symmetric spaces*, J. Analyse Math. **57** (1991), 250–281.

[8] C.A. Berenstein, D.-C. Chang, D. Pascuas, and L. Zalcman, *Variations on the theorem of Morera*, Contemp. Math. **137** (1992), 63–78.

[34] N. Garofalo and F. Segala, *Another step toward the solution of the Pompeiu problem in the plane*, Comm. Partial Differential Equations **18** (1993), 491–503.

4. The Supplementary Bibliography

In preparing the bibliographic supplement, I have aimed at including all papers dealing with, or related to, the Pompeiu Problem which have come to my attention and were not listed in [**Z2**]. Most of the items listed below appeared after the publication of [**Z2**]. Others were omitted on account of what now appears to have been an overly restrictive criterion of selectivity or by simple oversight. I welcome this opportunity to set matters right by listing them now.

References

[A] M.L. Agranovsky, *Perturbations of the spectrum in the Pompeiu problem*. Optimizatsiya, No. 48 (65) (1990), 47–57.

[AS] M.L. Agranovsky and A.M. Semenov, *Deformations of balls in Schiffer's conjecture for Riemannian symmetric spaces*, Israel J. Math. **95** (1996), 43–59.

[Ai] L. Aizenberg, *Variations on the theme of the Morera theorem and Pompeiu's problem*, Dokl. Akad. Nauk **337** (1994), 709–712. English transl.: Russian Acad. Sci. Dokl. Math. **50** (1995), 152–156.

[AiBW] L.A. Aizenberg, C.A. Berenstein, and L. Wertheim, *Mean-value characterization of pluriharmonic and separately harmonic functions*, Pacific J. Math. **175** (1996), 295–306.

[AiL1] L. Aizenberg and E. Liflyand, *Mean-value characterization of holomorphic and pluriharmonic functions*, Complex Variables Theory Appl. **32** (1997), 131–146.

[AiL2] L. Aizenberg and E. Liflyand, *Mean-value characterization of holomorphic and pluriharmonic functions II*, Complex Variables Theory Appl. **39** (1999), 381–390.

[Ar1] D.H Armitage, *The Pompeiu problem for spherical polygons*, Proc. Royal Irish Acad. **96**A (1996), 25–32.

[Ar2] D.H Armitage, *The Pompeiu property on the sphere*, New Zealand J. Math. **29** (2000), 11–18.

[BW] Y. Benyamini and Y. Weit, *Harmonic analysis of spherical functions on SU(1,1)*, Ann. Inst. Fourier (Grenoble) **42** (1992), 671–694.

[Be1] C.A Berenstein, *On the converse to Pompeiu's problem*, Notas e Comunicações de Matematica, (Universidade Federal de Pernambuco) **73** (1976).

[Be2] C.A. Berenstein, *El problema de Pompeiu*, Atas do Novo Coloquio Brasileiro de Matematica (Pocos de Caldas, 1973) vol. 1, 1977, 31–37.

[Be3] C.A. Berenstein, *The Pompeiu problem, what's new?* in: R. Deville et al. (eds.), Complex Analysis, Harmonic Analysis and Applications, Longman, Harlow, 1996, 1–11.

[BeKh] C. Berenstein and D. Khavinson, *Do solid tori have the Pompeiu property?*, Exposition. Math. **15** (1997), 87–93.

[BeP] C. Berenstein and D. Pascuas, *Morera and mean-value type theorems in the hyperbolic disk*, Israel J. Math. **86** (1994), 61–106.

[BeS] C.A. Berenstein and D. Struppa, *Complex analysis and convolution equations*, Current Problems in Mathematics, Fundamental Directions, vol. 54, 1989, 5–111 (Russian).

[BrW] A. Braun and Y. Weit, *On invariant subspaces of continuous vector valued functions*, J. Analyse Math. **41** (1982), 259–271.

[CafKS] L.A. Caffarelli, L. Karp and H. Shahgholian, *Regularity of a free boundary with application to the Pompeiu problem*, Ann. of Math. (2) **151** (2000), 269–292.

[CasP] C. Cascante and D. Pascuas, *Holomorphy tests based on Cauchy's integral formula*, Pacific J. Math. **171** (1995), 89–116.

[Cd1] M. Chamberland, *The Pompeiu problem and Schiffer's conjecture*, dissertation, University of Waterloo, 1994.

[Cd2] M. Chamberland, *Mean value integral equations and the Helmholtz equation*, Results Math. **30** (1996), 39–44.

[Cn1] T. Chatelain, *Quelques remarques sur des problèmes d'optimisation de valeurs propres*, Ulmer Seminare über Functionalanalysis und Differentialgleichungen **1** (1996), 112–121.

[Cn2] T. Chatelain, *A new approach to two overdetermined eigenvalue problems of Pompeiu type*, in: J. Blum et al. (eds.), Elasticité, Viscoélasticité et Contrôle Optimal (Lyon, 1995), ESAIM Proc., vol. 2, 1997, 235–242 (electronic).

[CnCH] T. Chatelain, M. Choulli and A. Henrot, *Some new ideas for a Schiffer's conjecture*, in: Modelling and Optimization of Distributed Parameter Systems (Warsaw, 1995), Chapman & Hall, 1996, 90–97.

[CnH] T. Chatelain and A. Henrot, *Some results about Schiffer's conjectures*, Inverse Problems **15** (1999), 647–658.

[CH] M. Choulli and A. Henrot, *Use of the domain derivative to prove symmetry results in partial differential equations*, Math. Nachr. **192** (1998), 91–103 .

[Da1] R. Dalmasso, *A note on the Schiffer conjecture*, Hokkaido Math. J. **28** (1999), 373–383.

[Da2] R. Dalmasso, *Le problème de Pompeiu*, Sémin. Théor. Spectr. Géom. **17**, Univ. Grenoble I, Saint-Martin-d'Hères (1999), 69–79.

[Da3] R. Dalmasso, *A new result on the Pompeiu problem*, Trans. Amer. Math. Soc. **352** (2000), 2723–2736.

[DeD] R.F. DeMar and P.J. Davis, *A complex Pompeiu problem*, Duke Math. J. **33** (1966), 91–101.

[E1] P. Ebenfelt, *Some results on the Pompeiu problem*, Ann. Acad. Sci. Fenn. Ser. A I Math. **18** (1993), 323–341.

[E2] P. Ebenfelt, *Singularities of the solution to a certain Cauchy problem and an application to the Pompeiu problem*, Duke Math. J. **71** (1993), 119–142.

[E3] P. Ebenfelt, *Propagation of singularities from singular and infinite points in certain complex-analytic Cauchy problems and an application to the Pompeiu problem*, Duke Math. J. **73** (1994), 561–582.

[EH1] M. El Harchaoui, *Inversion de la transformation de Pompéiu dans le disque hyperbolique (Cas de deux disques)*, Publ. Mat. **37** (1993), 133–164.

[EH2] M. El Harchaoui, *Inversion de la transformation de Pompéiu locale dans les espaces hyperboliques réel et complexe (Cas de deux boules)*, J. Analyse Math. **67** (1995), 1–37.

[Go] N. Garofalo, *A new result on the Pompeiu problem*, Rend. Sem. Mat. Univ. Politec. Torino, Special Issue (1989), 25–38.

[GoS] N. Garofalo and F. Segala, *Univalent functions and the Pompeiu problem*, Trans. Amer. Math. Soc. **346** (1994), 137–146.

[Gy] R. Gay, *Inversion de la transformation de Pompéiu locale*, Journées "Équations aux Dérivées Partielles" (Saint Jean de Monts, 1989), Exp. No. XVI, Ecole Polytech. Palaiseau, 1989, 14 pp.

[J] G. Johnsson, *The Cauchy problem in $\mathbb{C}^N$ for linear second order partial differential equations with data on a quadric surface*, Trans. Amer. Math. Soc. **344** (1994), 1–48.

[Ka] B. Kawohl, *Remarks on some old and current eigenvalue problems*, in: A. Alvino, E. Fabes, and G. Talenti (eds.), Partial Differential Equations of Elliptic Type (Cortona, 1992), Sympos. Math. vol. 35, Cambridge Univ. Press, 1994, 165–183.

[Kl] J.J. Klinkhammer, *On the Pompeiu problem in $\mathbf{R}^n$*, in: Proceedings of the Ashkelon Workshop on Complex Function Theory (May, 1996), IMCP vol. 11, 1997, 153–165.

[Ko1] T. Kobayashi, *Asymptotic behaviour of the null variety for a convex domain in a non-positively curved space form*, J. Fac. Sci. Univ. Tokyo Sect. IA Math. **36** (1989), 389–478.

[Ko2] T. Kobayashi, *Perturbation of domains in the Pompeiu problem*, Comm. Anal. Geom. **1** (1993), 515–541.

[Ko3] T. Kobayashi, *Convex domains and the Fourier transform on spaces of constant curvature*, UTMS **93-3**, University of Tokyo, 1993.

[Ko4] T. Kobayashi, *Bounded domains and the zero sets of Fourier transforms*, in: S. Gindikin and P. Michor (eds.), 75 Years of Radon Transform (Vienna, 1992), International Press, Cambridge MA, 1994, 223–239.

[L] H.T. Laquer, *The Pompeiu problem*, Amer. Math. Monthly **100** (1993), 461–467.

[Mal] S.A. Malyugin, *Functions with a constant integral on congruent cubes*, Mat. Zametki **24** (1978), 339–341. English transl.: Math. Notes **24** (1978), 682–683.

[Mas] P.A. Masharov, *A new Morera-type theorem on the disk*, Visnik Kharkiv. Univ. Ser. Mat. Prikl. Mat. Mekh. **475**:49 (2000), 126–132.

[Mo] R. Molzon, *Symmetry and overdetermined boundary value problems*, Forum Math. **3** (1991), 143–156.

[N] T. Nagasawa, *The Pompeiu and related problems and boundary behavior*, in: Variational Problems and Related Topics, Sûrikaisekikenkyûsho Kôkyûroku **1117** (1999), 28–35.

[O1] O.A. Ochakovska, *The support problem for functions with vanishing ball means*, Pratsi IM NAN Ukraine **31** (2000), 352–355.

[O2] O.A. Ochakovska, *A new uniqueness theorem for functions with vanishing integrals over balls of a fixed radius*, Pratsi IPMM NAN Ukraine, to appear.

[P] V.V. Proizvolov, *Integrals that are constant on congruent domains*, Mat. Zametki **21** (1977), 183–186; **24** (1978), 300. English transl.: Math. Notes **21** (1977), 103–105; **24** (1978), 661.

[Q1] E.T. Quinto, *Pompeiu transforms on geodesic spheres in real analytic manifolds*, Israel J. Math. **84** (1993), 353–363.

[Q2]   E.T. Quinto, *Radon transforms on curves in the plane*, in: E.T. Quinto, M. Cheney, and P. Kuchment (eds.), Tomography, Impedance Imaging, and Integral Geometry, (South Hadley, MA, 1993), Lectures in Appl. Math. vol. 30, 1994, 231–244.

[Rm1] A.G. Ramm, *The Pompeiu problem*, Appl. Anal. **64** (1997), 19–26.

[Rm2] A.G. Ramm, *Necessary and sufficient condition for a domain, which fails to have the Pompeiu property, to be a ball*, J. Inverse Ill-Posed Probl. **6** (1998), 165–171.

[Rn]   I.K Rana, *A Pompeiu problem for locally-finite measures on R^n*, Rend. Mat. (7) **3** (1983), 291–300.

[RwS] R. Rawat and A. Sitaram, *The injectivity of the Pompeiu transform and L^p-analogues of the Wiener-Tauberian theorem*, Israel J. Math. **91** (1995), 307–316.

[Se]   F. Segala, *Stability of the convex Pompeiu sets*, Ann. Mat. Pura. Appl. (4) **175** (1998), 295–306.

[Sz]   G. Szabo, *On functions having the same integral on congruent semidisks*, Ann. Univ. Sci. Budapest Sect. Comput. **3** (1982), 3–9.

[Ta]   S. Thangavelu, *Mean periodic functions on phase space and the Pompeiu problem with a twist*, Ann. Inst. Fourier (Grenoble) **45** (1995), 1007–1035.

[To]   K.W Thompson, *Additional results of Zalcman's Pompeiu problem*, Aequationes Math. **44** (1992), 42–47.

[ToSc] K.W. Thompson and T. Schonbek, *A problem of the Pompeiu type*, Amer. Math. Monthly **87** (1980), 32–36.

[Tr]   K. Trimeche, *Transmutation operators and mean-periodic functions associated with differential operators*, Math. Rep. **4** (1988), i–xiv and 1–282.

[U]   D.C. Ullrich, *More on the Pompeiu problem*, Amer. Math. Monthly **101** (1994), 165–168.

[V1]   V.V. Volchkov, *Functions with zero integrals with respect to certain sets*, Dokl. Akad. Nauk Ukrain. SSR Ser. A 1990, no. 8, 9–11.

[V2]   V.V. Volchkov, *Functions with zero integrals over cubes*, Ukraïn. Mat. Zh. **43** (1991), 859–863. English transl.: Ukrainian Math. J. **43** (1991), 806–810.

[V3]   V.V. Volchkov, *Mean-value theorems for some differential equations*, Dokl. Akad. Nauk Ukrain. SSR 1991, no. 6, 8–11.

[V4]   V.V. Volchkov, *Theorems on spherical averages for some differential equations*, Dokl. Akad. Nauk Ukraïni 1992, no. 5, 8–11.

[V5]   V.V. Volchkov, *On a problem of Zalcman and its generalizations*, Mat. Zametki **53**:2 (1993), 30–36. English transl.: Math. Notes **53** (1993), 134–138.

[V6]   V.V. Volchkov, *Morera-type theorems in domains with the weak cone condition*, Izv. Vyssh. Uchebn. Zaved. Mat. 1993, no. 10, 15–20.

[V7]   V.V. Volchkov, *On the Pompeiu problem and its generalizations*, Ukraïn. Mat. Zh. **45** (1993), 1444–1448. English transl.: Ukrainian Math. J. **45** (1993), 1623–1628.

[V8]   V.V. Volchkov, *New theorems on the mean for solutions of the Helmholtz equation*, Mat. Sb. **184**:7 (1993), 71–78. English transl.: Russian Acad. Sci. Sb. Math. **79** (1994), 281–286.

[V9]   V.V. Volchkov, *Problems of Pompeiu type on manifolds*, Dokl. Akad. Nauk Ukraïni 1993, no. 11, 9–13.

[V10] V.V. Volchkov, *Morera type theorems on the unit disc*, Anal. Math. **20** (1994), 49–63.

[V11] V.V. Volchkov, *New mean value theorems for polyanalytic functions*, Mat. Zametki **56**:3 (1994), 20–28. English transl.: Math. Notes **56** (1994), 889–895.

[V12] V.V. Volchkov, *Two-radius theorems for bounded domains in euclidean spaces*, Differentsial'nye Uravneniya **30** (1994), 1719–1724. English transl.: Differential Equations **30** (1994), 1587–1592.

[V13] V.V. Volchkov, *Approximation of functions on bounded domains in R^n by linear combinations of shifts*, Dokl. Akad. Nauk **334** (1993), 560–561. English transl.: Russian Acad. Sci. Dokl. Math. **49** (1994), 160–162.

[V14] V.V. Volchkov, *New two-radii theorems in the theory of harmonic functions*, Izv. Ross. Akad. Nauk Ser. Mat. **58**:1 (1994), 182–194. English transl.: Russian Acad. Sci. Izv. Math. **44** (1995), 181–192.

[V15] V.V. Volchkov, *Mean value theorems for a class of polynomials*, Sibirsk. Mat. Zh. **35** (1994), 737–745. English transl.: Siberian Math. J. **35** (1994), 656–663.

[V16] V.V. Volchkov, *A definitive version of the local two-radii theorem*, Mat. Sb. **186**:6 (1995), 15–34. English transl.: Sbornik: Math. **186** (1995), 783–802.

[V17] V.V. Volchkov, *Two-radii theorems on constant curvature spaces*, Dokl. Akad. Nauk **347** (1996), 300–302. English transl.: Doklady: Math. **53** (1996), 199–201.

[V18] V.V. Volchkov, *The final version of the mean value theorem for harmonic functions*, Mat. Zametki **59** (1996), 351–358. English transl.: Math. Notes **59** (1996), 247–252.

[V19] V.V. Volchkov, *Extremal cases of the Pompeiu problem*, Mat. Zametki **59** (1996), 671–680. English transl.: Math. Notes **59** (1996), 482–489.

[V20] V.V. Volchkov, *An extremum problem related to Morera's theorem*, Mat. Zametki **60** (1996), 804–809 . English transl.: Math. Notes **60** (1996), 606–610.

[V21] V.V. Volchkov, *Theorems on injectivity sets for Radon transforms over spheres*, Dokl. Akad. Nauk **354** (1997), 298–300. English transl.: Doklady: Math. **55** (1997), 359–361.

[V22] V.V. Volchkov, *Solution of the support problem for several function classes*, Mat. Sb. **188**:9 (1997), 13–30. English transl.: Sbornik: Math. **188** (1997), 1279–1294.

[V23] V.V. Volchkov, *Uniqueness theorems for some classes of functions with zero spherical means*, Mat. Zametki **62** (1997), 59–65. English transl.: Math. Notes **62** (1997), 50–55.

[V24] V.V. Volchkov, *Extremal problems on Pompeiu sets*, Mat. Sb. **189**:7 (1998), 3–22. English transl.: Sbornik: Math. **189**(1998), 955–976.

[V25] V.V. Volchkov, *Spherical means on Euclidean spaces*, Ukraïn. Mat. Zh. **50** (1998), 1310–1315. English transl.: Ukrainian Math. J. **50** (1998), 1496–1503.

[V26] V.V. Volchkov, *Injectivity sets of the Pompeiu transform*, Mat. Sb. **190**:11 (1999), 51–66. English transl.: Sbornik: Math. **190** (1999), 1607–1622.

[V27] V.V. Volchkov, *Injectivity sets for the Radon transform over a sphere*, Izv. Ross. Akad. Nauk Ser. Mat. **63**:3 (1999), 63–76. English transl.: Izvestiya: Math. **63** (1999), 481–493.

[V28] V.V. Volchkov, *About functions with zero integrals over parallelepipeds*, Dokl. Akad. Nauk **369** (1999), 444–445. English transl.: Doklady: Math. **60** (1999), 375-376.

[V29] V.V. Volchkov, *A new version of the two-radii theorem*, Heuristics and Didactics of Exact Sciences **10** (1999), 78.

[V30] V.V. Volchkov, *The Pompeiu Transform*, Donetsk State Univ., 1999.

[V31] V.V. Volchkov, *Extremal problems on Pompeiu sets. II*, Mat. Sb. **191**:5 (2000), 3–16. English transl.: Sbornik: Math. **191** (2000), 619-632.

[V32] V.V. Volchkov, *On polyhedra with the local Pompeiu property*, Dokl. Akad. Nauk **373** (2000), 448–450. English transl.: Doklady: Math. **62** (2000), 69-71.

[V33] V.V. Volchkov, *Functions with zero integrals over ellipsoids*, Dokl. Akad. Nauk **376** (2001), in press. English transl.: Doklady: Math.

[V34] V.V. Volchkov, *A definitive version of the local two-radii theorem on hyperbolic spaces*, Izv. Ross. Akad. Nauk Ser. Mat., to appear. English transl.: Izvestiya: Math.

[Vit1] Vit.V. Volchkov, *Theorems on spherical means in complex hyperbolic spaces*, Dopov. Nats. Akad. Nauk Ukr. Mat. Prirodozn. Tekh. Nauki 2000, no.4, 7–10.

[Vit2] Vit.V. Volchkov, *On functions with zero spherical means in complex hyperbolic spaces*, Mat. Zametki **68** (2000), 504–512. English transl.: Math. Notes **68** (2000).

[WCG] N.B. Willms, M. Chamberland and G.M.L. Gladwell, *A duality theorem for an overdetermined eigenvalue problem*, Z. Angew. Math. Phys. **46** (1995), 623–629.

[WG] N.B. Willms and G.M.L. Gladwell, *Saddle points and overdetermined problems for the Helmholtz equation*, Z. Angew Math. Phys. **45** (1994), 1–26.

[Z1] L. Zalcman, *Determining sets for functions and measures*, Real Analysis Exchange **11** (1985–86), 40–55.

[Z2] L. Zalcman, *A bibliographic survey of the Pompeiu problem*, in: B. Fuglede et al. (eds.), Approximation by Solutions of Partial Differential Equations, Kluwer Academic Publishers, Dordrecht, 1992, 185–194.

[ZaTri] V.P. Zastavnyi and R.M. Trigub, *Functions with a zero integral over sets that are congruent to a given set*, in: S.M. Nikolskii (ed.), Theory of Functions and Approximations, Part 2, Saratov Gos. Univ., Saratov, 1988, 69–71.

[Ze] D. Zeilberger, *Pompeiu's property on discrete space*, Proc. Nat. Acad. Sci. U.S.A. **75** (1978), 3555–3556.

DEPARTMENT OF MATHEMATICS, BAR-ILAN UNIVERSITY, RAMAT-GAN 52900, ISRAEL
E-mail address: `zalcman@macs.biu.ac.il`

II. Research Papers

Twistor Results for Integral Transforms

Toby Bailey and Michael Eastwood

ABSTRACT. We present a spectral sequence that describes the X-ray transform and similar integral transforms. It enables one to read off the kernel and range of these transforms.

Introduction

In its simplest guise, the X-ray transform associates to a real-valued function $f(p,q,r)$ on $\mathbb{R}^3$, its integral over the straight lines γ in $\mathbb{R}^3$. More precisely, the function

$$\hat{f}(\gamma) = \int_\gamma f \tag{1}$$

defined on the space of straight lines in $\mathbb{R}^3$ is called the X-ray transform of f. Of course, f is supposed to decay sufficiently fast at infinity that all these integrals make sense. We shall also suppose that f is smooth, i.e. infinitely differentiable. The range of this transform was found by John [22]. In fact, writing out (1) more explicitly:–

$$\hat{f}(w,x,y,z) = \int_{-\infty}^{\infty} f(w+xr, y+zr, r)\,dr \tag{2}$$

(for straight lines γ transverse to the (p,q)-plane) and differentiating under the integral sign, it is clear that

$$\frac{\partial^2 \hat{f}}{\partial w \partial z} - \frac{\partial^2 \hat{f}}{\partial x \partial y} = 0. \tag{3}$$

and John's result is that, save for appropriate decay conditions, this is the only restriction on $\hat{f}$.

It is, perhaps, not as well known as it should be, that the X-ray transform admits a convenient compactified formulation by regarding $\mathbb{R}^3$ as a standard affine coördinate chart on $\mathbb{RP}_3$. There is, however, a trick to this observation: $f(p,q,r)$

2000 *Mathematics Subject Classification.* Primary 53C65; Secondary 53C28.
Key words and phrases. Integral Geometry, Twistor Theory, X-ray Transform.
Support from the Australian Research Council is gratefully acknowledged.

on $\mathbb{R}^3$ should be regarded as a degree -2 homogeneous function $F(P,Q,R,S)$ on $\mathbb{R}^4 \setminus \{0\}$:–

$$(4) \qquad F(P,Q,R,S) = \frac{1}{S^2} f(\frac{P}{S}, \frac{Q}{S}, \frac{R}{S}) \quad \text{for } S \neq 0.$$

Equivalently, F may be regarded as a smooth section of an appropriate homogeneous line bundle on $\mathbb{RP}_3$. For details see [13, 14].

There are several advantages to this viewpoint. The decay conditions are now built into the geometry: F should extend as a smooth section to all of $\mathbb{RP}_3$. The symmetry group is enlarged to $SL(4, \mathbb{R})$. The space of straight lines in $\mathbb{R}^3$ is compactified to the Grassmannian $\text{Gr}_2(\mathbb{R}^4)$ of 2-planes in $\mathbb{R}^4$.

This compactified transform is the subject of this article. The aim is to give a simple formulation of this transform and its variants as a spectral sequence. From this spectral sequence one can simply read off the range and kernel of the transform. In particular, John's result [22] is immediate. There are generalisations to higher dimensions but, for simplicity, we shall stick to the dimensions described above.

All proofs will be omitted. They may be found in [1, 2, 8]. Suffice it to say that they are based on complex analysis and twistor theory. In fact, there is an integral formula due to Bateman [5]:–

$$\hat{f}(w,x,y,z) = \oint f((w+ix) + (iy+z)\zeta, (iy-z) + (w-ix)\zeta, \zeta) \, d\zeta$$

where f is a holomorphic function of 3 variables. It is clearly analogous to (2) with differentiation under the integral sign showing that $\hat{f}$ is harmonic:–

$$\frac{\partial^2 \hat{f}}{\partial w^2} + \frac{\partial^2 \hat{f}}{\partial x^2} + \frac{\partial^2 \hat{f}}{\partial y^2} + \frac{\partial^2 \hat{f}}{\partial z^2} = 0.$$

Twistor theory gives a proper interpretation of this formula [6]. It is this interpretation that leads, rather indirectly, to the spectral sequence for the X-ray transform. To present the results, however, complex analysis is not needed. There is some representation theory that cannot be avoided but it will be kept to a minimum.

Vector bundles on $\mathbb{RP}_3$

As already noted, we shall view the X-ray transform as acting on smooth sections of an appropriate line bundle on $\mathbb{RP}_3$. We shall need a notation for this line bundle and other vector bundles too. The notation we shall use comes from representation theory: see [9] for an informal discussion and [3, 4] for details. We shall consider the X-ray transform for smooth sections of the vector bundles

$$\stackrel{a}{\times}\!\!\!-\!\!\!\stackrel{b}{\bullet}\!\!\!-\!\!\!\stackrel{c}{\bullet} \quad \text{for integers } a,b,c \text{ with } b,c \geq 0.$$

The smooth sections

$$\Gamma(\mathbb{RP}_3, \stackrel{d}{\times}\!\!\!-\!\!\!\stackrel{0}{\bullet}\!\!\!-\!\!\!\stackrel{0}{\bullet})$$

correspond to smooth homogeneous functions on $\mathbb{R}^4 \setminus \{0\}$ of degree d. In general,

$$\stackrel{a}{\times}\!\!\!-\!\!\!\stackrel{b}{\bullet}\!\!\!-\!\!\!\stackrel{c}{\bullet} = \stackrel{a+2b-c}{\times}\!\!\!-\!\!\!\stackrel{0}{\bullet}\!\!\!-\!\!\!\stackrel{0}{\bullet} \otimes \stackrel{-2b+c}{\times}\!\!\!-\!\!\!\stackrel{b}{\bullet}\!\!\!-\!\!\!\stackrel{c}{\bullet}$$

and there is a geometric interpretation:–

$$\Gamma(\mathbb{RP}_3, \overset{-2b+c\ \ b\ \ c}{\times\!\!-\!\!\bullet\!\!-\!\!\bullet}) = \left\{ \begin{array}{c} \text{Smooth tensors } \omega_{ij\cdots k}{}^{\overbrace{lm\cdots n}^{c}} \text{ on } \mathbb{RP}_3 \\ \underbrace{}_{b} \\ \text{that are symmetric and trace-free.} \end{array} \right\}.$$

These conventions may look awkward at first but they are natural in representation theory: $\mathbb{RP}_3$ is viewed as a homogeneous space for $\mathrm{SL}(4,\mathbb{R})$ and the integers a,b,c specify the weight of the representation from which the corresponding bundle is induced. Here are some useful examples:–

$\overset{1\ \ 0\ \ 1}{\times\!\!-\!\!\bullet\!\!-\!\!\bullet}$ = the tangent bundle.

$\overset{-2\ \ 1\ \ 0}{\times\!\!-\!\!\bullet\!\!-\!\!\bullet}$ = the cotangent bundle.

$\overset{-3\ \ 0\ \ 1}{\times\!\!-\!\!\bullet\!\!-\!\!\bullet}$ = the bundle of 2-forms.

$\overset{-4\ \ 0\ \ 0}{\times\!\!-\!\!\bullet\!\!-\!\!\bullet}$ = the bundle of 3-forms or volume-forms.

$\overset{-4\ \ 2\ \ 0}{\times\!\!-\!\!\bullet\!\!-\!\!\bullet}$ = the bundle of symmetric covariant 2-tensors.

Vector bundles on $\mathrm{Gr}_2(\mathbb{R}^4)$

The vector bundles on $\mathrm{Gr}_2(\mathbb{R}^4)$ arising from the X-ray transform are denoted similarly:–

$\overset{a\ \ b\ \ c}{\bullet\!\!-\!\!\times\!\!-\!\!\bullet}$ for integers a,b,c with $a,c \geq 0$.

The notation is derived by viewing $\mathrm{Gr}_2(\mathbb{R}^4)$ as a homogeneous space for $\mathrm{SL}(4,\mathbb{R})$. As such, $\mathrm{Gr}_2(\mathbb{R}^4)$ admits an invariant pseudo-conformal structure. This allows a geometric interpretation of these bundles:–

$\overset{1\ \ 0\ \ 1}{\bullet\!\!-\!\!\times\!\!-\!\!\bullet}$ = the tangent bundle.

$\overset{1\ -2\ \ 1}{\bullet\!\!-\!\!\times\!\!-\!\!\bullet}$ = the cotangent bundle.

$\overset{2\ -3\ \ 0}{\bullet\!\!-\!\!\times\!\!-\!\!\bullet}$ = the bundle of self-dual 2-forms.

$\overset{0\ -3\ \ 2}{\bullet\!\!-\!\!\times\!\!-\!\!\bullet}$ = the bundle of anti-self-dual 2-forms.

$\overset{1\ \ 0\ \ 0}{\bullet\!\!-\!\!\times\!\!-\!\!\bullet}$ = a spin bundle.

$\overset{0\ \ 0\ \ 1}{\bullet\!\!-\!\!\times\!\!-\!\!\bullet}$ = the other spin bundle.

$\overset{0\ \ d\ \ 0}{\bullet\!\!-\!\!\times\!\!-\!\!\bullet}$ = the bundle of densities of weight d.

Strictly speaking, it is not this collection of bundles that we shall need but, rather, their 'odd' counterparts. The Grassmannian $\mathrm{Gr}_2(\mathbb{R}^4)$ is doubly covered by the Grassmannian $\mathrm{Gr}_2^+(\mathbb{R}^4)$ of oriented 2-planes in $\mathbb{R}^4$ just as $\mathbb{RP}_3$ is doubly covered by S^3. An ordinary smooth function on $\mathrm{Gr}_2(\mathbb{R}^4)$ lifts to an 'even' function on $\mathrm{Gr}_2^+(\mathbb{R}^4)$. The odd functions on $\mathrm{Gr}_2^+(\mathbb{R}^4)$ descend to sections of a corresponding line bundle on $\mathrm{Gr}_2(\mathbb{R}^4)$ but we shall simply write

$$\Gamma_{\mathrm{odd}}(\mathrm{Gr}_2(\mathbb{R}^4), \overset{0\ \ 0\ \ 0}{\bullet\!\!-\!\!\times\!\!-\!\!\bullet})$$

instead. Similarly, we shall write

$$\Gamma_{\mathrm{odd}}(\mathrm{Gr}_2(\mathbb{R}^4), \overset{a\ \ b\ \ c}{\bullet\!\!-\!\!\times\!\!-\!\!\bullet}) \quad \text{or just} \quad \Gamma_{\mathrm{odd}}(\overset{a\ \ b\ \ c}{\bullet\!\!-\!\!\times\!\!-\!\!\bullet})$$

for the odd sections of $\overset{a}{\bullet}\!\!\overset{b}{\times}\!\!\overset{c}{\bullet}$ on $\mathrm{Gr}_2^+(\mathbb{R}^4)$.

Representations of $\mathrm{SL}(4,\mathbb{R})$

Following [**4**], the real irreducible representations of $\mathrm{SL}(4,\mathbb{R})$ may be listed:–

$$\overset{a}{\bullet}\!\!\overset{b}{\bullet}\!\!\overset{c}{\bullet} \quad \text{for integers } a,b,c \text{ with } a,b,c \geq 0.$$

The defining action of $\mathrm{SL}(4,\mathbb{R})$ on $\mathbb{R}^4$ is $\overset{0}{\bullet}\!\!\overset{0}{\bullet}\!\!\overset{1}{\bullet}$ and, in general,

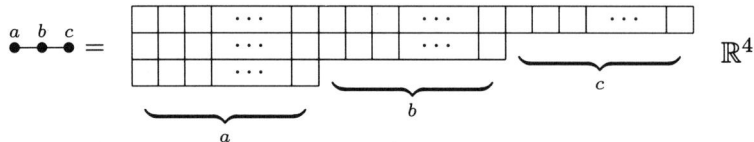

as a Young diagram [**12**].

The spectral sequence

For any bundle $\overset{a}{\times}\!\!\overset{b}{\bullet}\!\!\overset{c}{\bullet}$ on $\mathbb{RP}_3$, there is a spectral sequence

$E_1^{p,q} = 0$ unless $0 \leq p \leq 2$ and $0 \leq q \leq 1$

$E_1^{0,0} = \begin{cases} \Gamma_{\mathrm{odd}}(\overset{a}{\bullet}\!\!\overset{b}{\times}\!\!\overset{c}{\bullet}) & \text{if } a \geq 0 \\ 0 & \text{else,} \end{cases}$

$E_1^{0,1} = \begin{cases} \Gamma_{\mathrm{odd}}(\overset{-a-2}{\bullet}\!\!\overset{a+b+1}{\times}\!\!\overset{c}{\bullet}) & \text{if } a \leq -2 \\ 0 & \text{else,} \end{cases}$

$E_1^{1,0} = \begin{cases} \Gamma_{\mathrm{odd}}(\overset{a+b+1}{\bullet}\!\!\overset{-b-2}{\times}\!\!\overset{b+c+1}{\bullet}) & \text{if } a \geq -b-1 \\ 0 & \text{else,} \end{cases}$

$E_1^{1,1} = \begin{cases} \Gamma_{\mathrm{odd}}(\overset{-a-b-3}{\bullet}\!\!\overset{a}{\times}\!\!\overset{b+c+1}{\bullet}) & \text{if } a \leq -b-3 \\ 0 & \text{else,} \end{cases}$

$E_1^{2,0} = \begin{cases} \Gamma_{\mathrm{odd}}(\overset{a+b+c+2}{\bullet}\!\!\overset{-b-c-3}{\times}\!\!\overset{b}{\bullet}) & \text{if } a \geq -b-c-2 \\ 0 & \text{else,} \end{cases}$

$E_1^{2,1} = \begin{cases} \Gamma_{\mathrm{odd}}(\overset{-a-b-c-4}{\bullet}\!\!\overset{a}{\times}\!\!\overset{b}{\bullet}) & \text{if } a \leq -b-c-4 \\ 0 & \text{else,} \end{cases}$

It converges:–

$$E_1^{p,q} \Longrightarrow E_\infty^{p+q}$$

where $E_\infty^0 = 0$, $E_\infty^3 = 0$, and there are exact sequences

(5) $\quad \overset{a}{\bullet}\!\!\overset{b}{\bullet}\!\!\overset{c}{\bullet} \rightarrowtail \Gamma(\mathbb{RP}_3, \overset{a}{\times}\!\!\overset{b}{\bullet}\!\!\overset{c}{\bullet}) \to E_\infty^1 \twoheadrightarrow \overset{-a-2}{\bullet}\!\!\overset{a+b+1}{\bullet}\!\!\overset{c}{\bullet}$

$\qquad\qquad \updownarrow \qquad\qquad\qquad\qquad\qquad\qquad\qquad \updownarrow$

$\qquad\quad 0 \leq a \qquad\qquad\qquad\qquad\qquad\qquad -b-1 \leq a \leq -2$

and

$\overset{b}{\bullet}\!\!\overset{-a-b-3}{\bullet}\!\!\overset{a+b+c+2}{\bullet} \rightarrowtail E_\infty^2 \to \Gamma(\mathbb{RP}_3, \overset{a}{\times}\!\!\overset{b}{\bullet}\!\!\overset{c}{\bullet}) \twoheadrightarrow \overset{b}{\bullet}\!\!\overset{c}{\bullet}\!\!\overset{-a-b-c-4}{\bullet}.$

$\updownarrow \qquad\qquad\qquad\qquad\qquad\qquad\qquad\qquad\qquad \updownarrow$

$-b-c-2 \leq a \leq -b-3 \qquad\qquad\qquad\qquad\qquad a \leq -b-c-4$

Examples

The case $\overset{-2\ 0\ 0}{\times\!\!-\!\!\bullet\!\!-\!\!\bullet}$ The spectral sequence is

$$\begin{array}{|ccc|}\hline \Gamma_{\text{odd}}(\overset{0\ -1\ 0}{\bullet\!\!-\!\!\times\!\!-\!\!\bullet}) & 0 & 0 \\ \\ 0 & 0 & \Gamma_{\text{odd}}(\overset{0\ -3\ 0}{\bullet\!\!-\!\!\times\!\!-\!\!\bullet}) \\ \hline \end{array}$$

and the exact sequences for E_∞^1 and E_∞^2 collapse. The conclusion is an exact sequence:–

(6) $\quad \Gamma(\mathbb{RP}_3, \overset{-2\ 0\ 0}{\times\!\!-\!\!\bullet\!\!-\!\!\bullet}) \rightarrowtail \Gamma_{\text{odd}}(\overset{0\ -1\ 0}{\bullet\!\!-\!\!\times\!\!-\!\!\bullet}) \overset{\square}{\longrightarrow} \Gamma_{\text{odd}}(\overset{0\ -3\ 0}{\bullet\!\!-\!\!\times\!\!-\!\!\bullet}) \twoheadrightarrow \Gamma(\mathbb{RP}_3, \overset{-2\ 0\ 0}{\times\!\!-\!\!\bullet\!\!-\!\!\bullet}).$

The mapping $\square$ arises from the E_2 level of the spectral sequence. It is necessarily a linear differential operator invariant under the action of $\text{SL}(4,\mathbb{R})$. This information is sufficient to pin it down up to scale as the compactified extension of the ultrahyperbolic wave operator, namely the second order differential operator appearing as the left hand side of (3). In fact, the invariant linear differential operators between the bundles $\overset{a\ b\ c}{\bullet\!\!-\!\!\times\!\!-\!\!\bullet}$ on $\text{Gr}_2(\mathbb{R}^4)$ are classified: the result for $\text{Gr}_2(\mathbb{C}^4)$ is obtained in [**10, 21**] and there is no essential difference over the reals [**7**]. From (6) we deduce that the range of this simple X-ray transform is the kernel of $\square$ as in [**22**]. As a by-product, the cokernel of $\square$ is also identified with $\Gamma(\mathbb{RP}_3, \overset{-2\ 0\ 0}{\times\!\!-\!\!\bullet\!\!-\!\!\bullet})$. Since $\square$ is formally self-adjoint, this is not unexpected (cf. [**17**]).

The case $\overset{-3\ 0\ 0}{\times\!\!-\!\!\bullet\!\!-\!\!\bullet}$ The spectral sequence is

$$\begin{array}{|ccc|}\hline \Gamma_{\text{odd}}(\overset{1\ -2\ 0}{\bullet\!\!-\!\!\times\!\!-\!\!\bullet}) & \Gamma_{\text{odd}}(\overset{0\ -3\ 1}{\bullet\!\!-\!\!\times\!\!-\!\!\bullet}) & 0 \\ \\ 0 & 0 & 0 \\ \hline \end{array}$$

and the exact sequences for E_∞^1 and E_∞^2 again collapse. The conclusion is an exact sequence:–

$\Gamma(\mathbb{RP}_3, \overset{-3\ 0\ 0}{\times\!\!-\!\!\bullet\!\!-\!\!\bullet}) \rightarrowtail \Gamma_{\text{odd}}(\overset{1\ -2\ 0}{\bullet\!\!-\!\!\times\!\!-\!\!\bullet}) \overset{D}{\longrightarrow} \Gamma_{\text{odd}}(\overset{0\ -3\ 1}{\bullet\!\!-\!\!\times\!\!-\!\!\bullet}) \twoheadrightarrow \Gamma(\mathbb{RP}_3, \overset{-3\ 0\ 0}{\times\!\!-\!\!\bullet\!\!-\!\!\bullet}).$

Here, D is the Dirac operator. In particular, the range of the X-ray transform on $\Gamma(\mathbb{RP}_3, \overset{-3\ 0\ 0}{\times\!\!-\!\!\bullet\!\!-\!\!\bullet})$ is the kernel of D. As an explicit integral transform, we obtain, for a smooth function f on $\mathbb{R}^3$, a pair of functions

$$\hat{f}_1(w,x,y,z) = \int_{-\infty}^{\infty} f(w+xr, y+zr, r)\,dr$$

$$\hat{f}_2(w,x,y,z) = \int_{-\infty}^{\infty} rf(w+xr, y+zr, r)\,dr$$

and the Dirac equation says that

$$\frac{\partial \hat{f}_1}{\partial x} - \frac{\partial \hat{f}_2}{\partial w} = 0 \quad \text{and} \quad \frac{\partial \hat{f}_1}{\partial z} - \frac{\partial \hat{f}_2}{\partial y} = 0$$

as is evident by differentiating under the integral sign.

The case $\overset{-7\ \ 1\ \ 2}{\times\!\!-\!\!\bullet\!\!-\!\!\bullet}$ The spectral sequence is

$$\boxed{\begin{array}{ccc} \Gamma_{\text{odd}}(\overset{5\ -5\ \ 2}{\bullet\!\!-\!\!\times\!\!-\!\!\bullet}) & \Gamma_{\text{odd}}(\overset{3\ -7\ \ 4}{\bullet\!\!-\!\!\times\!\!-\!\!\bullet}) & \Gamma_{\text{odd}}(\overset{0\ -7\ \ 1}{\bullet\!\!-\!\!\times\!\!-\!\!\bullet}) \\ 0 & 0 & 0 \end{array}}$$

and the exact sequence for E^1_∞ collapses. We conclude that

$$\Gamma(\mathbb{RP}_3, \overset{-3\ \ 0\ \ 0}{\times\!\!-\!\!\bullet\!\!-\!\!\bullet}) = \ker \nabla^2 : \Gamma_{\text{odd}}(\overset{5\ -5\ \ 2}{\bullet\!\!-\!\!\times\!\!-\!\!\bullet}) \to \Gamma_{\text{odd}}(\overset{3\ -7\ \ 4}{\bullet\!\!-\!\!\times\!\!-\!\!\bullet})$$

where ∇^2 is an invariant differential operator (of second order).

The case $\overset{-2\ \ k+1\ \ 0}{\times\!\!-\!\!\bullet\!\!-\!\!\bullet}$ **for** $k \geq 0$ The spectral sequence is

$$\boxed{\begin{array}{ccc} \Gamma_{\text{odd}}(\overset{0\ \ k\ \ 0}{\bullet\!\!-\!\!\times\!\!-\!\!\bullet}) & 0 & 0 \\ 0 & \Gamma_{\text{odd}}(\overset{k\ -k-3\ \ k+2}{\bullet\!\!-\!\!\times\!\!-\!\!\bullet}) & \Gamma_{\text{odd}}(\overset{k+1\ -k-4\ \ k+1}{\bullet\!\!-\!\!\times\!\!-\!\!\bullet}) \end{array}}$$

and the exact sequence for E^1_∞ reads

$$0 \to \Gamma(\mathbb{RP}_3, \overset{-2\ \ k+1\ \ 0}{\times\!\!-\!\!\bullet\!\!-\!\!\bullet}) \to E^1_\infty \to \overset{0\ \ k\ \ 0}{\bullet\!\!-\!\!\bullet\!\!-\!\!\bullet} \to 0.$$

From here, a little diagram chasing (detailed in [**1**]) gives an X-ray transform and an exact sequence

$$\Gamma(\mathbb{RP}_3, \overset{0\ \ k\ \ 0}{\times\!\!-\!\!\bullet\!\!-\!\!\bullet}) \xrightarrow{\nabla} \Gamma(\mathbb{RP}_3, \overset{-2\ \ k+1\ \ 0}{\times\!\!-\!\!\bullet\!\!-\!\!\bullet}) \xrightarrow{\text{X-ray}} \Gamma_{\text{odd}}(\overset{0\ \ k\ \ 0}{\bullet\!\!-\!\!\times\!\!-\!\!\bullet})$$

where ∇ is an $SL(4,\mathbb{R})$-invariant first order linear differential operator on $\mathbb{RP}_3$. When $k = 0$, it is simply the exterior derivative from functions to 1-forms and we obtain the result of Michel [**24**], that a 1-form on $\mathbb{RP}_3$ whose integral over every great circle vanishes is closed. The case $k = 1$ is also due to Michel [**23**]. The case $k = 2$ is due to Estezet [**11**]—see [**15**] for an exposition and further discussion. The general case for tensors of compact support on $\mathbb{R}^3$ is due to Sharafutdinov [**25**].

The case $\overset{0\ \ 0\ \ 0}{\times\!\!-\!\!\bullet\!\!-\!\!\bullet}$ The spectral sequence is

$$\boxed{\begin{array}{ccc} 0 & 0 & 0 \\ \Gamma_{\text{odd}}(\overset{0\ \ 0\ \ 0}{\bullet\!\!-\!\!\times\!\!-\!\!\bullet}) & \Gamma_{\text{odd}}(\overset{1\ -2\ \ 1}{\bullet\!\!-\!\!\times\!\!-\!\!\bullet}) & \Gamma_{\text{odd}}(\overset{2\ -3\ \ 0}{\bullet\!\!-\!\!\times\!\!-\!\!\bullet}) \end{array}}$$

and the exact sequence for E^1_∞ reads

$$0 \to \underset{\parallel\ \mathbb{R}}{\overset{0\ \ 0\ \ 0}{\bullet\!\!-\!\!\bullet\!\!-\!\!\bullet}} \to \Gamma(\mathbb{RP}_3, \overset{0\ \ 0\ \ 0}{\times\!\!-\!\!\bullet\!\!-\!\!\bullet}) \to E^1_\infty \to 0.$$

We conclude that there is an X-ray transform from the space of smooth functions on $\mathbb{RP}_3$ to the quotient

$$\frac{\{\text{Smooth odd 1-forms } \Phi \text{ on } \mathrm{Gr}_2(\mathbb{R}^4) \text{ s.t. } d\Phi \text{ is anti-self-dual}\}}{\{\Phi \text{ of the form } df \text{ for some smooth odd function } f\}}$$

with kernel the constant functions on $\mathbb{RP}_3$. As explained in [**8**], this is a 'potential modulo gauge' representation of the closed odd anti-self-dual two-forms on $\mathrm{Gr}_2(\mathbb{R}^4)$

with zero 'charge'. There is an integral formula for this two-form:–

$$\left(\int_{-\infty}^{\infty} \frac{\partial^2 f}{\partial p^2}(w+xr, y+zr, r)\, dr\right) dw \wedge dx$$

$$+ \left(\int_{-\infty}^{\infty} \frac{\partial^2 f}{\partial p \partial q}(w+xr, y+zr, r)\, dr\right)(dw \wedge dz - dx \wedge dy)$$

$$+ \left(\int_{-\infty}^{\infty} \frac{\partial^2 f}{\partial q^2}(w+xr, y+zr, r)\, dr\right) dy \wedge dz.$$

Differentiation under the integral sign shows that it is closed. Though very much tied to our particular coördinate system, there is also an integral formula for the potential:–

$$\Phi = \left(\int_{-\infty}^{\infty} \frac{\partial f}{\partial p}(w+xr, y+zr, r)\, dr\right) dx$$

$$+ \left(\int_{-\infty}^{\infty} \frac{\partial f}{\partial q}(w+xr, y+zr, r)\, dr\right) dz.$$

Results on affine spaces

A somewhat condensed version of the material above was presented by the second author at the Summer Research Conference at Mount Holyoke College. In this section we answer some questions raised at the conference. We are grateful to Leon Ehrenpreis for asking how the moment conditions of the hyperplane Radon transform arise in the compactified setting and to Fulton Gonzalez for describing his 'plane-to-line' transform on $\mathbb{R}^3$.

The starting point of this article was to view $\mathbb{R}^3$ as a standard affine chart on $\mathbb{RP}_3$. If $[P, Q, R, S]$ are homogeneous coördinates on $\mathbb{RP}_3$, then $\mathbb{R}^3 \hookrightarrow \mathbb{RP}_3$ by $(p, q, r) \mapsto [p, q, r, 1]$. The bundle $\overset{-2\ \ 0\ \ 0}{\times\!\!-\!\!\bullet\!\!-\!\!\bullet}$ on $\mathbb{RP}_3$ may be trivialised on $\mathbb{R}^3$. Indeed, this is the effect of (4). It is then straightforward to interpret $\Gamma(\mathbb{RP}_3, \overset{-2\ \ 0\ \ 0}{\times\!\!-\!\!\bullet\!\!-\!\!\bullet})$ as smooth functions on $\mathbb{R}^3$ subject to certain decay conditions. More usual, however, is to ask about the X-ray transform of $\mathcal{S}(\mathbb{R}^3)$, the Schwartz space of rapidly decreasing functions. We may view $\mathcal{S}(\mathbb{R}^3) \hookrightarrow \Gamma(\overset{-2\ \ 0\ \ 0}{\times\!\!-\!\!\bullet\!\!-\!\!\bullet})$ as those sections that vanish to infinite order on the hyperplane at infinity. This interpretation may be carried over by the X-ray transform. To do this, let us first consider those sections that simply vanish at infinity. The hyperplane at infinity is defined by $S = 0$:–

$$\mathbb{RP}_2 \ni [P, Q, R] \longmapsto [P, Q, R, 0] \in \mathbb{RP}_3.$$

From this point of view, the lines in the hyperplane at infinity give rise to an embedded dual projective plane $\mathbb{RP}_2^* \hookrightarrow \mathrm{Gr}_2(\mathbb{R}^4)$. In classical twistor theory [20], this is referred to as a β-plane. Now the Funk transform, the compactified version of the classical Radon transform, provides an isomorphism

(7) $$\mathcal{F} : \Gamma(\mathbb{RP}_2, \overset{-2\ \ 0}{\times\!\!-\!\!\bullet}) \xrightarrow{\cong} \Gamma_{\mathrm{odd}}(\mathbb{RP}_2^*, \overset{0\ \ -1}{\bullet\!\!-\!\!\times}).$$

This isomorphism can also be proved by twistor methods [2]. The bundle $\overset{-2\ \ 0}{\times\!\!-\!\!\bullet}$ on $\mathbb{RP}_2$ is the restriction of $\overset{-2\ \ 0\ \ 0}{\times\!\!-\!\!\bullet\!\!-\!\!\bullet}$ on $\mathbb{RP}_3$ and $\overset{0\ \ -1}{\bullet\!\!-\!\!\times}$ on $\mathbb{RP}_2^*$ is the restriction of

$\stackrel{0\ -1\ 0}{\bullet\!-\!\times\!-\!\bullet}$ on $\mathrm{Gr}_2(\mathbb{R}^4)$. Moreover, the Radon transform is compatible with the X-ray transform. In fact, we can transform the evident exact sequence

$$0 \to \Gamma(\mathbb{RP}_3, \stackrel{-3\ 0\ 0}{\times\!-\!\bullet\!-\!\bullet}) \stackrel{\times S}{\longrightarrow} \Gamma(\mathbb{RP}_3, \stackrel{-2\ 0\ 0}{\times\!-\!\bullet\!-\!\bullet}) \to \Gamma(\mathbb{RP}_2, \stackrel{-2\ 0}{\times\!-\!\bullet}) \to 0$$

to obtain an isomorphic exact sequence

$$0 \to \{\theta \in \Gamma_{\mathrm{odd}}(\stackrel{1\ -2\ 0}{\bullet\!-\!\times\!-\!\bullet}) \text{ s.t. } D\psi = 0\} \to \{\phi \in \Gamma_{\mathrm{odd}}(\stackrel{0\ -1\ 0}{\bullet\!-\!\times\!-\!\bullet}) \text{ s.t. } \Box\phi = 0\}$$
$$\to \Gamma_{\mathrm{odd}}(\mathbb{RP}_2^*, \stackrel{0\ -1}{\bullet\!-\!\times}) \to 0.$$

In particular, we conclude that the X-ray transform gives an isomorphism

$$\{f \in \Gamma(\mathbb{RP}_3, \stackrel{-2\ 0\ 0}{\times\!-\!\bullet\!-\!\bullet}) \text{ s.t. } f|_{\mathbb{RP}_2} = 0\} \stackrel{\cong}{\longrightarrow}$$
$$\{\phi \in \Gamma_{\mathrm{odd}}(\mathrm{Gr}_2(\mathbb{R}^4), \stackrel{0\ -1\ 0}{\bullet\!-\!\times\!-\!\bullet}) \text{ s.t. } \Box\phi = 0 \text{ and } \phi|_{\mathbb{RP}_2^*} = 0\}.$$

Repeating this argument for $\Gamma(\mathbb{RP}_3, \stackrel{-3\ 0\ 0}{\times\!-\!\bullet\!-\!\bullet})$ and, generally, for $\Gamma(\mathbb{RP}_3, \stackrel{k\ 0\ 0}{\times\!-\!\bullet\!-\!\bullet})$ with $k \leq -2$, we conclude that the X-ray transform gives an isomorphism between $\mathcal{S}(\mathbb{R}^3)$ and those solutions of the ultrahyperbolic wave equation on $\mathrm{Gr}_2(\mathbb{R}^4)$ that vanish to infinite order on the β-plane $\mathbb{RP}_2^*$. As a decay condition for a smooth function on the space of lines in $\mathbb{R}^3$, this is rapid decrease on the space of parallel lines, uniform in the direction of the line. But this is the definition of Schwartz space for the space of lines in $\mathbb{R}^3$ and we have recovered John's result [22] precisely.

Notice that there are no moment or 'Cavalieri' conditions on the range of the X-ray transform. It is interesting to attempt the same reasoning as above for the hyperplane Radon transform to see where the moment conditions (as in [19], for example) arise. We have already seen the Radon transform on $\mathbb{RP}_2$ as (7). On $\mathbb{RP}_3$, there are compactified transforms

$$\mathcal{F} : \Gamma_{\mathrm{odd}}(\mathbb{RP}_3, \stackrel{-3\ 0\ 0}{\times\!-\!\bullet\!-\!\bullet}) \stackrel{\cong}{\longrightarrow} \Gamma_{\mathrm{odd}}(\mathbb{RP}_3^*, \stackrel{0\ 0\ -1}{\bullet\!-\!\bullet\!-\!\times})$$

and

$$\mathcal{F} : \Gamma_{\mathrm{odd}}(\mathbb{RP}_3, \stackrel{-4\ 0\ 0}{\times\!-\!\bullet\!-\!\bullet}) \stackrel{\cong}{\longrightarrow} \Gamma_{\mathrm{odd}}(\mathbb{RP}_3^*, \stackrel{0\ 0\ 0}{\bullet\!-\!\bullet\!-\!\times}).$$

Again, there is an exact sequence

$$0 \to \Gamma_{\mathrm{odd}}(\mathbb{RP}_3, \stackrel{-4\ 0\ 0}{\times\!-\!\bullet\!-\!\bullet}) \stackrel{\times S}{\longrightarrow} \Gamma_{\mathrm{odd}}(\mathbb{RP}_3, \stackrel{-3\ 0\ 0}{\times\!-\!\bullet\!-\!\bullet}) \to \Gamma_{\mathrm{odd}}(\mathbb{RP}_2, \stackrel{-3\ 0}{\times\!-\!\bullet}) \to 0$$

but, now, the hyperplane transform restricted to $\mathbb{RP}_2$ gives only a number up to sign by integration and so we cannot simply transform the whole sequence. Rather, we obtain

$$\partial/\partial Z : \Gamma_{\mathrm{odd}}(\mathbb{RP}_3^*, \stackrel{0\ 0\ 0}{\bullet\!-\!\bullet\!-\!\times}) \longrightarrow \Gamma_{\mathrm{odd}}(\mathbb{RP}_3^*, \stackrel{0\ 0\ -1}{\bullet\!-\!\bullet\!-\!\times})$$

as the transform of $g \mapsto Sg$ where $[W, X, Y, Z]$ are dual homogeneous coördinates to $[P, Q, R, S]$. So the question of whether $f \in \Gamma_{\mathrm{odd}}(\mathbb{RP}_3, \stackrel{-3\ 0\ 0}{\times\!-\!\bullet\!-\!\bullet})$ vanishes on the hyperplane at infinity is the question of whether its Radon transform $\mathcal{F}f$ can be written as $\partial g/\partial Z$ for some odd smooth function g on $\mathbb{RP}_3^*$. This question may be addressed directly. If $\mathcal{F}f = \partial g/\partial Z$, then

$$g(W, X, Y, Z) = g(W, X, Y, -Z) + \int_{-Z}^{Z} \mathcal{F}f(W, X, Y, t)dt.$$

Since $g(W,X,Y,Z)$ is homogeneous of degree zero, we may rewrite this equation as

$$g(W/Z, X/Z, Y/Z, 1) = g(W/Z, X/Z, Y/Z, -1) + \int_{-Z}^{Z} \mathcal{F}f(W,X,Y,t)dt.$$

Letting $Z \to \infty$ we obtain

$$g(0,0,0,1) = g(0,0,0,-1) + \int_{-\infty}^{\infty} \mathcal{F}f(W,X,Y,t)dt.$$

But $g(W,X,Y,Z)$ is also odd so

$$\int_{-\infty}^{\infty} \mathcal{F}f(W,X,Y,t)dt = 2g(0,0,0,1).$$

In particular, this integral is a constant, independent of (X,Y,Z). This is the first of the Cavalieri conditions. The higher conditions arise similarly in order that $\mathcal{F}f$ be in the range of $\partial^{k+1}/\partial Z^{k+1}$: $\Gamma_{\mathrm{odd}}(\mathbb{RP}_3^*, \overset{k\ \ 0\ \ 0}{\times\!\!-\!\!\bullet\!\!-\!\!\bullet}) \to \Gamma_{\mathrm{odd}}(\mathbb{RP}_3^*, \overset{-1\ 0\ \ 0}{\times\!\!-\!\!\bullet\!\!-\!\!\bullet})$. Specifically, we require that

$$\int_{-\infty}^{\infty} \mathcal{F}f(W,X,Y,t)t^k dt = \text{a homogeneous polynomial in } X, Y, Z \text{ of degree } k.$$

The argument is easily reversible: these necessary conditions are also sufficient.

Finally we consider the plane-to-line transform of [16]. This transform takes a smooth function f defined on the space of planes in $\mathbb{R}^3$ and yields a smooth function $\hat{f}$ on the space of lines in $\mathbb{R}^3$ by

$$\hat{f}(\gamma) = \int_{\gamma \subset P} f(P). \tag{8}$$

For each line γ, the planes P containing γ are parameterised by a circle and the integration in (8) is performed with respect to the translation and rotation-invariant measure. Notice that decay conditions are unnecessary: the integrals are over closed cycles as in the compactified X-ray transform. In fact, (8) may be seen in the compactified formulation: with $\mathbb{R}^3 \hookrightarrow \mathbb{RP}_3$ as usual, the planes in $\mathbb{R}^3$ compactify to projective planes in $\mathbb{RP}_3$ leaving only the plane at infinity missing. Therefore, the space of planes in $\mathbb{R}^3$ may be identified as the complement of a single point in $\mathbb{RP}_3^*$. Dually, we may as well start with $\mathbb{RP}_3 \setminus \{x\}$ and (8) becomes our previous compactified X-ray transform. The projective lines through x define an embedding $\mathbb{RP}_2 \hookrightarrow \mathrm{Gr}_2(\mathbb{R}^4)$ (known as an α-plane in twistor theory [20]). Thus, we obtain

$$\Gamma(\mathbb{RP}_3 \setminus \{x\}, \overset{-2\ \ 0\ \ 0}{\times\!\!-\!\!\bullet\!\!-\!\!\bullet}) \to \{\phi \in \Gamma_{\mathrm{odd}}(\mathrm{Gr}_2(\mathbb{R}^4) \setminus \mathbb{RP}_2, \overset{0\ -1\ \ 0}{\bullet\!\!-\!\!\times\!\!-\!\!\bullet}) \text{ s.t. } \Box\phi = 0\}$$

and in [16] it is shown that this is an isomorphism. To see this by twistor methods it is necessary to explain the origin of (5): all other aspects of the proof are unchanged. It involves Dolbeault cohomology:–

$$H^0(\mathbb{CP}_3, \mathcal{O}(\overset{a\ \ b\ \ c}{\times\!\!-\!\!\bullet\!\!-\!\!\bullet})) \hookrightarrow \Gamma(\mathbb{RP}_3, \overset{a\ \ b\ \ c}{\times\!\!-\!\!\bullet\!\!-\!\!\bullet}) \to E_\infty^1 \twoheadrightarrow H^1(\mathbb{CP}_3, \mathcal{O}(\overset{a\ \ b\ \ c}{\times\!\!-\!\!\bullet\!\!-\!\!\bullet}))$$

and (5) arises by using the Bott-Borel Weil Theorem (as, for example, in [4]) to compute the first and last terms. But Dolbeault cohomology extends

$$H^0(\mathbb{CP}_3 \setminus \{x\}, \mathcal{O}(\overset{-2\ \ 0\ \ 0}{\times\!\!-\!\!\bullet\!\!-\!\!\bullet})) = H^0(\mathbb{CP}_3, \mathcal{O}(\overset{-2\ \ 0\ \ 0}{\times\!\!-\!\!\bullet\!\!-\!\!\bullet})) = 0$$
$$H^1(\mathbb{CP}_3 \setminus \{x\}, \mathcal{O}(\overset{-2\ \ 0\ \ 0}{\times\!\!-\!\!\bullet\!\!-\!\!\bullet})) = H^1(\mathbb{CP}_3, \mathcal{O}(\overset{-2\ \ 0\ \ 0}{\times\!\!-\!\!\bullet\!\!-\!\!\bullet})) = 0$$

(as, for example, in [18]) and the twistor proof in [8] goes through.

References

[1] T.N. Bailey and M.G. Eastwood, Zero-energy fields on real projective space, Geom. Dedicata **67** (1997), 245–258.

[2] T.N. Bailey, M.G. Eastwood, A.G. Gover, and L.J. Mason, The Funk transform as a Penrose transform, Math. Proc. Camb. Phi. Soc. **125** (1999), 67–81.

[3] T.N. Bailey and M.A. Singer, Twistors, massless fields and the Penrose transform, in Twistors in Mathematics and Physics, Lond. Math. Soc. Lecture Notes in Math. **156** (1990) 299–338.

[4] R.J. Baston and M.G. Eastwood, The Penrose Transform: its Interaction with Representation Theory, Oxford University Press 1989.

[5] H. Bateman, The solution of partial differential equations by means of definite integrals, Proc. Lond. Math. Soc. **1(2)** (1904), 451–458.

[6] M.G. Eastwood, Introduction to Penrose transform, in The Penrose Transform and Analytic Cohomology in Representation Theory, Contemp. Math. **154** (1993), 71–75.

[7] M.G. Eastwood, On the weights of conformally invariant operators, in Further Advances in Twistor Theory vol. II, Pitman Research Notes in Math. **232** (1995), 114–119.

[8] M.G. Eastwood, Complex methods in real integral geometry, in Proceedings of the Sixteenth Winter School on Geometry and Physics, Srní, Suppl. Rendi. Circ. Mat. Palermo **46** (1997), 55–71.

[9] M.G. Eastwood, Variations on the de Rham complex, Notices Amer. Math. Soc. **46** (1999) 1368–1376.

[10] M.G. Eastwood and J.W. Rice, Conformally invariant differential operators on Minkowski space and their curved analogues, Commun. Math. Phys. **109** (1987) 207–288.

[11] P. Estezet, Tenseurs symétriques à énergie nulle sur les variétés à courbure constante, Thèse de Doctorat de Troisième Cycle, Université de Grenoble I, 1988.

[12] W. Fulton and J. Harris, Representation Theory, a First Course, Graduate Texts in Mathematics vol. 129, Springer 1991.

[13] I.M. Gelfand, S.G. Gindikin, and M.I. Graev, Integral geometry in affine and projective spaces (Russian), Current Problems in Math. vol. 16, Akad. Nauk SSSR (1980) 53–226.

[14] S.G. Gindikin, Real integral geometry and complex analysis, in Integral geometry, Radon transforms and complex analysis, Springer Lecture Notes in Math. **1684** (1998) 70–98.

[15] H. Goldschmidt, The Radon transform for symmetric forms on real projective spaces, in Integral Geometry and Tomography, Contemp. Math. **113** (1990), 81–96.

[16] F.B. Gonzalez, John's Equation and the plane-to-line transform on $\mathbb{R}^3$, in Harmonic Analysis and Integral Geometry, Research Notes in Math. vol. 422, Chapman and Hall/CRC 2000.

[17] E. Grinberg, Euclidean Radon transforms: ranges and restrictions, in Integral Geometry, Contemp. Math. **63** (1987), 109–133.

[18] R.C. Gunning, Lectures on Complex Analytic Varieties: the Local Parametrization Theorem, Princeton University Press 1970.

[19] S. Helgason, Geometric Analysis on Symmetric Spaces, Math. Surveys and Monographs vol. 39, Amer. Math. Soc. 1994.

[20] S.A. Huggett and K.P. Tod, An Introduction to Twistor Theory, London Mathematical Society Student Texts vol. 4, Cambridge University Press 1985.

[21] H.P. Jakobsen, Conformal covariants. Publ. Res. Inst. Math. Sci. **22** (1986), 345–364.

[22] F. John, The ultrahyperbolic differential equation with four independent variables, Duke Math. Jour. **4** (1938), 300–322.

[23] R. Michel, Problèmes d'analyse géométriques liés à la conjecture de Blaschke, Bull. Soc. Math. France **101** (1973), 17–69.

[24] R. Michel, Sur quelques problèmes de géométrie globale des géodésiques, Bol. Soc. Bras. Mat. **9** (1978), 19–38.

[25] V.A. Sharafutdinov, Integral Geometry of Tensor Fields, VSP Utrecht 1994.

DEPARTMENT OF MATHEMATICS, UNIVERSITY OF EDINBURGH, JAMES CLERK MAXWELL BUILDING, THE KING'S BUILDINGS, MAYFIELD ROAD, EDINBURGH EH9 3JZ, SCOTLAND
E-mail address: tnb@mathematics.edinburgh.ac.uk

DEPARTMENT OF PURE MATHEMATICS, UNIVERSITY OF ADELAIDE, SOUTH AUSTRALIA 5005
E-mail address: meastwoo@maths.adelaide.edu.au

Contemporary Mathematics
Volume **278**, 2001

Injectivity for a weighted vectorial Radon transform

Jan Boman

ABSTRACT. We prove injectivity and support theorems for weighted vectorial Radon transforms acting on divergence free vector fields assuming the weight is positive and real analytic. By means of a counterexample we show that the same statements are not true if the weight function is only assumed to be positive and C^∞.

1. Introduction. A weighted plane Radon transform can be defined by

$$R_\rho f(L) = \int_L f \rho_L \, ds, \quad L \in \mathcal{L},$$

where f is an integrable function with compact support, $\mathcal{L}$ denotes the set of (oriented) lines L in $\mathbf{R}^2$, $\rho_L(x)$ is a positive weight function defined for $x \in L$, $L \in \mathcal{L}$, and ds is arc length measure on the line L. Here we shall consider a weighted vectorial Radon transform $\mathcal{V}_\rho$ defined by

$$(1.1) \qquad \mathcal{V}_\rho f(L) = \int_L \langle f, e_L \rangle \rho_L ds = (R_\rho \langle f, e_L \rangle)(L), \quad L \in \mathcal{L},$$

where f is a vector field with compact support in the plane, e_L is the unit vector in the direction of L, and $\langle \cdot, \cdot \rangle$ is the scalar product in $\mathbf{R}^2$. The weight function ρ_L will be allowed to depend on the orientation of the line L.

In this note we shall prove that a *divergence free* vector field f with compact support can be recovered from $\mathcal{V}_\rho f$, if the weight $\rho_L(x)$ is positive and real analytic as a function of (x, L). We also show that the corresponding statement is not true if the weight $\rho_L(x)$ is only assumed to be positive and C^∞ (Theorem 3).

Denote the set of compactly supported integrable functions or vector fields in $\mathbf{R}^2$ by $L_c^1(\mathbf{R}^2)$ and the set of distributions with compact support by $\mathcal{E}'(\mathbf{R}^2)$.

THEOREM 1. *Assume that the weight $(x, L) \mapsto \rho_L(x)$ is positive and real analytic. Let f be a vector field in $L_c^1(\mathbf{R}^2)$ (or $\mathcal{E}'(\mathbf{R}^2)$) satisfying $\mathrm{div}\, f = \partial_1 f_1 + \partial_2 f_2 = 0$ in the sense of the theory of distributions, and assume that*

$$(1.2) \qquad \mathcal{V}_\rho f(L) = 0 \quad \text{for all lines} \quad L \in \mathcal{L}.$$

2000 *Mathematics Subject Classification.* 44A12.

© 2001 American Mathematical Society

Then $f = 0$.

The theorem is applicable to flow in a bounded region with tangental direction along the boundary. For if D is a bounded region with C^1 boundary and the vector field $f \in C^1(\bar{D})$ is divergence free in the interior of D, has vanishing normal component along the boundary of D, and vanishes outside D, then div $f = 0$ in $\mathbf{R}^2$ in the sense of the theory of distributions.

The idea of the proof of Theorem 1 is as follows. If ρ is constant, it is well known that $\mathcal{V}_\rho f = 0$ implies $-\partial_2 f_1 + \partial_1 f_2 = 0$ (see e.g. [**SSLP**]). Together with div $f = 0$ this implies that f is harmonic, and since f has compact support it must be identically zero. We shall denote the scalar quantity $-\partial_2 f_1 + \partial_1 f_2$ by rot f. In the more general case when ρ is smooth and positive but not necessarily constant we shall prove that $\mathcal{V}_\rho f = 0$ implies $Qf = 0$ where Q is a pseudodifferential operator with leading term rot. This means that $Qf = \text{div } f = 0$ is an elliptic system of pseudodifferential operators. Such a system can only have smooth solutions, and if ρ is also real analytic it can only have real analytic solutions. And if f is real analytic and has compact support it must be identically zero.

Using more refined arguments — description of local regularity of f in terms of the analytic wave front set — we also prove the following local uniqueness theorem, similar to those given in for instance [**BQ1**], [**BQ2**], [**Bo1**], [**Q1**]; see also Quinto's article [**Q2**] in these proceedings.

THEOREM 2. *Assume that the weight $\rho_L(x)$ and the vector field f satisfy the assumptions of Theorem 1 except (1.2). Let $\mathcal{L}_0$ be an open connected set of lines in $\mathcal{L}$ such that at least one line in $\mathcal{L}_0$ is disjoint from the support of f. Assume that $\mathcal{V}_\rho f(L) = 0$ for all lines $L \in \mathcal{L}_0$. Then $f = 0$ on the union of all lines in $\mathcal{L}_0$.*

As a special case we get the traditional kind of support theorem:

COROLLARY. *Assume that the weight $\rho_L(x)$ and the vector field f satisfy the assumptions of Theorem 1 except (1.2), and let K be a convex, compact subset of $\mathbf{R}^2$. Assume that $\mathcal{V}_\rho f(L) = 0$ for all lines L not intersecting K. Then $f = 0$ in the complement of K.*

Using the methods in [**Bo2**] we prove the following theorem, which shows that the assumption in Theorem 1 and Theorem 2 that $\rho_L(x)$ is real analytic cannot be weakened to C^∞.

THEOREM 3. *There exists a positive weight function $(x, L) \mapsto \rho_L(x) \in C^\infty$ and a C^∞ divergence free vector field f with compact support, not identically zero, such that*

(1.3) $$\int_L \langle f, e_L \rangle \rho_L ds = 0 \quad \text{for all lines } L \in \mathcal{L}.$$

As a general reference for pseudodifferential operators we suggest [**H**], ch. 18. Analytic pseudodifferential operators were introduced by Boutet de Monvel and Krée, and invertibility of elliptic analytic symbols was proved in [**BK**]. See also [**Bj**], where the microlocal version of the same theorem is treated, which we shall need in Section 3. For a survey on vector tomography see [**SS**]. Some algorithms and computer experiments are described in [**S**].

I wish to thank Kent Stråhlén for useful comments on an earlier version of this note.

2. The pseudodifferential operator $\mathcal{V}^*_{1/\rho}\mathcal{V}_\rho$. Denoting the line $\langle \omega, x\rangle = p$ by $L(\omega, p)$ for $(\omega, p) \in S^1 \times \mathbf{R}$, writing $\rho(x, \omega) = \rho_{L(\omega,p)}(x)$ for $p = \langle \omega, x\rangle$, and using the convention that the line $L(\omega, p)$ is oriented in the direction $\omega^\perp = (-\omega_2, \omega_1)$ we can write

$$\mathcal{V}_\rho f(\omega, p) = \int_{L(\omega,p)} \langle \omega^\perp, f(x)\rangle \rho(x,\omega) ds = (R_\rho \langle \omega^\perp, f\rangle)(\omega, p), \quad (\omega, p) \in S^1 \times \mathbf{R}.$$

For a weight function $\sigma(x,\omega) \geq 0$ we define the weighted adjoint vectorial Radon transform $\mathcal{V}^*_\sigma$ by

$$(2.1) \qquad \mathcal{V}^*_\sigma \phi(x) = \int_{S^1} \omega^\perp \phi(\omega, \langle \omega, x\rangle) \sigma(x,\omega) d\omega,$$

where $d\omega$ is arc length measure on the circle S^1. Note that $\mathcal{V}^*_\sigma$ takes scalar valued functions on $\mathcal{L} \simeq S^1 \times \mathbf{R}$ into vector valued functions on $\mathbf{R}^2$. The operator $\mathcal{V}^*_\rho$ is the adjoint of $\mathcal{V}_\rho$ in the sense that

$$\langle \mathcal{V}_\rho f, \phi\rangle_{L^2(S^1\times\mathbf{R})} = \langle f, \mathcal{V}^*_\rho \phi\rangle_{L^2(\mathbf{R}^2)};$$

here $\langle \cdot, \cdot\rangle_E$ denotes the usual inner product in the space E, and we note that we have vector valued functions in the right hand side. When ρ is constant equal to 1 we write just $\mathcal{V}$ and $\mathcal{V}^*$, respectively.

The operator $\sqrt{-\Delta}$ is defined for sufficiently rapidly decaying smooth functions h by

$$(\sqrt{-\Delta}h)\widehat{\,}(\xi) = |\xi|\hat{h}(\xi).$$

Here the Fourier transform of h is defined by $\hat{h}(\xi) = \int h(x)e^{-i\langle x,\xi\rangle} dx$. The operator Q mentioned in the introduction can now be defined by

$$(2.2) \qquad Qf = \sqrt{-\Delta}\,\mathrm{rot}\,\mathcal{V}^*_{1/\rho}\mathcal{V}_\rho f.$$

PROPOSITION 1. *Assume $(x,\omega) \mapsto \rho(x,\omega)$ is C^∞ and positive. Then Q is a pseudodifferential operator with polyhomogeneous symbol in the standard class, and*

$$(2.3) \qquad Q = 4\pi\,\mathrm{rot} + \quad \textit{lower order terms.}$$

If, in addition, $(x,\omega) \mapsto \rho(x,\omega)$ is real analytic, then Q is an analytic pseudodifferential operator.

The special case when ρ is constant equal to 1 is of particular interest:

$$(2.4) \qquad \sqrt{-\Delta}\,\mathrm{rot}\,\mathcal{V}^*\mathcal{V} = 4\pi\,\mathrm{rot}.$$

We shall see that Proposition 1 is an immediate consequence of (2.4) and

LEMMA 1. *Assume $\rho(x,\omega)$ is C^∞ and positive. Then $\mathcal{V}^*_{1/\rho}\mathcal{V}_\rho$ is a pseudodifferential operator and*

$$(2.5) \qquad \mathcal{V}^*_{1/\rho}\mathcal{V}_\rho = \mathcal{V}^*\mathcal{V} + \quad \textit{lower order terms.}$$

If, in addition, $(x,\omega) \mapsto \rho(x,\omega)$ is real analytic, then $\mathcal{V}^*_{1/\rho}\mathcal{V}_\rho$ is an analytic pseudodifferential operator.

PROOF. Let $\chi(x)$ be a function in $C_0^\infty(\mathbf{R}^2)$ equal to 1 in a neighborhood of a ball $D_R = \{x \in \mathbf{R}^2; |x| \leq R\}$ containing the support of f. By the definitions of $\mathcal{V}_\rho$ and $\mathcal{V}_{1/\rho}^*$ we can then write

$$\mathcal{V}_{1/\rho}^* \mathcal{V}_\rho f(x) = \int_{S^1} \frac{1}{\rho(x,\omega)} \int_{-\infty}^\infty \omega^\perp \langle \omega^\perp, f(x+t\omega^\perp)\rangle \rho(x+t\omega^\perp, \omega) \chi(x+t\omega^\perp) dt d\omega.$$

Changing variable $t\omega^\perp = z$ and observing that $\omega = \pm z^\perp/|z|$ depending on the sign of t and that $dtd\omega = dz/|z|$ we obtain

$$\mathcal{V}_{1/\rho}^* \mathcal{V}_\rho f(x) = \int_{\mathbf{R}^2} z\langle z, f(x+z)\rangle G(x,z) |z|^{-3} dz,$$

where

(2.6) $\qquad G(x,z) = \chi(x+z)\Big(\dfrac{\rho(x+z, z^\perp/|z|)}{\rho(x, z^\perp/|z|)} + \dfrac{\rho(x+z, -z^\perp/|z|)}{\rho(x, -z^\perp/|z|)}\Big).$

Viewing $f(x)$ as a column vector we can write $\mathcal{V}_{1/\rho}^* \mathcal{V}_\rho f(x) = \int K(x,z) f(x+z) dz$, where the kernel $K(x,z)$ is the 2 by 2 matrix

(2.7) $\qquad K(x,z) = G(x,z) \dfrac{1}{|z|^3} \begin{pmatrix} z_1^2 & z_1 z_2 \\ z_1 z_2 & z_2^2 \end{pmatrix}.$

Since $\rho(x,\omega)$ is C^∞ on $S^1 \times \mathbf{R}^2$ it is clear that $G(x,z)$ and $K(x,z)$ are C^∞ when $|z| \neq 0$ and that $G(x,z)$ is bounded near $z=(0,0)$, hence $K(x,z)$ is integrable as a function of z for fixed x. Let $p(x,\xi)$ be the Fourier transform of $K(x,z)$ with respect to z. Then $p \in C^\infty$ and (note that $K(x,z)$ is an even function of z)

$$\mathcal{V}_{1/\rho}^* \mathcal{V}_\rho f(x) = (2\pi)^{-2} \int_{\mathbf{R}^2} e^{i\langle x,\xi\rangle} p(x,\xi) \hat{f}(\xi) d\xi.$$

It is easily seen from the properties of K that $p(x,\xi)$ is a symbol in the standard class. The asymptotic expansion of $p(x,\xi)$ as $|\xi| \to \infty$ is obtained from the Taylor expansion of $K(x,z)$ at $z=0$. To obtain the leading term we expand $\rho(x+z,\omega)$ around $z=0$

$$\rho(x+z,\omega) = \rho(x,\omega) + \langle z, \psi_\omega(x,z)\rangle$$

with $\psi_\omega(x,z)$ smooth in all variables. Inserting this expression into (2.6) we obtain $G(x,z) = 2 + \mathcal{O}(|z|)$ as $|z| \to 0$ uniformly with respect to x for $|x| \leq R$, hence for such x we have $K = K_0 + K_1$, where

(2.8) $\qquad K_0(x,z) = K_0(z) = \dfrac{2}{|z|^3} \begin{pmatrix} z_1^2 & z_1 z_2 \\ z_1 z_2 & z_2^2 \end{pmatrix},$

and $K_1(x,z)$ is bounded. K_1 must therefore be the kernel of an operator of order at most -2. To finish the proof of (2.5) it is enough to observe that K must be equal to K_0 if ρ is constant, hence K_0 must be the kernel of $\mathcal{V}^*\mathcal{V}$.

Assume finally that $\rho(x,\omega)$ is real analytic and positive. Then $K(x,z)$ is real analytic for $|x| < R$ and z in some pointed neighborhood D_ε^0 of the origin, but this is not sufficient for $\mathcal{V}_{1/\rho}^* \mathcal{V}_\rho$ to be an analytic pseudodifferential operator. However, $K(x,z)$ has a stronger property: it can be written $K(x,z) = a(z/|z|, x, z)/|z|$ for $(x,z) \in D_R \times D_\varepsilon^0$, where $a(\omega, x, z)$ is real analytic on $S^1 \times D_R \times D_\varepsilon^0$. By standard estimates it follows from here that the complete symbol of $\mathcal{V}_{1/\rho}^* \mathcal{V}_\rho$ satisfies the estimates required for an analytic symbol (see [**BK**] or [**Bj**]). The proof of Lemma 1 is complete.

For the proof of Lemma 1 it was not necessary to compute the symbol of $\mathcal{V}^*\mathcal{V}$, but it is interesting to do this. This symbol must be the Fourier transform of the matrix $K_0(z)$. The Fourier transform (in the distribution sense) of $z_1^2/|z|^3$ is $2\pi\xi_2^2/|\xi|^3$, and the Fourier transform of $z_1z_2/|z|^3$ is $-2\pi\xi_1\xi_2/|\xi|^3$ (see Lemma 2), hence

$$(2.9) \qquad p_0(\xi) = \hat{K}_0(\xi) = \frac{4\pi}{|\xi|^3}\begin{pmatrix} \xi_2^2 & -\xi_1\xi_2 \\ -\xi_1\xi_2 & \xi_1^2 \end{pmatrix}.$$

Multiplying this matrix to the left with the row-vector $(-i\xi_2, i\xi_1)$ — the symbol of rot — gives $4\pi|\xi|^{-1}(-i\xi_2, i\xi_1)$; thus we have proved (2.4). Observe moreover that the matrix in (2.9) can be written

$$\begin{pmatrix} |\xi|^2 & 0 \\ 0 & |\xi|^2 \end{pmatrix} - \begin{pmatrix} \xi_1^2 & \xi_1\xi_2 \\ \xi_1\xi_2 & \xi_2^2 \end{pmatrix} = |\xi|^2\begin{pmatrix} 1 & 0 \\ 0 & 1 \end{pmatrix} - \begin{pmatrix} \xi_1 \\ \xi_2 \end{pmatrix}(\xi_1 \quad \xi_2).$$

Taking into account (2.9) we find that $\mathcal{V}^*\mathcal{V}$ satisfies the identity

$$(-\Delta)^{3/2}\mathcal{V}^*\mathcal{V}f = -4\pi\Delta f + 4\pi\operatorname{grad}\operatorname{div} f.$$

PROOF OF PROPOSITION 1. By Lemma 1 we can write $\mathcal{V}^*_{1/\rho}\mathcal{V}_\rho = \mathcal{V}^*\mathcal{V} + H$, where H is a pseudodifferential operator of lower order than that of $\mathcal{V}^*\mathcal{V}$. Inserting this expression in (2.2) and using (2.4) we obtain

$$Q = \sqrt{-\Delta}\operatorname{rot}(\mathcal{V}^*\mathcal{V} + H) = 4\pi\operatorname{rot} + \sqrt{-\Delta}\operatorname{rot} H,$$

and since the order of H is at most -2, the order of the second term can be at most 0, which completes the proof.

END OF PROOF OF THEOREM 1. As was already briefly indicated above, the proof goes as follows. Let f be a vector field with compact support and assume that $\mathcal{V}_\rho f = 0$. Then $Qf = 0$, if the operator Q is defined by (2.2). It is now clear from Proposition 1 that the system $Qf = \operatorname{div} f = 0$ is an elliptic system of analytic pseudodifferential operators, hence f must be real analytic. And since f has compact support it must be identically zero.

3. The support theorem.

As in [**BQ1**] and several later papers (see Section 1) we shall reason as in Hörmander's proof of Holmgren's uniqueness theorem, [**H**], Theorem 8.6.5. This means that the essential parts of the proof are a microlocal regularity theorem for solutions of the equation $\mathcal{V}_\rho f = 0$ (Proposition 2), and a unique continuation theorem for functions (distributions) that are microlocally real analytic in a certain cotangent direction: if the distribution f and the hypersurface S satisfy $WF_A(f) \cap N^*(S) = \emptyset$, then f is determined in a full neighborhood of S if it is given in a one-sided neighborhood of S ([**H**], Theorem 8.5.6); here $WF_A(f)$ denotes the analytic wave front set of f and $N^*(S)$ denotes the conormal manifold of the set S, that is, the set of $(x, \xi) \in T^*(\mathbf{R}^n)$ such that $x \in S$ and ξ is perpendicular to the tangent plane to S at x.

PROPOSITION 2. *Let f be an integrable function (distribution) with compact support such that* div $f = 0$ *in the sense of the theory of distributions, and $\mathcal{V}_\rho f(L) = 0$ for all lines L in some neighborhood of L_0. Assume that the weight function $\rho(x,\omega)$ is real analytic and positive. Then*

$$WF_A(f) \cap N^*(L_0) = \emptyset.$$

For the proof we shall need the following lemma.

LEMMA 2. *Let $k(x)$ be a function on $\mathbf{R}^2$ which is even, homogeneous of degree -1, and C^∞ outside the origin. Then the Fourier transform of k (as an element of $\mathcal{S}'$) is given by*

(3.1) $$\hat{k}(\xi) = 2\pi k(\xi^\perp).$$

PROOF. Write k as a sum of two even and homogeneous functions supported in the cones $|x_2| \leq 2|x_1|$ and $x_2| \geq |x_1|/2$, respectively. It is sufficient to prove the formula for one of those functions, say the first one, which we shall also denote by k. Then k can be written $k(x_1, x_2) = |x_1|^{-1} h(x_2/x_1)$ for some function $h \in C_0^\infty(\mathbf{R})$. By the definition of the Fourier transform we have $\langle \hat{k}, \varphi \rangle = \langle k, \hat{\varphi} \rangle$ for all test functions φ in the Schwartz class $\mathcal{S}(\mathbf{R}^2)$; here $\langle \cdot, \cdot \rangle$ denotes the pairing between distributions and test functions. It is enough to use test functions of the form $\varphi(\xi) = \varphi_1(\xi_1)\varphi_2(\xi_2)$. Then we can write $\langle k, \hat{\varphi} \rangle$ as a repeated integral

(3.2) $$\int |x_1|^{-1} \int h(x_2/x_1) \hat{\varphi}_2(x_2) dx_2 \, \hat{\varphi}_1(x_1) dx_1.$$

The inner integral can be written $|x_1| \int \hat{h}(\xi_2 x_1) \varphi_2(\xi_2) d\xi_2$ if $x_1 \neq 0$. Inserting this into (3.2) and changing the order of integration gives

$$\langle \hat{k}, \varphi \rangle = \int \varphi_2(\xi_2) \int \hat{h}(\xi_2 x_1) \hat{\varphi}_1(x_1) dx_1 d\xi_2$$

$$= 2\pi \int \varphi_2(\xi_2) |\xi_2|^{-1} \int h(-\xi_1/\xi_2) \varphi_1(\xi_1) d\xi_1 d\xi_2.$$

The last expression was obtained by an application of Parseval's formula to the inner integral in the middle term. Thus we have shown that $\hat{k}(\xi) = 2\pi |\xi_2|^{-1} h(-\xi_1/\xi_2)$, which proves (3.1) for this k and hence completes the proof of the lemma.

PROOF OF PROPOSITION 2. Let $(x^0, \omega^0) \in N^*(L_0)$. Take an even function $\psi \in C^\infty(S^1)$ with $\psi(\omega^0) \neq 0$ and supp ψ contained in a small neighborhood of $\pm\omega^0$. By means of the localized adjoint Radon transform $\mathcal{V}^*_{\psi/\rho}$ we define an operator T by $Tf = \mathcal{V}^*_{\psi/\rho} \mathcal{V}_\rho f$. By the assumption we have $\mathcal{V}_\rho f(\omega, p) = 0$ for ω close to ω^0 and p close to $\langle x^0, \omega^0 \rangle$; hence $Tf(x) = 0$ for x close to x^0, if the support of ψ is sufficiently small. In analogy with the proof of Theorem 1 we shall consider the system

(3.3) $$\sqrt{-\Delta}\operatorname{rot} Tf = 0, \quad \operatorname{div} f = 0,$$

and we shall prove that this system is microlocally elliptic at (x^0, ω^0). Let us first recall what this means. The system (3.3) can be written $Pf = 0$, where $P = (P_{ij})$ is a 2×2 matrix of first order pseudodifferential operators. The principal symbol $p^0_{ij}(x, \xi)$ of P_{ij} is homogeneous of degree 1 in ξ. By definition the system is microlocally elliptic at (x^0, ξ^0) if the determinant of the matrix $p^0(x^0, \xi^0) = (p_{ij}^0(x^0, \xi^0))$ is different from zero.

We may choose coordinates so that $\omega^0 = (0, 1)$. In our case $P_{21} = \partial_1$ and $P_{22} = \partial_2$, so $p_{21}^0 = i\xi_1$ and $p_{22}^0 = i\xi_2$. It follows that the determinant of $p^0(x^0, \omega^0)$ is equal to $ip_{11}^0(x^0, \omega^0)$. If T is the 2×2 matrix T_{ij}, then by the definition of P we have $P_{11} = \sqrt{-\Delta}(-\partial_2 T_{11} + \partial_1 T_{21})$. Denoting the principal symbol of T_{ij} by t_{ij}^0 we conclude that

$$p_{11}^0(x, \xi) = |\xi|(-i\xi_2 t_{11}^0(x, \xi) + i\xi_1 t_{21}^0(x, \xi)),$$

so $p_{11}^0(x^0, \omega^0) = -it_{11}^0(x^0, \omega^0)$. It is therefore enough to show that $t_{11}^0(x^0, \omega^0) \neq 0$.

Reasoning as in the proof of Theorem 1 we easily see that the kernel $M(x, z)$ of T can be written as in (2.7) with $G(x, z)$ replaced by $G(x, z)\psi(z^\perp/|z|)$. Taylor expanding $\rho(x + z, \omega)$ as before we obtain $M = M_0 + M_1$ where M_1 is bounded and $M_0(x, z) = M_0(z) = K_0(z)\psi(z^\perp/|z|)$, $K_0(z)$ given by (2.8). It follows that the principal symbol $t_{11}^0(x, \xi) = t_{11}^0(\xi)$ is equal to the Fourier transform of the function $2z_1^2|z|^{-3}\psi(z^\perp/|z|)$, hence by Lemma 2

$$t_{11}^0(\xi) = 4\pi\xi_2^2|\xi|^{-3}\psi(\xi/|\xi|),$$

so $t_{11}^0(\omega^0) \neq 0$ and the proof is complete.

PROOF OF THEOREM 2. The proof is the same as that of e.g. Theorem 2.2 in [**BQ1**], so we will be very brief here. It is enough to show that an arbitrary line $L_1 \in \mathcal{L}_0$ is disjoint from supp f. Let L_0 be a line in $\mathcal{L}_0$ not intersecting the support of f. Connect L_0 and L_1 with a continuous curve of lines $L_t \in \mathcal{L}_0$, $0 \leq t \leq 1$. If L_1 intersects supp f there must be a $t^* \in (0, 1]$ such that L_t is disjoint from supp f for $t < t^*$ and L_{t^*} intersects supp f. But this is impossible by virtue of Proposition 2 and Theorem 8.5.6 in [**H**] (see the paragraph preceding Proposition 2). The proof is complete.

4. The counterexample. For the proof of Theorem 3 we shall closely follow [**Bo2**]. To avoid lengthy repetitions we shall be rather brief and refer to [**Bo2**] for most of the steps.

PROOF OF THEOREM 3. We shall choose $f = (\nabla h)^\perp = (-\partial_2 h, \partial_1 h)$ for some scalar valued smooth function h with compact support. Then clearly div $f = 0$, and $\langle f, e_L \rangle = \langle (\nabla h)^\perp, \omega^\perp \rangle = \langle \omega, \nabla h \rangle = D_\omega h$, where D_ω denotes directional derivative with respect to x in the direction ω. Hence (1.3) takes the form

(4.1) $$\int_{L(\omega,p)} D_\omega h(\cdot)\rho(\cdot, \omega)ds = (R_\rho D_\omega h)(\omega, p) = 0, \quad (\omega, p) \in S^1 \times \mathbf{R}.$$

As in [**Bo2**] we choose h first and then construct ρ such that (4.1) holds. Take a function $\phi \in C^\infty(\mathbf{R})$ such that, supp $\phi \subset [4/5, 6/5]$, $\phi = 1$ on $[9/10, 11/10]$, and choose $h = \sum_{k=0}^\infty h_k/k!$, where h_0 will be defined later, and

$$h_k(x) = \phi(2^k(1-r))\cos 6^k\theta, \quad k = 1, 2, \ldots,$$

with $x = (r\cos\theta, r\sin\theta)$.

For lines intersecting a compact part of the open unit disk the construction of ρ is trivial as soon as we have verified that $D_\omega h$ assumes positive as well as negative values on the line $L(\omega, p)$ whenever that line intersects the interior of the disk (see Lemma 2 in [**Bo2**]). Fix a line $L(\omega, p)$ with $\varepsilon < |p| < 1 - \varepsilon$, where $\varepsilon > 0$. We claim that $D_\omega h_k$ changes sign on $L(\omega, p)$ if k is sufficiently large. Indeed, writing $\omega = (\cos\varphi, \sin\varphi)$ and applying the operator

(4.2) $\langle \omega, \nabla \rangle = \omega_1 \partial_1 + \omega_2 \partial_2 = \cos\varphi\, \partial_1 + \sin\varphi\, \partial_2 = \cos(\varphi - \theta)\, \partial_r + \sin(\varphi - \theta)\, r^{-1} \partial_\theta$

to h_k it is easily seen that the term containing $\partial_\theta h_k$ determines the sign of $D_\omega h_k$ on $L(\omega, p)$ if k is large enough; moreover, since $|p| > \varepsilon$ that term certainly changes sign on $L(\omega, p)$ if k is large enough. For lines passing close to the origin we shall need the term h_0. Choose $h_0 = h_{01} + h_{02}$, where $h_{01}(x) = \phi(4r)\cos 2\theta$ and $h_{02}(x) = \phi(8r)\sin 2\theta$. For continuity reasons it is enough to prove that $D_\omega h_0$ changes sign on the line $L(\omega, 0)$ through the origin for every ω. The function h_0 is even, $h_0(-x) = h_0(x)$, and hence $D_\omega h_0$ is odd, so it is enough to find one point on $L(\omega, 0)$ where $D_\omega h_0 \neq 0$. And it is readily verified that $D_\omega h_{01}$ and $D_\omega h_{02}$ vanish identically on $L(\omega, 0)$ precisely when $\sin 2\varphi = 0$ and $\cos 2\varphi = 0$, respectively; this proves the statement.

For the remaining lines, i.e., for $|\langle \omega, x \rangle| > p_0$ with some p_0, $1/2 < p_0 < 1$, to be chosen later, we choose

(4.3) $\rho(x,\omega) = 1 - \sum_{k=4}^{\infty} A_k(\omega, p) \langle \omega, \nabla h_k \rangle$ for $\langle \omega, x \rangle = p$, where

$$A_k(\omega, p) = k! \psi_{k-3}(|p|) \int_{L(\omega,p)} \langle \omega, \nabla h \rangle ds \bigg/ \int_{L(\omega,p)} \langle \omega, \nabla h_k \rangle^2 ds.$$

Here $\{\psi_k\}_{k=1}^{\infty}$ is a partition of unity on the interval $(1/2, 1)$ such that $\operatorname{supp} \psi_k \subset [1 - 2^{-k+1}, 1 - 2^{-k-1}]$ and $\sup |\partial^m \psi_k| \leq C_m 2^{km}$ for all m and k. It is obvious that (4.1) holds for $|p| > p_0$ and that ρ is C^∞ except possibly at points corresponding to lines tangent to the unit circle. When we have proved that the series (4.3) is uniformly convergent, it will also be clear that we can choose $p_0 < 1$ such that $\rho(x, \omega)$ is positive for $|\langle \omega, x \rangle| > p_0$.

To verify that $\rho \in C^\infty$ everywhere we shall prove that all derivatives of ρ tend to zero as $|\langle \omega, x \rangle| = |p|$ tends to 1. Setting

$$G(\omega, p) = \int_{L(\omega, p)} \langle \omega, \nabla h \rangle ds, \quad H_k(\omega, p) = \int_{L(\omega, p)} \langle \omega, \nabla h_k \rangle^2 ds, \quad k = 1, 2, \ldots,$$

we can write

(4.4) $\rho(x,\omega) = 1 - \sum_{k=4}^{\infty} k!\, G(\omega, p) \frac{\psi_{k-3}(|p|)}{H_k(\omega, p)} \langle \omega, \nabla h_k \rangle$ for $\langle \omega, x \rangle = p$.

We are going to estimate the derivatives of each factor in an arbitrary term in the sum above. To estimate the derivatives of $\langle \omega, \nabla h_k \rangle$ is easy; as in Lemma 1 (A) in [**Bo2**] we can prove

(4.5) $|\partial^\alpha \langle \omega, \nabla h_k \rangle| \leq C_\alpha 6^{|\alpha|k}, \quad k = 1, 2, \ldots;$

here ∂^α is an arbitrary derivative with respect to φ, x_1, and x_2. Similarly, using (4.2) we easily obtain the analogues of Lemma 1 (C) and (3.2) in [**Bo2**]:

(4.6) $|\partial^\beta H_k(\omega, p)| \leq C_\beta 6^{|\beta|k}, \quad k = 1, 2, \ldots,$

(4.7) $|\partial^\beta G(\omega, p)| \leq C_{m,\beta} 2^{-mk}/k!, \quad |p| \geq 1 - 2^{-k}, \quad k, m = 1, 2, \ldots,$

β being an arbitrary derivative with respect to φ and p. As in [**Bo2**] the proof of (4.7) depends on a sufficiently large number of partial integrations.

The only estimate that requires a different argument from the ones given in [**Bo2**] is the estimate from below of $H_k(\omega, p)$. If $\psi_{k-3}(|p|) \neq 0$, then $|p| \leq 1 - 4 \cdot 2^{-k}$, so the line $L(\omega, p)$ certainly intersects the ring where $\phi(2^k(1 - |x|)) = 1$ in two intervals of length at least $2^{-k}/5$. Hence

$$\left(\int_{L(\omega,p)} |\partial_\theta h_k|^2 ds\right)^{1/2} \geq c_1 6^k \cdot 2^{-k/2}, \quad \text{and} \quad \left(\int_{L(\omega,p)} |\partial_r h_k|^2 ds\right)^{1/2} \leq C\, 2^k.$$

If $h_k(x) \neq 0$ and $x \in L(\omega, p)$ we have $|x| > 1 - 2 \cdot 2^{-k}$ and

$$|\langle x, \omega \rangle| = |x||\cos(\varphi - \theta)| = |p| \leq 1 - 4 \cdot 2^{-k},$$

hence $|\cos(\varphi - \theta)| < 1 - 2^{-k}$, which gives $|\sin(\varphi - \theta)| > 2^{-k/2}$. Applying (4.2) to h_k we therefore obtain

(4.8) $$H_k(\omega, p)^{1/2} \geq 2^{-k/2} c_1 6^k 2^{-k/2} - C\, 2^k > c_2 > 0,$$

if $\psi_{k-3}(|p|) \neq 0$ and k is sufficiently large. Combining (4.6) and (4.8) we now obtain

(4.9) $$\left|\partial^\beta \frac{\psi_{k-3}(|p|)}{H_k(\omega, p)}\right| \leq C_\beta 6^{|\beta|k}.$$

Using Leibnitz' formula and the estimates (4.5), (4.7), and (4.9) we can now conclude that the series (4.4) is uniformly convergent after an arbitrary number of differentiations with respect to x_1, x_2, and φ, hence $\rho \in C^\infty$. This completes the proof of Theorem 3.

References

[Bj] Björk, J.-E., *Rings of differential operators*, North-Holland Publ. Comp., Amsterdam, Oxford, New York, 1979.

[BK] Boutet de Monvel, L., and Krée, P., *Pseudo-differential operators and Gevrey classes*, Ann. Inst. Fourier **17** (1967), 295-323.

[Bo1] Boman, J., *Holmgren's uniqueness theorem and support theorems for real analytic Radon transforms*, Contemp. Math. **140** (1992), 23-30.

[Bo2] Boman, J., *An example of non-uniqueness for a generalized Radon transform*, J. Anal. Math. **61** (1993), 395-401.

[BQ1] Boman, J., and Quinto, E. T., *Support theorems for real-analytic Radon transforms*, Duke Math. J. **55** (1987), 943-948.

[BQ2] Boman, J., and Quinto, E. T., *Support theorems for real-analytic Radon transforms on line complexes in three-space*, Trans. Amer. Math. Soc. **335** (1993), 877-890.

[H] Hörmander, L., *The analysis of linear partial differential operators, Vol. 1, 3.*, Springer-Verlag, Berlin, Heidelberg, New York, Tokyo, 1983.

[N] Natterer, F., *The mathematics of computerized tomography*, Wiley&Sons, New York, Brisbane, Toronto, Singapore, 1986.

[S] Stråhlén, K., *Studies of vector tomography*, thesis, Lund Institute of Technology, Doctoral Theses in Mathematical Sciences 1999:8.

[SS] Sparr, G., and Stråhlén, K., *Vector field tomography, an overview*, to appear.

[SSLP] Sparr, G., Stråhlén, K., Lindström, K., and Persson, H. W., *Doppler tomography for vector fields*, Inverse Problems **11** (1995), 1051-1061.

[Q1] Quinto, E. T., *Radon transforms satisfying the Bolker assumption*, in "75 years of Radon Transform" (S. Gindikin and P. Michor, ed.), 263-270, International Press, Boston, 1995.

[Q2] Quinto, E. T., *Radon transforms, differential equations, and microlocal analysis*, in these proceedings.

DEPARTMENT OF MATHEMATICS, STOCKHOLM UNIVERSITY, S-10691 STOCKHOLM, SWEDEN
E-mail address: jabo@matematik.su.se

Shape reconstruction in 2D from limited-view multifrequency electromagnetic data

Oliver Dorn, Eric L. Miller, and Carey M. Rappaport

ABSTRACT. In geophysical applications it is often the case that electromagnetic data are available for a number of frequencies but only in a strongly limited-view geometry. In these applications, it might be expected that the highest frequency data alone already give the reconstructions with the highest possible resolution, and the question arises how the additional information given by the lower frequency data should be incorporated into the reconstruction process. In the present paper we consider a shape reconstruction problem in electromagnetic cross-borehole tomography and demonstrate a possible way of using the lower frequency data as a stabilizing and regularizing component in the reconstructions. We employ an iterative shape reconstruction routine which was introduced earlier by the authors and which makes use of an adjoint field inversion technique as well as a level set representation of the shapes. We present numerical experiments which show that stable reconstructions of nontrivial shapes can be achieved from noisy limited-view electromagnetic data by properly incorporating the available lower frequency information.

1. Introduction

In a recent paper [12] the authors have introduced a two-step shape reconstruction method for electromagnetic tomography which uses adjoint fields and level sets. The first step of the method employs a source-type reconstruction scheme and yields a good first guess for the shapes. This preprocessing step is designed to be fast and stable and, in particular, to be able to handle the strong nonlinearity in the inverse problem which is due to the high contrasts of the parameter perturbations. The second step is iterative and uses the outcome of the preprocessing step as an initialization. In this second step a level set representation for the shapes

---

1991 *Mathematics Subject Classification*. Primary 35R30, 86A22.

Key words and phrases. Electromagnetic tomography, shape reconstruction, inverse problems.

The work of OD has been supported in part under a grant from the U. S. Dept. of Energy DE-FG07-97ID3566, and under the NSERC Research Grant 84306.

The work of ELM has been supported in part by CenSSIS, the Center for Subsurface Sensing and Imaging Systems, under the Engineering Research Centers Program of the National Science Foundation (award number EEC-9986821), by a grant from the U. S. Dept. of Energy DE-FG07-97ID3566, and by A CAREER Grant from the National Science Foundation MIP-9623721 .

The work of CMR has been supported in part by CenSSIS, the Center for Subsurface Sensing and Imaging Systems, under the Engineering Research Centers Program of the National Science Foundation (award number EEC-9986821), and by The Army Research Office, Multidisciplinary University Research Initiative Grant No. DAAG55-97-0013.

is employed in order to be able to keep track of topological changes of the shapes which often occur during the iteration process.

The aim is to retrieve an unknown number of penetrable objects (inclusions) imbedded in an inhomogeneous background medium from observations of electromagnetic (EM) fields. The EM fields in our application are in the frequency band of 5 to 30 MHz. The wavelengths of these fields are typically between 2-15 m in moist soil, where the relative dielectric constant is typically around 20 [**32**]. We are in particular interested in the imaging and monitoring of pollutant plumes at environmental cleanup sites given cross-borehole EM data, where the distances of the boreholes are not much larger than 10-20 m.

We assume that the known conductivity distribution is positive but small everywhere. The electromagnetic characteristics (permittivity and conductivity) of the background medium as well as of the material forming the inclusions are assumed to be known, but the main topological information concerning the *number*, *sizes*, *shapes*, and *locations* of the inclusions is missing and has to be reconstructed from the EM data. Inside the pollutant plumes, the permittivity is assumed to be constant with a known value which is typically much smaller (or larger) than the background permittivity, and the background permittivity itself is arbitrary but also known. No topological constraints are made on the shapes of the plumes. For example, they are allowed to be multiply connected, and to enclose 'cavities' or 'holes' filled with background material.

For alternative approaches to solving shape recovery problems in various applications we refer to [**16, 19, 25, 29**] and the references therein.

The paper is organized as follows. In section 2, we will formulate the shape reconstruction problem which we want to solve. Section 3 gives a short overview of the iterative shape reconstruction routine which uses a level set representation of the shapes and adjoint fields for the inversion task. The source-type preprocessing routine is outlined in section 4, and numerical experiments are presented and discussed in section 5. Finally, section 6 summarizes the results of the paper and indicates some interesting directions for future research.

2. The inverse problem

2.1. The Helmholtz Equation.
We consider the 2D Helmholtz Equation

(2.1) $$\Delta u + k^2(x)u = q(x) \quad \text{in } \mathbf{R}^2,$$

with complex wavenumber

(2.2) $$k^2(x) = \omega^2 \mu_0 \epsilon_0 \left[\epsilon(x) + i \frac{\sigma(x)}{\omega \epsilon_0} \right].$$

Here, $i^2 = -1$, ω denotes the angular frequency $\omega = 2\pi f$, μ_0 is the magnetic permeability in free space $\mu_0 = 4\pi \times 10^{-7}$ Henrys per meter, ϵ_0 is the dielectric permittivity in free space $\epsilon_0 = 8.854 \times 10^{-12}$ Farads per meter, ϵ is the relative dielectric permittivity (dimensionless), and σ is the electric conductivity in Siemens per meter. The form of (2.2) corresponds to time-harmonic line sources $\tilde{q}(x,t)$ which have a time-dependence $\tilde{q}(x,t) = q(x)e^{-i\omega t}$. For these sources we require that there exists a radius $r_0 > 0$ such that $\mathrm{supp}(q) \subset\subset B_{r_0}(0)$, where $B_r(x) = \{y \in \mathbf{R}, |x-y| < r\}$ denotes the open ball centered in x with radius $r > 0$. For simplicity we assume throughout the paper that we can find a ball $B_R(0)$ with $R > r$ such that the complex wavenumber $k^2(x)$ is constant with value k_0^2 in $\mathbf{R}^2 \backslash B_R(0)$, and that for this k_0 the field u generated by (2.1) satisfies the Sommerfeld radiation condition

$$\lim_{r \to \infty} \sqrt{r} \left(\frac{\partial u}{\partial r} - i k_0 u \right) = 0 \tag{2.3}$$

with $r = |x|$ where the limit is assumed to hold uniformly in all directions $x/|x|$. With this assumption, the problem (2.1)-(2.3) possesses a uniquely determined solution u in $\mathbf{R}^2$ [6].

Furthermore we will consider in this paper only the case that the conductivity is positive everywhere, $\sigma > 0$ in $\mathbf{R}^2$, and that it is small. Typical values in our geophysical examples will be $\sigma \approx 10^{-3} - 10^{-4}$ Siemens per meter or less [32].

We want to introduce some notation here which will be useful in the following. We denote the wavenumber $k^2(x)$ in short form by

$$k^2(x) = \kappa(x) = a\epsilon(x) + ib\sigma(x), \quad a = \omega^2 \mu_0 \epsilon_0, \quad b = \omega \mu_0. \tag{2.4}$$

The frequencies are assumed to be positive, $\omega > 0$, such that $a, b > 0$.

2.2. Formulation of the shape reconstruction problem. We assume that we are given p different source distributions q_j, $j = 1, \ldots, p$. For each of these sources, data are gathered at the detector positions x_d, $d = 1, \ldots, D_j$, for various frequencies f_k, $k = 1, \ldots, K$. The total number of receivers D_j, as well as their positions x_d, might vary with the source q_j. We assume, for simplicity in the notation, that these positions do not depend on the frequency f_k. This restriction is, however, not necessary for the derivation of the inversion method. We require that there exists a radius $r_1 > 0$ such that all receiver positions are inside the ball of radius r_1, i.e. $x_{jd} \in B_{r_1}(0)$ for all $d = 1, \ldots, D_j$, $j = 1, \ldots, p$.

In the application of *EM cross-borehole tomography*, the sources and receivers are typically situated in some boreholes, and the permittivity distribution ϵ (and/or the conductivity distribution σ) between these boreholes has to be recovered from the gathered data. In the 2D geometry considered here, typical sources are time-harmonic line sources which can be modelled in (2.1) by

$$q_j(x) = J_j \delta(x - x_j), \quad j = 1, \ldots, p, \tag{2.5}$$

where x_j denotes the 2D coordinates of the j-th line source, $j = 1, \ldots, p$, and the complex number J_j is the strength of the source. We will use these sources in our numerical experiments in section 5.

For a given source q_j and a given frequency f_k we collect a set of data $\tilde{G}_{jk}$ which is described by

$$\tilde{G}_{jk} = \left(\tilde{u}_{jk}(x_{j1}), \ldots, \tilde{u}_{jk}(x_{jd}), \ldots, \tilde{u}_{jk}(x_{jD_j}) \right)^T \in Z_j \tag{2.6}$$

with $Z_j = \mathbb{C}^{D_j}$ being the data space corresponding to a single experiment using one source and one frequency only. In (2.6), the fields $\tilde{u}_{jk}$ solve (2.1)-(2.3) with the correct permittivity distribution $\tilde{\epsilon}(x)$, i.e.

$$\Delta \tilde{u}_{jk} + [a_k \tilde{\epsilon}(x) + ib_k \sigma(x)] \tilde{u}_{jk} = q_j(x) \quad \text{in } \mathbf{R}^2 \tag{2.7}$$

with

$$a_k = \omega_k^2 \mu_0 \epsilon_0, \quad b_k = \omega_k \mu_0, \quad \omega_k = 2\pi f_k. \tag{2.8}$$

More generally, we define for a given source q_j the measurement operator M_j acting on solutions u_{jk} of (2.1) by

(2.9) $$M_j u_{jk} = \left(u_{jk}(x_{j1}), \ldots, u_{jk}(x_{jd}), \ldots, u_{jk}(x_{jD_j})\right)^T \in Z_j.$$

With this notation, (2.6) is written as

(2.10) $$\tilde{G}_{jk} = M_j \tilde{u}_{jk}, \qquad j = 1, \ldots, p, \; k = 1, \ldots, K.$$

We gather these data sets $\tilde{G}_{j,k}$ for all sources q_j, $j = 1, \ldots, p$, and all frequencies f_k, $k = 1, \ldots, K$, and the aim is to recover from this collection of data sets

(2.11) $$\tilde{G} = (\tilde{G}_{1,1}, \ldots, \tilde{G}_{p,K})^T$$

the unknown parameter distribution $\tilde{\epsilon}(x)$ in the domain of interest.

DEFINITION 2.1. Let us assume that we are given a constant $\hat{\epsilon} > 0$, an open ball $B_r(0) \subset \mathbf{R}^2$ with $r > \max(r_0, r_1) > 0$, and a bounded function $\epsilon_b : \mathbf{R}^2 \to \mathbf{R}$. We call a pair (Ω, ϵ), which consists of a compact domain $\Omega \subset\subset B_r(0)$ and a bounded function $\epsilon : \mathbf{R}^2 \to \mathbf{R}$, *admissible* if we have

(2.12) $$\epsilon|_\Omega = \hat{\epsilon}, \quad \epsilon|_{\mathbf{R}^2 \setminus \Omega} = \epsilon_b|_{\mathbf{R}^2 \setminus \Omega}.$$

In other words, a pair (Ω, ϵ) is *admissible* if ϵ is equal to a preassigned constant value $\hat{\epsilon}$ inside of Ω, and equal to the preassigned *background permittivity* ϵ_b outside of Ω. The domain Ω is called the *scattering domain*.

REMARK 2.2. For an admissible pair (Ω, ϵ), and for given $\hat{\epsilon}$, ϵ_b, the permittivity ϵ is uniquely determined by Ω.

Formulation of the shape reconstruction problem. Given a constant $\hat{\epsilon} > 0$, a background distribution ϵ_b, and some data $\tilde{G}$ as in (2.11). Find a shape $\tilde{\Omega}$ such that the corresponding admissible pair $(\tilde{\Omega}, \tilde{\epsilon})$ reproduces the data, i.e. (2.10) holds with $\tilde{u}_{jk}$ given by (2.7) for $j = 1, \ldots, p$, $k = 1, \ldots, K$.

3. A level set method for shape reconstruction

The *level set method* was originally developed by Osher and Sethian for describing the motion of curves and surfaces [23, 27]. Since then, it has found applications in a variety of quite different situations. Examples are image enhancement, computer vision, interface problems, crystal growth, or etching and deposition in the microchip fabrication. For an overview we refer to [28].

The idea of using a level set representation as part of a solution scheme for inverse problems involving obstacles was first suggested by Santosa in [26]. In that paper, two linear inverse problems, a deconvolution problem and the problem of reconstructing a diffraction screen, are solved by employing an optimization approach as well as a time evolution approach using the level set technique for describing the shapes. Santosa also outlines in that paper how the two presented methods can be generalized to nonlinear shape recovery problems.

More recently, a related method was applied to a nonlinear shape recovery problem by Litman *et al.* in [18]. In that work, an inverse transmission problem in free space is solved by a controled evolution of a level set function. This evolution is governed by a Hamilton-Jacobi type equation, whose velocity function has to be determined properly in order to minimize a given cost functional.

The approach developed here is not based on a Hamilton-Jacobi type equation. We follow an optimization approach, and employ a very specific inversion routine (an adjoint field technique) for solving it. This has the advantage that we do not have to propagate the level set function explicitly by computing a numerical Hamiltonian. Instead, our inversion routine provides us in each step with an update that has to be applied directly to the most recent level set function. Doing so, we automatically 'propagate' the level set function until the method converges.

3.1. Level set representation of the domains Ω. Assume that we are given a domain $\Omega \subset\subset B_r(0)$. The characteristic function $\chi_\Omega : \mathbf{R}^2 \to \{0,1\}$ is defined in the usual way as

$$(3.1) \qquad \chi_\Omega(x) = \begin{cases} 1 &, \ x \in \Omega \\ 0 &, \ x \in \mathbf{R}^2 \backslash \Omega. \end{cases}$$

DEFINITION 3.1. We call a function $\phi : \mathbf{R}^2 \to \mathbf{R}$ a *level set representation of* Ω if

$$(3.2) \qquad \chi_\Omega(x) = \Psi_\phi(x) \qquad \text{on } \mathbf{R}^2$$

where $\Psi_\phi : \mathbf{R}^2 \to \{0,1\}$ is defined as

$$(3.3) \qquad \Psi_\phi(x) = \begin{cases} 1 &, \ \phi(x) \leq 0 \\ 0 &, \ \phi(x) > 0. \end{cases}$$

For each function $\phi : \mathbf{R}^2 \to \mathbf{R}$ there is a domain Ω associated with ϕ by (3.2),(3.3) which we call $\Omega[\phi]$. It is clear that different functions $\phi_1, \phi_2, \phi_1 \neq \phi_2$, can be associated with the same domain $\Omega[\phi_1] = \Omega[\phi_2]$, but that different domains cannot have the same level set representation. Therefore, we can use the level set representation for unambiguously specifying a domain Ω by any one of its associated level set functions.

DEFINITION 3.2. We call a triple (Ω, ϵ, ϕ), which consists of a domain $\Omega \subset\subset B_r(0)$ and bounded functions $\epsilon, \phi : \mathbf{R}^2 \to \mathbf{R}$, *admissible* if the pair (Ω, ϵ) is admissible in the sense of definition 2.1, and ϕ is a valid level set representation of Ω.

REMARK 3.3. For an admissible triple (Ω, ϵ, ϕ), and for given $\hat{\epsilon}$, ϵ_b, the pair (Ω, ϵ) is uniquely determined by ϕ.

We use this definition to reformulate our shape reconstruction problem.

Level set formulation of the shape reconstruction problem. Given a constant $\hat{\epsilon} > 0$, a background distribution ϵ_b, and some data $\tilde{G}$ as in (2.11). Find a level set function $\tilde{\phi}$ such that the corresponding admissible triple $(\tilde{\Omega}, \tilde{\epsilon}, \tilde{\phi})$ reproduces the data, i.e. (2.10) holds with $\tilde{u}_{jk}$ given by (2.7) for $j = 1, \ldots, p$, $k = 1, \ldots, K$.

3.2. Operators. Throughout this paper we will denote the space of parameters ϵ_s by F, and the space of level set functions ϕ describing the shapes Ω by Φ. For simplicity, both spaces are assumed to be L_2-Hilbert spaces equipped with suitable inner products. The space of data $\tilde{G}_{jk}$ corresponding to the source position q_j and frequency f_k is given by $Z_j = \mathbf{C}^{D_j}$, where D_j is the number of receivers correponding to the source q_j.

We consider the following *decomposition of ϵ in $\mathbf{R}^2$*

(3.4) \hspace{2em} (i) \hspace{1em} $\epsilon = \epsilon_b + \epsilon_s$ \hspace{0.5em} in $\mathbf{R}^2$

(3.5) \hspace{2em} (ii) \hspace{1em} $\text{supp}(\epsilon_s) \subset\subset B_r(0)$.

It says that the permittivity distribution ϵ is decomposed into the background distribution ϵ_b and the perturbation ϵ_s which is assumed to have compact support and which we will refer to as the *scattering permittivity* in the following.

Given a constant $\hat{\epsilon}$ and a bounded function $\epsilon_b : \mathbf{R}^2 \to \mathbf{R}$, there is with each level set function $\phi \in \Phi$ a uniquely determined scattering permittivity $\Lambda(\phi)$ associated by putting

(3.6) \hspace{2em} $\Lambda(\phi)(x) = \begin{cases} \hat{\epsilon} - \epsilon_b(x) &, \phi(x) \leq 0 \\ 0 &, \phi(x) > 0. \end{cases}$

With (3.3) we can write this also as

(3.7) \hspace{2em} $\Lambda(\phi)(x) = \Psi_\phi(x)(\hat{\epsilon} - \epsilon_b(x))$ \hspace{1em}, \hspace{1em} $x \in \mathbf{R}^2$.

Notice that the operator Λ is chosen such that the triple (Ω, ϵ, ϕ) with $\epsilon = \epsilon_b + \Lambda(\phi)$ and domain $\Omega[\phi]$ forms an admissible triple (Ω, ϵ, ϕ) in the sense of definition 3.2.

We see that, for (Ω, ϵ, ϕ) an admissible triple, $\Lambda(\phi)$ is just the scattering permittivity ϵ_s as defined in (3.4),(3.5)

(3.8) \hspace{2em} $\Lambda(\phi)(x) = \epsilon_s(x) = \chi_\Omega(x)(\hat{\epsilon} - \epsilon_b(x))$ \hspace{1em}, \hspace{1em} $x \in \mathbf{R}^2$.

Let us assume now that we are given a background permittivity ϵ_b and that we have collected some data $\tilde{G}_{jk}$ which correspond to the 'true' permittivity distribution

(3.9) \hspace{2em} $\tilde{\epsilon} = \epsilon_b + \tilde{\epsilon}_s$,

where $\tilde{\epsilon}_s$ is the 'true' scattering permittivity. The *residual operators* R_{jk} map for a source position q_j and a frequency f_k a given scattering permittivity ϵ_s to the corresponding mismatch in the data

(3.10) \hspace{2em} $R_{jk} : F \longrightarrow Z_j$ \hspace{1em}, \hspace{1em} $R_{jk}(\epsilon_s) = M_j u_{jk} - \tilde{G}_{jk}$

where u_{jk} solves

(3.11) \hspace{2em} $\Delta u_{jk} + \left[a_k(\epsilon_b + \epsilon_s)(x) + i b_k \sigma(x)\right] u_{jk} = q_j$

and M_j is the measurement operator defined in (2.9). From (2.10) we see that for the 'true' scattering permittivity the residuals vanish,

(3.12) \hspace{2em} $R_{jk}(\tilde{\epsilon}_s) = 0$ \hspace{1em} for $j = 1, \ldots, p,\ k = 1, \ldots, K$,

if the data are noise-free.

The *forward operators* T_{jk} which map a given level set function $\phi \in \Phi$ into the corresponding mismatch in the data are defined by

(3.13) \hspace{2em} $T_{jk} : \Phi \longrightarrow Z_j$ \hspace{1em}, \hspace{1em} $T_{jk}(\phi) = R_{jk}(\Lambda(\phi))$

for $j = 1, \ldots, p,\ k = 1, \ldots, K$. The goal is to find a level set function $\tilde{\phi} \in \Phi$ such that

(3.14) \hspace{2em} $T_{jk}(\tilde{\phi}) = 0$ \hspace{1em} for $j = 1, \ldots, p,\ k = 1, \ldots, K$.

We mention that all three operators Λ, R_{jk} and T_{jk} are nonlinear.

We will also need an expression for the linearized operators $T'_{jk}[\phi]$. As it is described in [12], it is convenient to use an approximation to these operators which is easier to handle in the formal derivation of the adjoint scheme employed here. The approximated linearized operator is given by

(3.15) $$T'_{jk}[\phi] : \Phi \longrightarrow Z_j \quad , \quad T'_{jk}[\phi]\delta\phi = M_j v_{jk}$$

where v_{jk} solves the linearized equation

(3.16) $$\Delta v_{jk} + \left[a_k(\epsilon_b + \epsilon_s)(x) + ib_k\sigma(x)\right] v_{jk} = -a_k \delta\epsilon_s(x) u_{jk}(x),$$

with u_{jk} a solution of (3.11) and

(3.17) $$\delta\epsilon_s(x) = -[\hat{\epsilon} - \epsilon_b(x)] \frac{\delta\phi(x)}{|\nabla\phi(x)|} C_\rho(\Gamma) \chi_{B_\rho(\Gamma)}(x).$$

Here, $\chi_{B_\rho(\Gamma)}(x)$ is the characteristic function supported on the finite width neighborhood $B_\rho(\Gamma) = \cup_{y \in \Gamma} B_\rho(y)$, which is introduced in [12] as an approximation to the surface Dirac delta function concentrated on $\Gamma = \partial\Omega[\phi]$. We have $C_\rho(\Gamma) = L(\Gamma)/\text{Vol}(B_\rho(\Gamma))$ where $L(\Gamma) = \int_{B_r(0)} \hat{\delta}_\Gamma(x) dx$ is the length of Γ, and $\text{Vol}(B_\rho(\Gamma)) = \int_{B_r(0)} \chi_{B_\rho(\Gamma)}(x) dx$ is the volume of $B_\rho(\Gamma)$. For a very small ρ we will get a very large weight $C_\rho(\Gamma)$, whereas for increasing ρ this weight $C_\rho(\Gamma)$ decreases accordingly.

3.3. A nonlinear Kaczmarz-type approach. The algorithm works in a 'single-step fashion' as follows. Instead of using the data (2.11) for all sources and all frequencies simultaneously, we only use the data for one source and one frequency at a time while updating the linearized residual operator after each determination of the corresponding incremental correction $\delta\phi$.

To be more specific, let us assume that we are given a level set function $\phi^{(n)}(x)$ and a scattering permittivity $\epsilon_s^{(n)}(x)$ such that $(\Omega^{(n)}, \epsilon_b + \epsilon_s^{(n)}, \phi^{(n)})$ forms an admissible triple in the sense of definition 3.2. Using a data set $\tilde{G}_{jk}$ corresponding to the fixed source position q_j and the frequency f_k, we want to find an update $\delta\phi^{(n)}$ to $\phi^{(n)}$ such that for the admissible triple

(3.18) $$\left(\Omega^{(n+1)}, \epsilon_b + \epsilon_s^{(n+1)}, \phi^{(n+1)}\right) :=$$
$$\left(\Omega[\phi^{(n)} + \delta\phi^{(n)}], \epsilon_b + \Lambda(\phi^{(n)} + \delta\phi^{(n)}), \phi^{(n)} + \delta\phi^{(n)}\right)$$

the residuals in the data corresponding to this source and this frequency vanish

(3.19) $$T_{jk}(\phi^{(n+1)}) = T_{jk}(\phi^{(n)} + \delta\phi^{(n)}) = 0.$$

Applying a Newton-type approach, we get from (3.19) a correction $\delta\phi^{(n)}$ for $\phi^{(n)}$ by solving

(3.20) $$T'_{jk}[\phi^{(n)}]\delta\phi^{(n)} = -T_{jk}(\phi^{(n)}) = -\left(M_j u_{jk} - \tilde{G}_{jk}\right)$$

where u_{jk} satisfies (3.11) with $\epsilon_s = \Lambda(\phi^{(n)})$

(3.21) $$\Delta u_{jk} + \left[a_k(\epsilon_b + \Lambda(\phi^{(n)}))(x) + ib_k\sigma(x)\right] u_{jk} = q_j(x)$$

and

(3.22) $$\epsilon_b(x) + \Lambda(\phi^{(n)})(x) = \begin{cases} \hat{\epsilon} &, \quad x \in \Omega[\phi^{(n)}] \\ \epsilon_b(x) &, \quad x \in \mathbf{R}^2 \setminus \Omega[\phi^{(n)}]. \end{cases}$$

Since we have only few data given for one source and one frequency, equation (3.20) usually will have many solutions (in the absence of noise), such that we have to pick one according to some criterion. We choose to take that solution which minimizes the energy norm of $\delta\phi^{(n)}$

(3.23) $\quad$ Min $\|\delta\phi^{(n)}\|_2 \quad$ subject to $\quad T'_{jk}(\phi^{(n)})\delta\phi^{(n)} = -\left(M_j u_{jk} - \tilde{G}_{jk}\right)$.

This solution can be formulated explicitly. It is

(3.24) $\quad \delta\phi^{(n)}_{\text{MN}} = -T'_{jk}[\phi^{(n)}]^* \left(T'_{jk}[\phi^{(n)}]T'_{jk}[\phi^{(n)}]^*\right)^{-1} \left(M_j u_{jk} - \tilde{G}_{jk}\right),$

where $T'_{jk}[\phi^{(n)}]^*$ denotes the adjoint operator to $T'_{jk}[\phi^{(n)}]$.

After correcting ϕ by $\phi \to \phi + \delta\phi_{jk}$, where $\delta\phi_{jk}$ is given by (3.24), we use the updated residual equation (3.20) to compute the next correction $\delta\phi_{j'k'}$. Doing this for one equation after the other, until each of the sources q_j and each of the frequencies f_k has been considered exactly once, will yield one complete sweep of the algorithm. This procedure is similar to the Kaczmarz method for solving linear systems, or the algebraic reconstruction technique (ART) in x-ray tomography [20] and the simultaneous iterative reconstruction technique (SIRT) as presented in [9]. Related approaches have also been employed in ultrasound tomography by Natterer and Wübbeling [21], in more general bilinear inverse problems by Natterer [22], in optical tomography by Dorn [10], and in 3D-electromagnetic induction tomography (EMIT) by Dorn et alii [11]. Because of the similarity of the derived method to the ART scheme in x-ray tomography, we will often refer to it in short as 'levelART'.

3.4. The adjoint linearized operator $T'_{jk}[\phi^{(n)}]^*$. In order to calculate the minimal norm solution (3.24), we will need a practically useful expression for the adjoint of the linearized operator $T'_{jk}[\phi]$. We will present such an expression in this section. The calculation of the action of this operator will typically require to solve an adjoint Helmholtz problem. This explains the name 'adjoint field method' which is often used for the general inversion method employed here.

Let $\hat{C}_{jk}$ be defined as $\hat{C}_{jk} = T'_{jk}[\phi^{(n)}]T'_{jk}[\phi^{(n)}]^*$, or some approximation to it as it was derived in [12]. In order to calculate a correction $\delta\phi^{(n)}_{\text{MN}}$ by (3.24) we have to apply the operator $T'_{jk}[\phi^{(n)}]^*$ to the vector

(3.25) $\quad\quad\quad\quad\quad \zeta := \hat{C}_{jk}^{-1}\left(M_j u_{jk} - \tilde{G}_{jk}\right).$

An explicit formula for $T'_{jk}[\phi^{(n)}]^*\zeta$ is given next.

(3.26) $\quad T'_{jk}[\phi^{(n)}]^*\zeta = \dfrac{[\hat{\epsilon} - \epsilon_b(x)]}{a_k|\nabla\phi^{(n)}(x)|} \operatorname{Re}\left(\overline{u_{jk}(x)z_{jk}(x)}\right) C_\rho(\Gamma)\chi_{B_\rho(\Gamma)}(x),$

where u_{jk} solves

(3.27) $\quad\quad\quad\quad\quad \Delta u_{jk} + \kappa_k(x)u_{jk} = q_j(x),$

and z_{jk} solves the 'adjoint equation'

(3.28) $\quad\quad\quad\quad\quad \Delta z_{jk} + \kappa_k(x)z_{jk} = \sum_{d=1}^{D_j} \bar{\zeta}_d \delta(x - x_{jd})$

with

(3.29) $\quad\quad\quad\quad\quad \kappa_k(x) = a_k\left[\epsilon_b(x) + \Lambda(\phi^{(n)})(x)\right] + ib_k\sigma(x)$

and a_k, b_k defined as in (2.8).

Table 1: *The 'levelART' algorithm.*
Preparation step.
- Calculate $\hat{C}_{jk}$ and $D_{jk} = \hat{C}_{jk}^{-1}$ for each source q_j, $j = 1, \ldots, p$, and each frequency f_k, $k = 1, \ldots, K$, and store in memory for later use.
- Build groups of frequencies $G_m = \{f_1, \ldots, f_{K_m}\}$, $m = 1, \ldots, M$.

Initialization.

$n = 0$;

$(\Omega^{(0)}, \epsilon^{(0)}, \phi^{(0)})$ given from low-frequency STAF.

Reconstruction loop.

    FOR $m = 1 : M$ march over frequency groups G_m
        FOR $i = 1 : I_m$ perform I_m sweeps for frequency group G_m
            FOR $k = 1 : K_m$ march over frequencies in G_m
              FOR $j = 1 : p$ This loop needs only one LR-factorization
                  $\zeta_{jk} = D_{jk}(M_j u_{jk} - \tilde{G}_{jk})$; u_{jk} solves (3.27) with $\epsilon^{(n)}$, q_j, f_k
                  $\delta\phi_{jk} = -\frac{\hat{\epsilon}-\epsilon_b(x)}{a_k}\text{Re}(\overline{u_{jk} z_{jk}})\chi_{B_\rho(\Gamma)}$; z_{jk} solves (3.28) with $\epsilon^{(n)}$ and ζ_{jk}
              END
            $\delta\phi^{(n)}(x) = \sum_{j=1}^p \delta\phi_{jk}(x)$;
            $\phi^{(n+1)} = C_{\text{LS}}^{(n)}(\phi^{(n)} + \eta\frac{C_\rho(\Gamma)}{c_1}\delta\phi^{(n)})$;   update level set function
            Optional step: 'curve shortening by diffusion'. See [12].
            $\epsilon^{(n+1)} = \epsilon_b + \Lambda(\phi^{(n+1)})$; $n = n+1$;   Reinitialization $n \to n+1$
        END
    END   alternatively, some stopping criteria can be used here
    END
$(\Omega^{(N)}, \epsilon^{(N)}, \phi^{(N)}) = (\Omega[\phi^{(n)}], \epsilon_b + \Lambda(\phi^{(n)}), \phi^{(n)})$;   Final reconstruction.

3.5. Implementation: The levelART algorithm. In the levelART algorithm described in Table 1, η is a relaxation parameter for the update of the level set function which is determined empirically. The constant $C_\rho(\Gamma)$ could be calculated explicitly for the actual curve $\Gamma^{(n)}$, or it could be approximated by some value corresponding to a simple geometrical object (to give an example, in case of a single circle it would be $C_\rho(\Gamma) = (2\rho)^{-1}$). In our numerical experiments so far, however, it is simply considered as part of η. The same holds true for the constant c_1 which has been introduced in [12] as an approximation to $|\nabla\phi|$. The constant ρ is in our numerical experiments chosen between 30-40 cm, which corresponds to 2-3 grid cells. The scaling factor $C_{\text{LS}}^{(n)}$ is determined after each update to keep the global minimum (or maximum) of the level set function at a constant value. For more details see [12].

4. A simple source-type method (STAF)

The second method which we investigate in this paper, and which is mainly designed to provide us with a good first guess for the levelART algorithm, is the 'source-type adjoint field' (STAF) inversion method. Roughly speaking, the general idea of source-type reconstruction methods in inverse scattering is to split a given nonlinear inverse scattering problem into two subproblems. The first one is linear, and tries to recover a virtual 'equivalent source' in the medium which would be able to fit the data if applied with the known background distribution. This equivalent source is related to the unknown scattering permittivity by a nonlinear 'constitutive' relation. Therefore, in the second part of the algorithm, a nonlinear inverse problem has to be solved to derive the scattering permittivity from the recovered equivalent source distribution.

This idea is not at all new. It has been applied for example in the Source-Type Integral Equation (STIE) method of Habashy et al. [14], or in the method presented by Chew et al. in [5]. More recently, similar ideas have been applied by Abdullah et al. [1], Caorsi et al. [3], and van den Berg et al. [30, 31].

All of these approaches have in common that they use the source-type method as a stand-alone inversion scheme. Such a method has the advantage that it is not as sensitive to strong nonlinearities in the inverse problem as for example perturbation methods or the Born or Rytov approximation are [8, 13, 17].

On the other hand, interpreting the inverse scattering problem as an inverse source problem is not without drawbacks. For example, the existence of so-called 'non-radiating sources' or 'invisible sources' gives rise to a nonuniqueness in the inverse source problem, which is difficult to deal with when solving the nonlinear part [1, 7, 14]. Moreover, it is not clear at all how to combine properly the information corresponding to different experiments, since each experiment creates its own 'equivalent sources' and its own 'invisible sources'. For more information about possible applications, advantages and drawbacks of the source-type scheme as a stand-alone inversion tool we refer to [1, 2, 3, 5, 7, 14, 30, 31].

Our approach is different from those mentioned above. We only want to find a good approximation to the shapes of the objects, and a corresponding initial level set function which can be used to start the levelART algorithm. In particular, we can make use of our prior information about the permittivity distribution. This will allow us to circumvent most of the problems of source-type schemes which have been mentioned above.

We will now describe this method, which we will call the Source-Type Adjoint Field (STAF) method, in more details.

4.1. Formulation as an inverse scattering problem. To start with we formulate the following inverse scattering problem.

Inverse Scattering Problem. Let us assume that we are given a bounded function $\epsilon_b : \mathbf{R}^2 \to \mathbf{R}$, and some data $\tilde{G}$ as in (2.11). Find a bounded function $\tilde{\epsilon}_s : \mathbf{R}^2 \to \mathbf{R}$ with $\mathrm{supp}(\epsilon_s) \subset\subset B_r(0)$ such that $\tilde{\epsilon} = \epsilon_b + \tilde{\epsilon}_s$ reproduces the data, i.e. (2.10) holds with $\tilde{u}_{jk}$ given by (2.7) for $j = 1, \ldots, p$, $k = 1, \ldots, K$.

For a fixed frequency f_k and a source q_j, let $\tilde{u}_{jk}$ be the solution of

$$(4.1) \qquad \Delta \tilde{u}_{jk} + \left(a_k(\epsilon_b + \tilde{\epsilon}_s)(x) + ib_k\sigma(x)\right) \tilde{u}_{jk} = q_j(x),$$

and let u_{jk} be the solution of the 'unperturbed' equation

$$(4.2) \qquad \Delta u_{jk} + (a_k\epsilon_b(x) + ib_k\sigma(x)) u_{jk} = q_j(x).$$

Define
$$\tilde{v}_{jk} := u_{jk} - \tilde{u}_{jk}. \tag{4.3}$$
Subtraction of (4.2) from (4.1) shows that $\tilde{v}_{jk}$ solves
$$\Delta \tilde{v}_{jk} + (a_k \epsilon_{\mathrm{b}}(x) + ib_k \sigma(x)) \, \tilde{v}_{jk} = \tilde{Q}^s_{jk}(x), \tag{4.4}$$
where the 'scattering source' $\tilde{Q}^s_{jk}(x)$ is defined as
$$\tilde{Q}^s_{jk}(x) = a_k \tilde{\epsilon}_{\mathrm{s}}(x) \tilde{u}_{jk}(x). \tag{4.5}$$
Denoting by Y the Hilbert space of sources, we introduce a 'source type' forward operator A_{jk} by putting
$$A_{jk} : Y \longrightarrow Z_j \quad , \quad A_{jk} Q^s_{jk} = M_j v_{jk} \tag{4.6}$$
where M_j is the measurement operator defined in (2.9), and v_{jk} solves
$$\Delta v_{jk} + (a_k \epsilon_{\mathrm{b}}(x) + ib_k \sigma(x)) \, v_{jk} = Q^s_{jk}(x). \tag{4.7}$$
The operator A_{jk} is linear.

Let us assume now that we apply the 'correct' scattering source $\tilde{Q}^s_{jk}(x)$ defined by (4.5) as argument of A_{jk}. Then we know from (2.10), (3.9), (3.10) that
$$A_{jk} \tilde{Q}^s_{jk} = M_j \tilde{v}_{jk} = M_j(u_{jk} - \tilde{u}_{jk}) = M_j u_{jk} - \tilde{G}_{jk} = R_{jk}(0). \tag{4.8}$$
The vectors $R_{jk}(0)$ are easily computed by solving a forward problem on the background distribution (4.2). Therefore, all we have to do to get back the scattering source $\tilde{Q}^s_{jk}$ from the data $\tilde{G}_{jk}$ is to solve (4.8) for $\tilde{Q}^s_{jk}$. Doing so amounts to solving an ill-posed but *linear* inverse problem.

Once we have recovered $\tilde{Q}^s_{jk}(x)$, we want to get back $\tilde{\epsilon}_{\mathrm{s}}(x)$ by using the constitutive relation (4.5). This second part of the inversion scheme can be interpreted as solving a *nonlinear* inverse problem since $\tilde{u}_{jk}(x)$ depends on $\tilde{\epsilon}_{\mathrm{s}}(x)$.

Notice that $\tilde{Q}^s_{jk}(x)$ varies with different sources and frequencies, but that $\tilde{\epsilon}_{\mathrm{s}}(x)$ is the same for all sources and all frequencies (if we neglect dispersion). We will make use of this observation when we try to solve the nonlinear part (4.5). In the following, we describe the method which we will use to recover the scattering source $\tilde{Q}^s_{jk}(x)$ from a given data set $\tilde{G}_{jk}$ for a fixed source q_j and a fixed frequency f_k.

4.2. Looking for a scattering source. Since for a fixed (primary) source position and a fixed frequency we have only few data given to recover $\tilde{Q}^s_{jk}$, and since we have to take into account that also 'non-radiating' and 'invisible' sources have been generated in the experiment, we assume that there will be many solutions (in absence of noise) of (4.8). To pick one, we are looking for the solution with minimal norm
$$\text{Min} \; \|Q^s_{jk}\|_Y \quad \text{subject to} \quad A_{jk} Q^s_{jk} = R_{jk}(0). \tag{4.9}$$
It is given by
$$Q^s_{jk,\mathrm{MN}} = A^*_{jk} \left(A_{jk} A^*_{jk} \right)^{-1} R_{jk}(0), \tag{4.10}$$
where A^*_{jk} denotes the adjoint operator to A_{jk}. In our numerical experiments presented in this paper, we will simply replace the operator $A_{jk} A^*_{jk}$ by (a constant times) the identity which amounts to applying strong regularization.

The following theorem, which is proven in [12], tells us how to calculate the action of A^*_{jk} on a vector $\zeta \in Z_j$ in an efficient way.

THEOREM 4.1. *Let* $\zeta = (\zeta_1, \ldots, \zeta_{D_j})^T \in Z_j$ *and let* x_{jd}, $d = 1, \ldots, D_j$ *be the detector positions corresponding to the source* q_j. *Then,* $A^*_{jk}\zeta$ *is given by*

$$(4.11) \qquad A^*_{jk}\zeta = \overline{z_{jk}}\, \chi_{B_r(0)},$$

where z_{jk} *solves*

$$(4.12) \qquad \Delta z_{jk} + (a_k \epsilon_b + i b_k \sigma)\, z_{jk} = \sum_{d=1}^{D_j} \overline{\zeta_d} \delta(x - x_{jd}).$$

4.3. Recovery of the scattering permittivity. After we have found a scattering source Q^s_{jk} which satisfies (4.9), we want to use the constitutive relation

$$(4.13) \qquad Q^s_{jk}(x) = a_k \tilde{\epsilon}_s(x) \tilde{u}_{jk}(x),$$

which holds for the 'correct' scattering source $\tilde{Q}^s_{jk}$ according to (4.5), to find an approximation for $\tilde{\epsilon}_s(x)$.

Let $\tilde{u}_{jk}$ be a solution of (4.1) and u_{jk} a solution of (4.2). We decompose Q^s_{jk}, $\tilde{u}_{jk}$, u_{jk} and $\tilde{\epsilon}_s$ into amplitude and phase

$$(4.14) \qquad Q^s_{jk}(x) = |Q^s_{jk}(x)|\, e^{ir(x)}, \qquad \tilde{u}_{jk} = |\tilde{u}_{jk}|\, e^{i\tilde{s}(x)},$$

$$(4.15) \qquad u_{jk} = |u_{jk}|\, e^{is(x)}, \qquad \epsilon_s(x) = |\epsilon_s(x)|\, e^{it(x)},$$

where we have omitted the subscripts jk in the argument functions r, $\tilde{s}$, s, and t for simplicity in the notation. Making use of the fact that $\tilde{\epsilon}_s(x) \in \mathbf{R}$ we see that

$$(4.16) \qquad t(x) \in \{0, \pi\} \quad \text{for all } x \in \mathbf{R}^2.$$

With (4.14),(4.15) equation (4.9) decomposes into two equations, one for the amplitude and one for the phase. They are

$$(4.17) \qquad |Q^s_{jk}(x)| = a_k\, |\tilde{\epsilon}_s(x)|\, |\tilde{u}_{jk}|,$$
$$(4.18) \qquad r(x) = \tilde{s}(x) + t(x).$$

The observation in our numerical experiments is that, although $s(x)$ and $\tilde{s}(x)$ might be quite different from each other for large perturbations $\tilde{\epsilon}_s(x)$, the amplitudes $|u_{jk}(x)|$ and $|\tilde{u}_{jk}(x)|$ do not differ too much from each other in the scattering region. Therefore, in our applications it is a reasonable approximation to assume that

$$(4.19) \qquad |\tilde{u}_{jk}(x)| \approx |u_{jk}(x)| \quad \text{in } B_r(0).$$

With this approximation, (4.17) yields the following estimate for $|\tilde{\epsilon}_s(x)|$

$$(4.20) \qquad |\tilde{\epsilon}_s^{(jk)}(x)| \approx \frac{|Q^s_{jk}(x)|}{a_k |u_{jk}(x)|} \quad \text{in } B_r(0).$$

We have added the indices j and k on the left hand side of (4.20) to indicate that we have used only the data $\tilde{G}_{jk}$ corresponding to source q_j and frequency f_k for its determination.

Notice that the step (4.19), (4.20) is *nonlinear* since taking the amplitude of a complex number is a nonlinear operation. Therefore, the approach presented here is quite different from the usual Born approximation which approximates $\tilde{u}_{jk}$ by u_{jk}.

For the purposes of the present paper, the determination of $|\tilde{\epsilon}_s(x)|$ is already sufficient in order to get a good first guess for the shapes of the unknown inclusions,

Table 2: *The STAF algorithm.*
Select a frequency f_k.
> FOR $j = 1 : p$ This loop needs only one LR-factorization
> $R_{jk}(0) = M_j u_{jk} - \tilde{G}_{jk},$ u_{jk} solves (4.2)
> $Q_{jk}^s = A_{jk}^* R_{jk}(0) = \overline{z_{jk}} \chi_{B_r(0)},$ z_{jk} solves (4.12) with $\zeta_{jk} = R_{jk}(0)$
> $|\tilde{\epsilon}_s^{(jk)}(x)| = \frac{|Q_{jk}^s(x)|}{a_k |u_{jk}(x)|}$
> END

$|\tilde{\epsilon}_s(x)| = \frac{1}{p} \sum_{j=1}^{p} |\tilde{\epsilon}_s^{(jk)}(x)|,$ $x \in B_r(0)$

Fix a constant $C_l > 0$. The reconstructed shape is $\{x \in B_r(0) : |\tilde{\epsilon}_s(x)| \geq C_l\}$.

since we can now make use of our prior information about the correct value of $\hat{\epsilon}$ in (2.12).

In source-type reconstruction schemes it is generally not clear what the best way is to combine information corresponding to different source positions q_j, $j = 1, \ldots, p$, in the reconstruction process. We have chosen to use the following formula for this purpose:

$$(4.21) \qquad |\tilde{\epsilon}_s^{(k)}(x)| = \frac{1}{p} \sum_{j=1}^{p} |\tilde{\epsilon}_s^{(jk)}(x)|.$$

We mention that the reconstruction formula (4.21) was chosen here mainly because of its apparent similarity to the backprojection technique without filtering in Computerized Tomography (CT), with the lines replaced by more complicated patterns given by $|\tilde{\epsilon}_s^{(jk)}(x)|$. It should be emphasized, however, that the mathematical derivation of the reconstruction scheme applied here is completely different from the derivation of the backprojection scheme in CT. Whereas the latter one admits the interpretation as an adjoint scheme for approximately inverting the linear Radon transform [15], the derivation of (4.21) involves a nonlinear step (taking the absolute values of the fields), and tries to approximately invert a *nonlinear* forward operator.

4.4. Implementation: The STAF algorithm. The Source Type Adjoint Field (STAF) scheme, which we have employed in our numerical experiments presented in section 5, is in brief algorithmic form described in Table 2.

5. Numerical Experiments

5.1. Discretization of the computational domain. In our numerical experiments, we use a Finite-Differences Frequency Domain (FDFD) code written in MATLAB for solving (2.1)-(2.3). The code uses appropriately designed perfectly matched layers (PML) to avoid reflections at the artificial computational boundaries [24].

The physical domain is partitioned into 100×100 elementary cells (pixels) in the first numerical example, and into 180×110 elementary cells in the second example. Each of these grid cells has a physical size of approximately 0.14×0.14 m^2, such that the total computational domain in the first example covers an area of 14×14 m^2, and in the second example of 15×25 m^2. The eight layers which are closest to the boundaries of the computational domain are used as a PML.

We will refer to the first numerical example as the 'full-view' situation, and to the second numerical example as the 'limited-view', 'cross-borehole' or 'geophysical' situation. This terminology is motivated by the source and receiver geometries used, which are as follows.

In the full-view example, we have 64 sources and 64 receivers given which surround the domain of interest. Each source position is at the same time a receiver position and vice versa. The distance of two adjacent sources or receivers from each other is four pixels or 55 cm. The area enclosed by these sources and receivers has a size of 10×10 m^2.

In the two limited-view examples, 74 sources and receivers are positioned equally spaced in two boreholes. The distance of two adjacent sources or receivers from each other is again 4 pixels or 55 cm, and the distance of the two boreholes from each other is about 10 m.

We mention that, in all of our numerical examples, the regions beyond the source and receiver positions are part of the inversion problem, too. This means, the area which has to be recovered from the data is the whole area situated between the PML boundaries. In some of our numerical experiments, artifacts can be observed developping in the outer areas during the early stages of the reconstruction process.

We apply time-harmonic dipole sources of the form (2.5) with frequencies of $f = 5, 10, 15, 20, 25$, or 30 MHz. In our examples, this corresponds to wavelengths in the background medium between 2 meters for $f = 30$ MHz and 13 meters for $f = 5$ MHz. The size of an individual grid cell is chosen such that each of these wavelengths is sampled by at least 16 pixels in order to avoid numerical artifacts due to undersampling.

The data in our numerical examples are generated by running the FDFD forward modelling code on the correct permittivity and conductivity distributions. Using the same forward code for creating the data and for doing the reconstruction is usually called 'inverse crime'. Therefore, to make sure that the situations we model in our experiments are as realistic as possible, we have tested the forward modelling code thoroughly, and add Gaussian noise with signal-to-noise ratios of up to 10 dB to the real and imaginary parts of the generated data. In an earlier paper [**12**] we have also investigated the effect of more systematic errors in the data to the reconstruction process, for example the effect of an additional perturbation in the conductivity distribution. These results let us expect that both, the STAF and the levelART reconstruction algorithm, are relatively stable with respect to measurement noise if appropriate regularization is applied (see [**12**]). This, of course, has to be verified in future work by using real data and the corresponding 3D generalizations of these two methods.

5.2. A full-view example. Our first numerical example demonstrates the performance of the STAF algorithm as well as the levelART algorithm in a situation where a relatively complicated shape has to be recovered from data which correspond to receiver positions completely surrounding the domain of interest. All data are noise-free in this idealized example. The geometry is shown in Figure 1. The positions of the sources and receivers are indicated by dots in the Figure. The background medium in this example consists of a homogeneous conductivity distribution $\sigma_b = 3.0 \times 10^{-4}$ Siemens/m, and a homogeneous permittivity distribution $\epsilon_b = 20$. Inside the object, the permittivity is $\hat{\epsilon} = 15$, having a moderate contrast

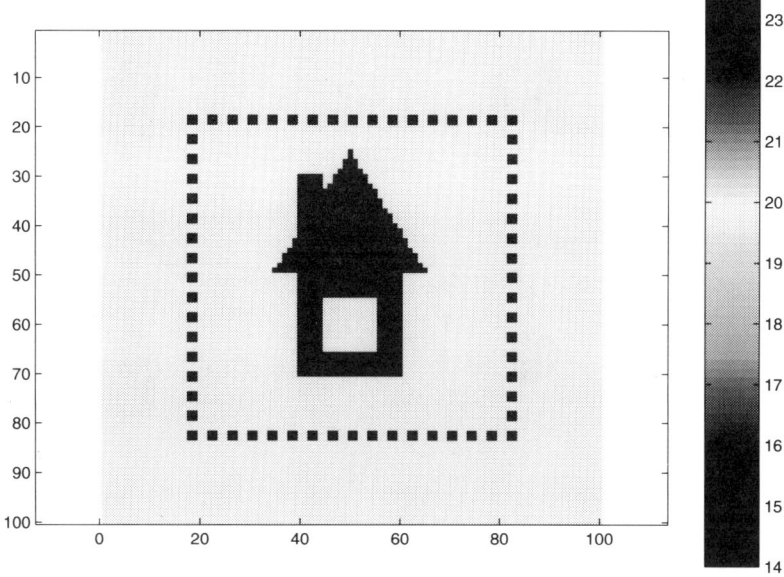

FIGURE 1. Original object for the example with full view. The dots in the figure indicate the source and receiver positions. The permittivity in the background is $\epsilon_b = 20$, and in the object $\hat{\epsilon} = 15$.

to the background distribution. Notice that an interesting feature of this geometry is the 'hole' in the body of the object which is difficult to reconstruct.

First we apply the STAF reconstruction scheme by using the data corresponding to one fixed frequency only. We do this for six different frequencies, namely 5, 10, 15, 20, 25 and 30 MHz. The constant C_l in the reconstruction scheme (see Table 2) which determines the shapes is chosen to be 0.7 times the maximal value of the reconstructed function $|\tilde{\epsilon}_s(x)|$. Figure 2 shows the different reconstructions. As it was expected the resolution of the reconstructions increases if we go to higher frequencies. The lowest frequency (5 MHz) only gives us a large blop at the correct position of the sought object. At $f = 15$ MHz the triangular shape of the roof of the house becomes visible, and the hole in the bottom of the house is correctly recovered at a frequency of $f = 25$ MHz. The highest frequencies ($f = 25$ MHz and $f = 30$ MHz) give us a quite decent reconstruction of the correct shape of the unknown object.

Figure 3 demonstrates how the levelART algorithm performs in this situation when we use the lowest frequency (5 MHz) STAF reconstruction as an initial guess. We use three groups of frequencies for this reconstruction (see Table 1), each of them containing only one single frequency, namely $G_1 = \{10 \text{ MHz}\}$, $G_2 = \{20 \text{ MHz}\}$, and $G_3 = \{30 \text{ MHz}\}$. We start the reconstruction with the lowest frequency $f = 10$ MHz and perform $I_1 = 30$ sweeps as described in Table 1. Continuing from that result, we perform $I_2 = 30$ sweeps with $f = 20$ MHz, and finally $I_3 = 30$ sweeps with $f = 30$ MHz. We believe that marching in this way over the frequencies, starting with the lower frequencies and continuing with the higher frequencies, is helpful in dealing with the strong nonlinearity (high contrast) of the problem, since the lower

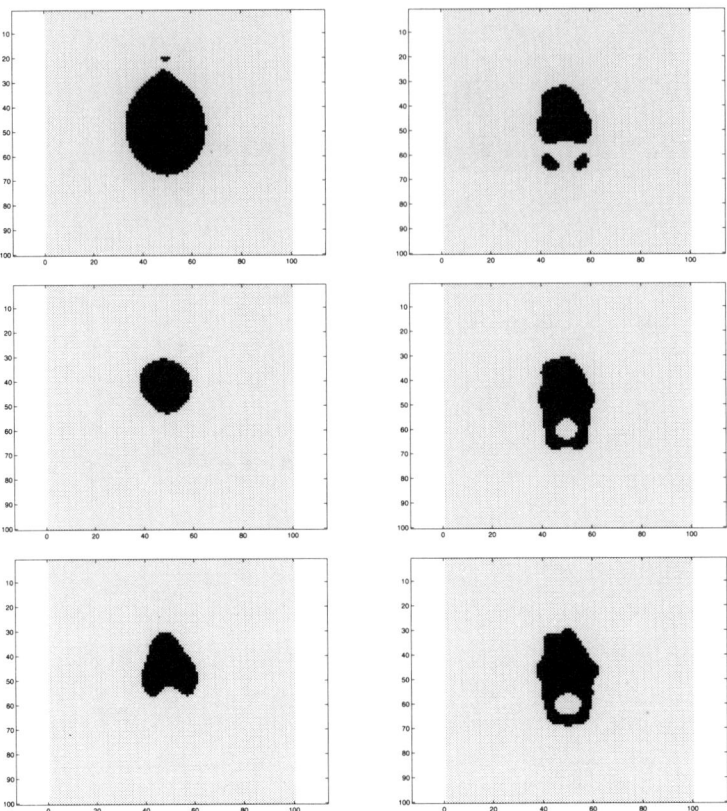

FIGURE 2. Different STAF reconstructions for the full-view example using data with a fixed frequency. Left column from top to bottom: $f = 5$ MHz, $f = 10$ MHz, $f = 15$ MHz; Right column from top to bottom: $f = 20$ MHz, $f = 25$ MHz, $f = 30$ MHz. The data are noise-free.

frequencies seem to be less sensitive to the high contrast in the parameters than the higher frequencies are. This is in agreement with related observations reported earlier by Chen [4].

Comparing the final result of the levelART algorithm with the STAF reconstructions in Figure 2, we conclude that the iterative levelART scheme can be used to improve each of the STAF reconstructions.

5.3. A cross-borehole situation with multiple objects. In our second numerical example, we consider a more realistic situation which is typical for geophysical applications. Comparable situations occur for example when we wish to monitor pollutant plumes at environmental cleanup sites from cross-borehole EM data. The geometry of this second example is shown in Figure 4.

The background permittivity distribution in this example consists of four tilted layers with values of $\epsilon_b = 21$ in the top layer, and then continuing downwards with 20, 19, and again 21 for the deepest layer. The conductivity distribution σ_b is homogeneous with a value of $\sigma_b = 3.0 \times 10^{-4}$ S/m everywhere.

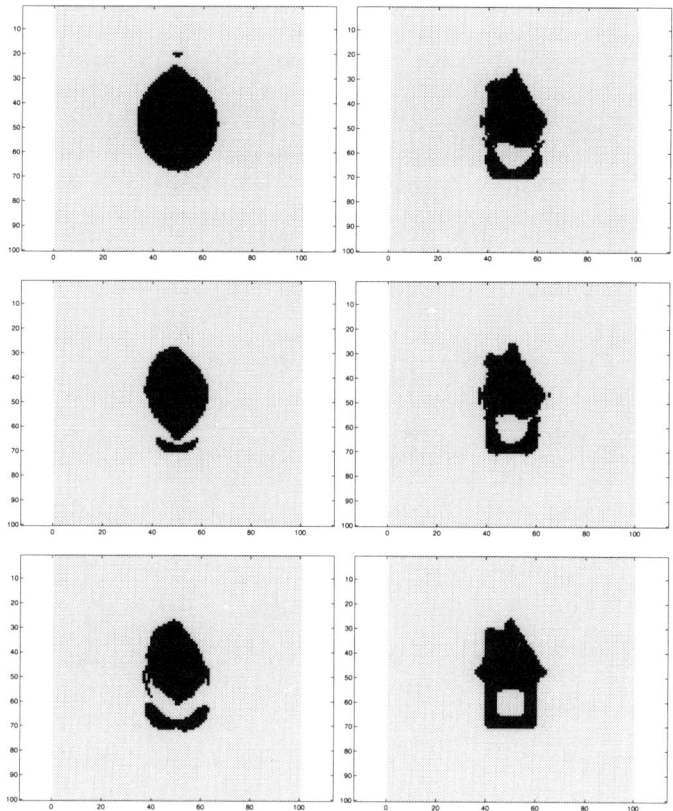

FIGURE 3. The levelART process: Evolution of permittivity $\epsilon^{(n)}$. Left column from top to bottom: STAF reconstruction of $\epsilon^{(0)}$ for 5 MHz (top left); This is the starting guess for the following reconstruction using levelART. After 10 steps of levelART with 10 MHz; After 30 steps with 10 MHz; Right column from top to bottom: After 10 more steps with 20 MHz; After 30 steps with 20 MHz; Final reconstruction after 30 steps of levelART with 30 MHz (bottom right). The algorithm used noise-free data.

Embedded in this background are three compact inclusions as it can be seen in Figure 4. The permittivity inside these inclusions is $\hat{\epsilon} = 5$, having a high contrast to the background values. The three inclusions are oriented such that there are two 'channels' of background material between them, one of them in the vertical and one in the horizontal direction. The difficulty in this example is to separate the three inclusions from each other from the limited-view data. In particular, the reconstruction of the vertical channel is critical since we expect that the resolution in the horizontal direction will suffer from the missing data.

As in the full-view example, we first apply the STAF resonstruction scheme by using only the data corresponding to one fixed frequency. We do this for each of the six frequencies 5, 10, 15, 20, 25, and 30 MHz. Again, the constant C_l in the

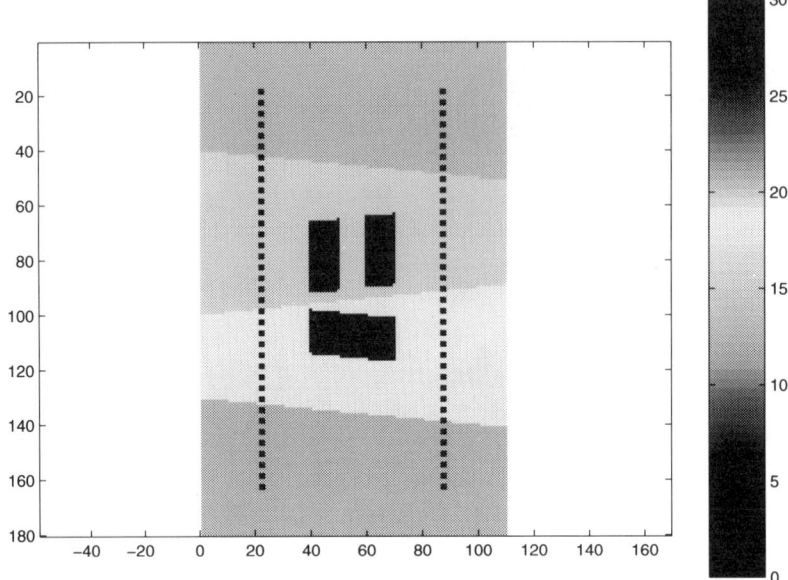

FIGURE 4. True permittivity distribution in the cross-borehole example. The dots in the figure indicate the source and receiver positions. The permittivity in the background layers is (from top to bottom) $\epsilon_b = 21, 20, 19,$ and 21. Inside the object it is $\hat{\epsilon} = 5$.

reconstruction scheme is chosen to be 0.7 times the maximal value of the reconstructed function $|\tilde{\epsilon}_s(x)|$. Figure 5 shows the different reconstructions in the case where the data are noise-free, and Figure 6 shows the corresponding reconstructions in the case where the data are contaminated by additive Gaussian noise with a signal-to-noise ratio (SNR) of 10 dB.

We observe in the Figures 5 and 6 that, in contrast to our first example, the STAF reconstructions do *not* improve with increasing frequency in this limited-view situation. The reconstruction corresponding to the lowest frequency (5 MHz) is similar to the reconstruction which we get in the full-view situation, which means that the lower frequencies are less sensitive to the large amount of missing data. When we increase the frequency, the reconstructions look like deteriorated images which are oscillatory in the horizontal direction, due to the missing data on the top and on the bottom of the reconstruction domain. In the case of noisy data in Figure 6, the higher frequency reconstructions look even worse due to additional blurring which is caused by the noise in the data. In short, the figures show that the sensitivity of the STAF reconstructions to limited view and noise in the data increases with increasing frequency.

Combining the observations so far we conclude that low frequency data are fairly insensitive to noise in the data and to limited view, but the corresponding reconstructions have low resolution. On the other hand, the high frequency data yield in the ideal situation of complete data sets high resolution reconstructions, but these reconstructions deteriorate dramatically when only noisy limited-view data are available.

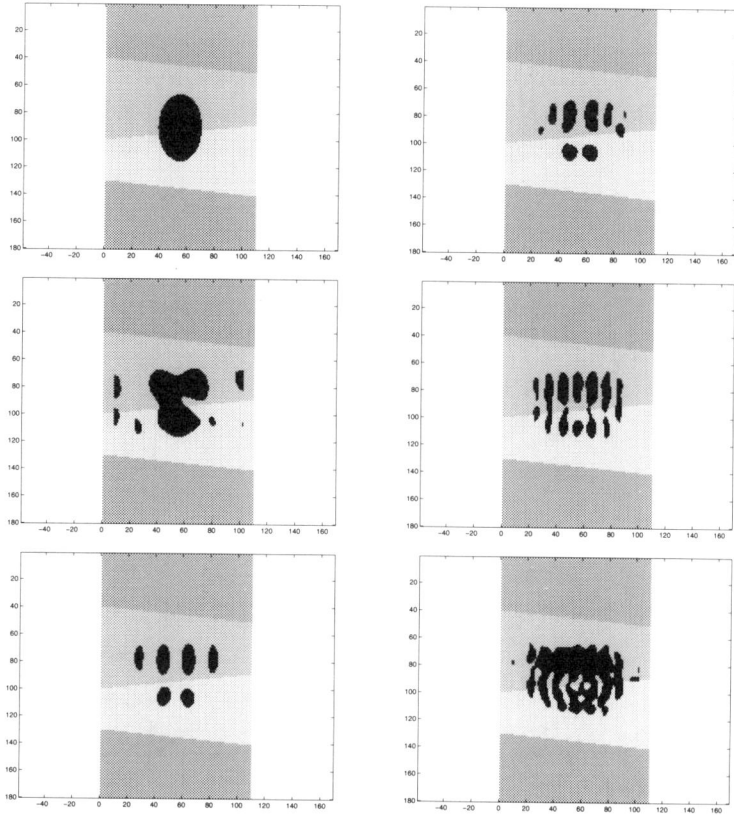

FIGURE 5. Different STAF reconstructions for the cross-borehole example using limited-view data with a fixed frequency. Left column from top to bottom: $f = 5$ MHz, $f = 10$ MHz, $f = 15$ MHz; Right column from top to bottom: $f = 20$ MHz, $f = 25$ MHz, $f = 30$ MHz. The data are noise-free.

In the levelART approach we try to make use of the favourable features of the low frequency data as well as of the high frequency data, but in a way such that the unwanted features of each of them are compensated for by the other one. In order to achieve this goal, we define two (or in general even more) groups of frequencies (see Table 1), namely $G_1 = \{15 \text{ MHz}\}$, and $G_2 = \{20, 25, \text{and } 30 \text{ MHz}\}$. As initial guess we use the STAF reconstruction of $f = 5$ MHz. We make this choice because of our observation that the lower frequency reconstructions are less sensitive to the limited-view geometry and to the noise in the data. Now we perform $I_1 = 20$ sweeps of the levelART algorithm with the frequency 15 MHz, corresponding to the first frequency group. Since we have already a good initial guess available, the nonlinearity in the problem has been reduced and it causes no problems to use that frequency in this stage of the reconstruction process.

We see in Figure 7 that after 20 sweeps with 15 MHz we almost have the three reconstructed objects separated from each other. However, it can be observed that it becomes more and more difficult for the algorithm to improve the reconstruction with only the data corresponding to this frequency. Therefore, we continue with

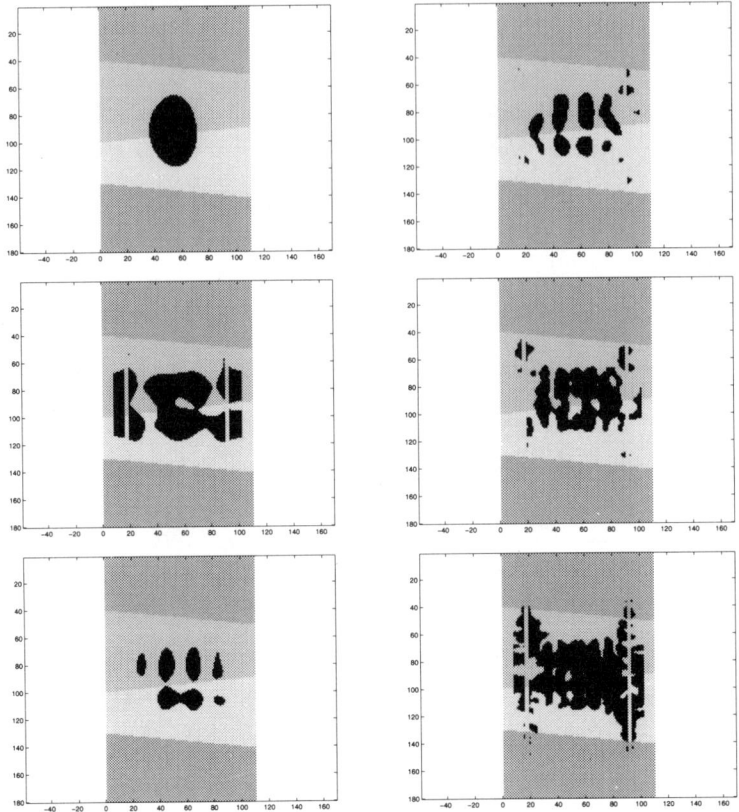

FIGURE 6. Different STAF reconstructions for the cross-borehole example using limited-view data with a fixed frequency. Left column from top to bottom: $f = 5$ MHz, $f = 10$ MHz, $f = 15$ MHz; Right column from top to bottom: $f = 20$ MHz, $f = 25$ MHz, $f = 30$ MHz. The data are noisy with a signal-to-noise ratio of 10 dB.

the second frequency group, and perform $I_2 = 30$ sweeps with the three frequencies 20, 25, and 30 MHz. Each of these sweeps uses first the lowest frequency (20 MHz) data for calculating one update for the shape, then uses the next higher frequency (25 MHz) data for the next update, and finally the highest frequency (30 MHz) data for the final update of the given sweep. The succeeding sweep starts again with the lowest frequency (20 MHz) data of that group G_2.

We believe that, in this combination, the updates from the lower frequency data in the given group act as regularizing and stabilizing elements in the combined reconstruction process, and compensate for the deteriorating effects due to the missing data and the noise in the highest frequencies. The resolution of the final reconstruction nevertheless seems to be mainly determined by the highest frequency data in the group, as it can be seen in Figure 7.

The regularizing effect of lower frequency data becomes even more visible in the following experiment. Here, we want to find out whether it would be sufficient just to run the levelART algorithm with data corresponding to only one of the

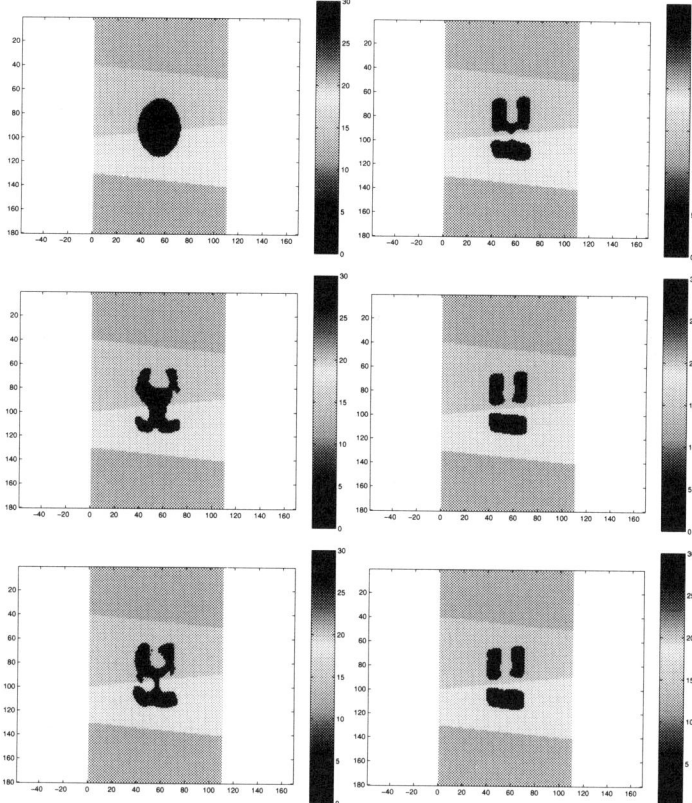

FIGURE 7. The levelART process: Evolution of permittivity $\epsilon^{(n)}$. Left column from top to bottom: STAF reconstruction $\epsilon^{(0)}$ for 5 MHz (top left); This is the starting guess for the following reconstruction using levelART. After 10 steps of levelART with 15 MHz; After 20 steps with 15 MHz; Right column from top to bottom: After 10 more sweeps with 20+25+30 MHz; After 20 sweeps with 20+25+30 MHz; Final reconstruction after 30 sweeps of levelART with 20 + 25 + 30 MHz (bottom right). The algorithm used noisy data with a SNR of 10 dB and curve shortening by diffusion.

higher frequencies and using the STAF reconstruction with 5 MHz as an initial guess. We use the same geometry as in the above cross-borehole experiment, but do not add any noise to the data. Starting with the STAF result for 5 MHz as displayed in Figure 8, we perform for each of the frequencies 15 MHz, 20 MHz, 25 MHz, and 30 MHz 110 sweeps with levelART using only that single frequency. No additional regularization (as for example 'curve shortening by diffusion' [**12**]) is applied. The results we get are displayed in Figure 8. Then, to compare with, we do an experiment where we first run 40 steps of levelART using the data with frequency 15 MHz, and then run 25 additional sweeps using the data with frequencies 20 MHz, 25 MHz, and 30 MHz repeatedly in a cyclic order. In this way, all experiments (single-frequency and multi-frequency) are comparable with each other in terms of

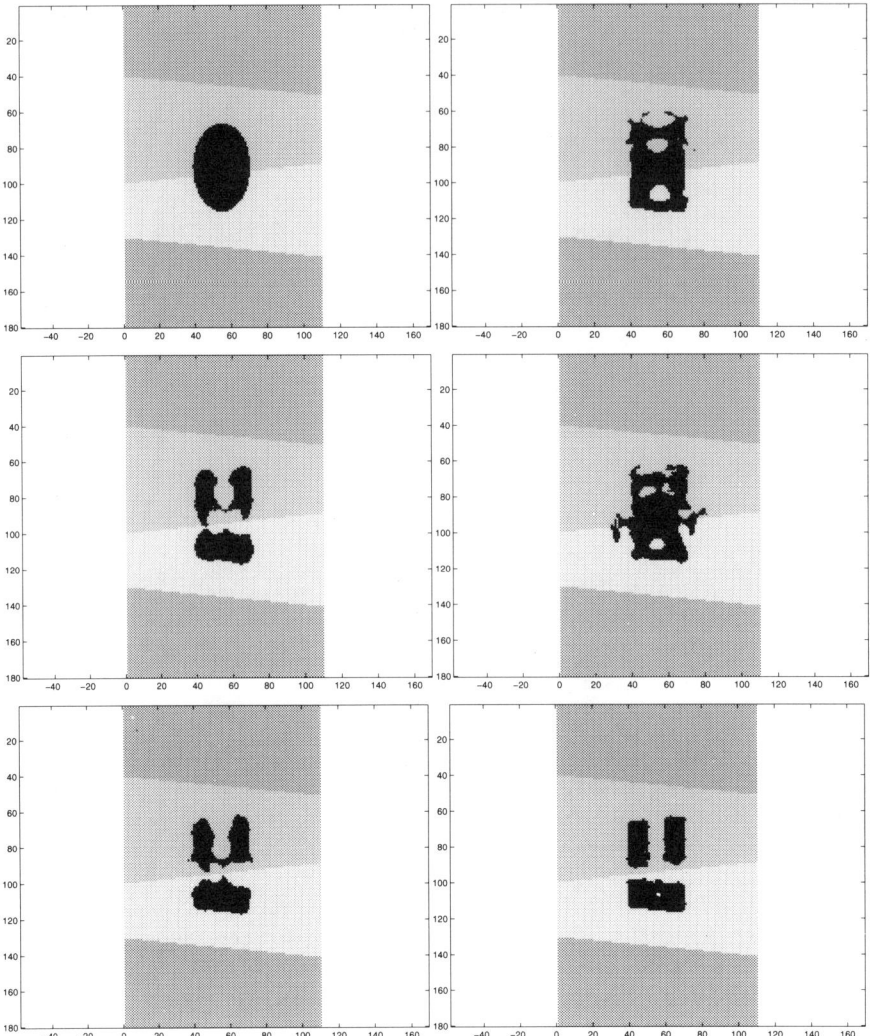

FIGURE 8. The levelART algorithm: Final reconstructions of permittivity ϵ. Left column from top to bottom: STAF reconstruction $\epsilon^{(0)}$ for 5 MHz (top left), which is the starting guess for each of the following reconstructions: final reconstruction after 110 iteration steps using only 15 MHz data; using only 20 MHz data. Right column from top to bottom: final reconstruction after 110 steps using only 25 MHz data; using only 30 MHz data; bottom right: final reconstruction after first using 40 steps with 15 MHz data, and then 25 more sweeps using 20, 25 and 30 MHz data in a cyclic order. The data are noise-free and no additional regularization is applied.

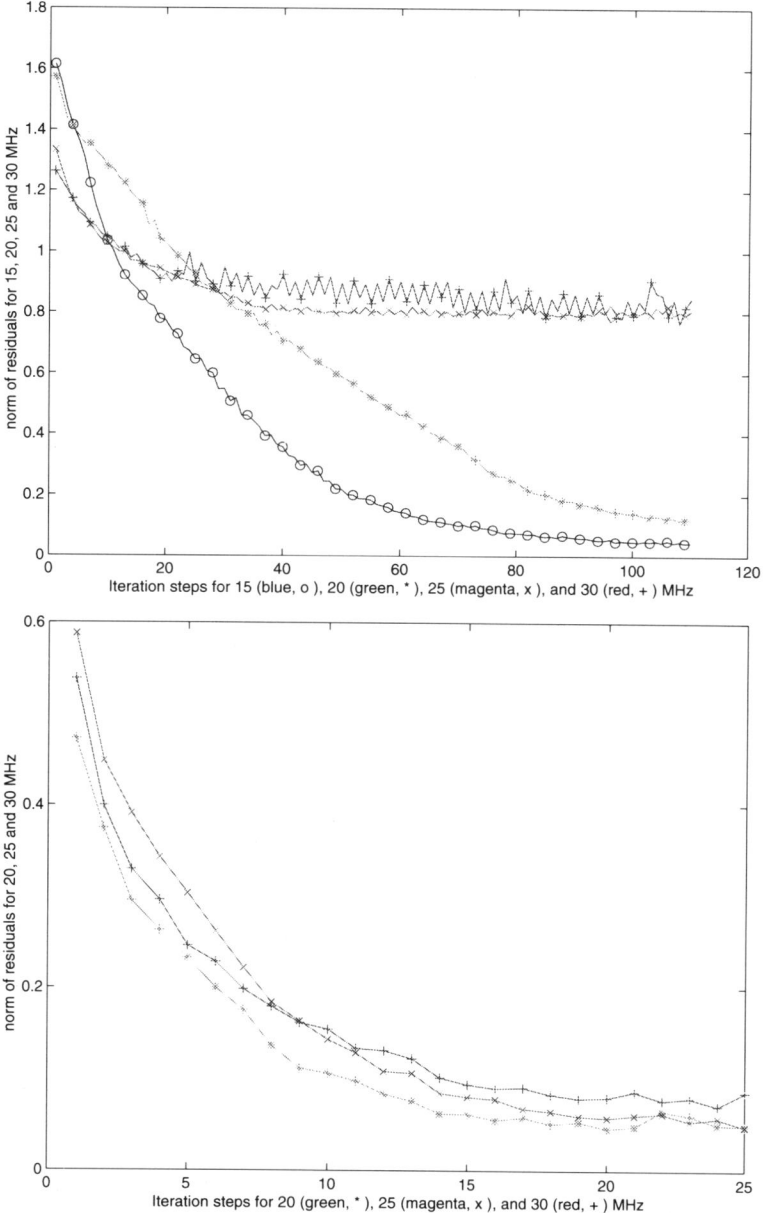

FIGURE 9. Top: norms of residuals in the reconstructions which use only one single frequency (110 iterations). Bottom: norms of the residuals in the reconstruction which starts with the frequency 15 MHz (40 iterations, not shown in the figure), and then continues in a cyclic order with 20, 25 and 30 MHz data, using each of these exactly 25 times.

total computational cost. The result of this last experiment is also displayed in Figure 8.

The results show that the reconstructions using only the data corresponding to one single frequency are quite useful as long as this frequency is not too high. The reconstructions for 15 MHz and 20 MHz data clearly approximate the correct geometry of the three objects, but these objects are not yet completely separated from each other. In addition, the resolution in these reconstructions is still not very good. Increasing the frequency should improve the resolution, but the images corresponding to 25 MHz and 30 MHz in Figure 8 show that instead the algorithm 'breaks down' in the sense that the reconstructions get trapped far away from the correct ones. The combined reconstruction in the bottom right image of Figure 8, on the other hand, behaves perfectly stable and shows an improved resolution due to the incorporated higher frequency data.

In Figure 9 we display the evolution of the norms of the residuals during the reconstruction processes. When using only data corresponding to one low frequency (15 MHz and 20 MHz), the residuals decrease continuously and the reconstruction is performed stably. However, when higher frequency data (25 MHz and 30 MHz) are used instead, the residuals get 'trapped' on their way and start oscillating around some intermediate value. On the other hand, when using a combination of lower and higher frequency data, all residuals (including those corresponding to higher frequencies) decrease continuously until the reconstruction task is completed.

We want to mention that other choices for building frequency groups are certainly possible, and might even yield better results than those presented here. So far it is not clear whether there is an 'optimal' way of selecting these groups.

6. Summary and future directions

We have investigated the use of multifrequency electromagnetic data for shape reconstruction in a limited-view source-receiver geometry. We have compared the results of two recently introduced shape reconstruction methods, namely a very simple one-step source type method (STAF) on the one hand, and an iterative method (levelART) on the other hand which uses adjoint fields and level sets for the inversion. It was shown that, in the full view situation, the STAF algorithm yields low resolution reconstructions if applied to the data corresponding to the lower frequencies, and decent higher resolution reconstructions if applied to the higher frequency data. The reconstructions from the lower frequency data are fairly insensitive to noise and missing data in a limited-view geometry, whereas the higher frequency reconstructions deteriorate dramatically in that situation.

The levelART algorithm, if applied with a low-frequency STAF reconstruction as an initial guess, is able to yield improved reconstructions in all cases. If only the data corresponding to one single frequency are used for levelART in the strongly limited-view geometry, stable reconstructions (with lower resolution) are possible for the lower frequencies. However, in contrast to the full-view situation, the algorithm breaks down when going to limited-view data with higher frequencies.

On the other hand, when applying levelART with multifrequency electromagnetic data, it can make use of the lower frequency data for stabilizing and regularizing the reconstruction process, and for overcoming the problems caused by the limited view, the noisy data, and the high contrasts in the perturbations. In this way, stable reconstructions with a good resolution are possible even in the limited-view geometry. In particular, in all cases which we have considered so far, the resolution of the final reconstructions seems to be mainly determined by the highest frequency data which are used for the inversion. Some additional regularization (for example 'curve shortening by diffusion') might be useful in order to increase the smoothness of the reconstructed boundaries. This feature of the levelART algorithm is discussed elsewhere [12].

So far we have only considered a somehow idealized 2D situation. However, we believe that the generalization of the method to a more realistic 3D situation is possible and doable. This is an interesting problem for future research.

References

[1] H. Abdullah and A. K. Louis, *The approximate inverse for solving an inverse scattering problem for acoustic waves in an inhomogeneous medium*, Inverse Problems **15**, 1999, 1213–1229.

[2] T. S. Angell, W. Jiang and R. E. Kleinman, *A distributed source method for inverse acoustic scattering*, Inverse Problems **13**, 1997, 531–46.

[3] S. Caorsi and G. L. Gragnani, *Inverse-scattering method for dielectric objects based on the reconstruction of the nonmeasurable equivalent current density*, Radio Science **34** (1), 1999, 1–8.

[4] Y. Chen, *Inverse scattering via Heisenberg's uncertainty principle*, Inverse Problems **13**, 1997, 253–282.

[5] W. C. Chew, Y. M. Wang, G. Otto, D. Lesselier, and J. Ch. Bolomey, *On the inverse source method of solving inverse scattering problems*, Inverse Problems **10**, 1994, 547–553.

[6] D. L. Colton and R. Kress, *Inverse Acoustic and Electromagnetic Scattering Theory*, 2nd edn, Berlin, Springer, 1998.

[7] A. J. Devaney and G. C. Sherman, *Nonuniqueness in inverse source and scattering problems*, IEEE Trans. Antennas Propag. **30**, 1984, 1034–1037.

[8] A. J. Devaney, *Reconstructive tomography with diffracting wavefields*, Inverse Problems **2**, 1986, 161–83.

[9] K. A. Dines and R. J. Lytle, *Computerized geophysical tomography*, Proc. IEEE **67**, 1979, 1065–1073

[10] O. Dorn, *A transport-backtransport method for optical tomography*, Inverse Problems **14**, 1998, 1107–1130

[11] O. Dorn, H. Bertete-Aguirre, J. G. Berryman and G. C. Papanicolaou, *A nonlinear inversion method for 3D electromagnetic imaging using adjoint fields*, Inverse Problems **15**, 1999, 1523–1558.

[12] O. Dorn, E. L. Miller and C. M. Rappaport, *A shape reconstruction method for electromagnetic tomography using adjoint fields and level sets*, Inverse Problems **16** (5), 2000, 1119–1156.

[13] T. M. Habashy, R. W. Groom and B. R. Spies, *Beyond the Born and Rytov Approximations: A Nonlinear Approach to Electromagnetic Scattering*, Journal of Geophysical Research **98** (B2), 1993, 1759–1775.

[14] T. M. Habashy, M. L. Oristaglio and A. T. de Hoop, *Simultaneous nonlinear reconstruction of two-dimensional permittivity and conductivity*, Radio Science **29** (4), 1994, 1101–1118.

[15] S. Helgason, *The Radon Transform*, Birkhauser, 1980.

[16] F. Hettlich and W. Rundell, *Recovery of the support of a source term in an elliptic differential equation*, Inverse Problems **13**, 1997, 959–976.

[17] K. T. Ladas and A. J. Devaney, *Iterative methods in geophysical diffraction tomography*, Inverse Problems **8**, 1992, 119–132.

[18] A. Litman, D. Lesselier and F. Santosa *Reconstruction of a two-dimensional binary obstacle by controlled evolution of a level-set*, Inverse Problems **14**, 1998, 685–706.

[19] Eric L. Miller, Misha Kilmer, and Carey Rappaport, *A New Shape-Based Method for Object Localization and Characterization from Scattered Field Data*, IEEE Trans. Geoscience and Remote Sensing, special issue on Computational Wave Issues in Remote Sensing, Imaging and Target Identification, Propagation, and Inverse Scattering, vol. 38 (4), July 2000, 1682–1696.

[20] F. Natterer, *The Mathematics of Computerized Tomography*, Stuttgart, Teubner, 1986.

[21] F. Natterer and F. Wübbeling, *A propagation-backpropagation method for ultrasound tomography*, Inverse Problems **11**, 1995, 1225–1232.

[22] F. Natterer, *Numerical Solution of Bilinear Inverse Problems*, Preprints "Angewandte Mathematik und Informatik" 19/96-N, 1996, Münster. This preprint is electronically available at http://wwwmath.uni-muenster.de/math/num/Preprints/1997/natterer_3.

[23] S. Osher and J. Sethian, *Fronts propagation with curvature dependent speed: Algorithms based on Hamilton-Jacobi formulations*, Journal of Computational Physics **56**, 1988, 12–49.

[24] Carey M. Rappaport, Misha Kilmer, and Eric L. Miller, *Accuracy Considerations in Using the PML ABC with FDFD Helmholtz Equation Computation*, International Journal of Numerical Modeling, special issue on the PML, vol. 13, 2000, 471–482.

[25] C. Rozier, D. Lesselier, D. Angell and R. E. Kleinman, *Shape retrieval of an obstacle immersed in shallow water from single frequency fields using a complete family method*, Inverse Problems **13**, 1996, 487–508.

[26] F. Santosa, *A Level-Set Approach for Inverse Problems Involving Obstacles*, ESAIM: Control, Optimization and Calculus of Variations **1**, 1996, 17–33.

[27] J. A. Sethian, *Numerical algorithms for propagating interfaces: Hamilton-Jacobi equations and conservation laws*, J. Diff. Geom. **31**, 1990, 131–61.

[28] J. A. Sethian, *Level Set Methods and Fast Marching Methods*, 2nd ed., Cambridge University Press, 1999.

[29] L. Souriau, B. Duchene, D. Lesselier and R. E. Kleinman, *Modified gradient approach to inverse scattering for binary objects in stratified media*, Inverse Problems **12**, 1996, 463–481.

[30] P. M. Van den Berg and R. E. Kleinman, *A contrast source inversion method*, Inverse Problems **13**, 1997, 1607–20.

[31] P. M. Van den Berg, A. L. van Broekhoven and A. Abubakar, *Extended contrast source inversion*, Inverse Problems **15**, 1999, 1325–1344.

[32] A. von Hippel, *Dielectric Materials and Applications*, Wiley, New York, 1953, 3–4.

DEPT OF COMPUTER SCIENCE, UNIVERSITY OF BRITISH COLUMBIA, VANCOUVER, BRITISH COLUMBIA, V6T 1Z4, CANADA
E-mail address: `dorn@cs.ubc.ca`

CENTER FOR SUBSURFACE SENSING AND IMAGING SYSTEMS, NORTHEASTERN UNIVERSITY, BOSTON, MA 02115, USA
E-mail address: `elmiller@ece.neu.edu`

CENTER FOR SUBSURFACE SENSING AND IMAGING SYSTEMS, NORTHEASTERN UNIVERSITY, BOSTON, MA 02115, USA
E-mail address: `rappaport@neu.edu`

Three Problems at Mount Holyoke

Leon Ehrenpreis

ABSTRACT. We discuss problems involving a parametric version of the Radon transform, a Pompeiu-like problem for tangent spheres, and a Morera type criterion for detecting holomorphic functions by moment conditions.

Formulation of the problems.

Of course many interesting problems were posed at the Mount Holyoke Conference; this paper represents my contribution to three of them.

To introduce the first problem we shall use the notation of parametric Radon Transform as formulated in [4]. This differs somewhat from the original description of F. John and promulgated by Helgason, Gelfand, etc.

If Ω is a set with a given measure, e.g. $\Omega \in \mathbb{R}^n$ then δ_Ω (the δ function of Ω) is the distribution which is defined by integration over Ω with respect to the given measure. The passage from Ω to δ_Ω allows us to replace geometric objects by analytic ones. By abuse of language we sometimes refer to δ_Ω as Ω.

Let f be a nice function (smooth and small at infinity) on $\mathbb{R}^n$. For any $n \times l$ matrix A and any vector $b \in \mathbb{R}^n$ we form the l plane parametric Radon transform

$$(1) \qquad F(A, b) = \int f(At + b)\, dt,$$

where t is a variable in $\mathbb{R}^l$. We write

$$(2) \qquad F = \mathbf{R}_p^l f.$$

Equation (1) makes good sense when $l \leq n$ and the matrix A has rank l. But when rank $A < l$ the integral generally diverges. Nevertheless it is proven in [4] that F is locally integrable in A, b when f is suitably small at infinity and smooth provided that $l \leq n - 1$.

John observed that F (formally) satisfies a system of differential equations, which we call the *John equations*:

$$(3) \qquad \frac{\partial^2 F}{\partial a_j^i \partial b_k} = \frac{\partial^2 F}{\partial a_k^i \partial b_j}.$$

1991 *Mathematics Subject Classification.* 44A12, 42A38, 30E05.

© 2001 American Mathematical Society

(a_j^i is the entry of A in the i column and j row.) It is shown in [4] that when F is interpreted as a distribution (3) holds when $l \leq n-2$, in particular, for $n = 3, l = 1$ which is John's original case.

It is important to note that (3) is valid if we replace dt in (1) by any other suitable measure. This means that (3) is not the whole story–we need some way of knowing that the measure is dt. Since dt is the Haar measure on $\mathbb{R}^l$ we have

(4) $$F(A, b + At^0) = F(A, b)$$

for any $t^0 \in \mathbb{R}^l$. In infinitesimal form (4) becomes

(5) $$a^I \cdot \nabla_b F = 0$$

where a^I is the I column of A.

We refer to (4) or (5) as the *invariance equations*.

It is shown in [4] that for suitable classes of functions f the John equations and the invariance equations characterize the range of $\mathbf{R}_p^l$ when $l \leq n-2$. This is essentially John's result when $l = 1, n = 3$. The case of general n, l with $l \leq n-2$ has been the result of much research; we shall not describe these results as they are not in the direction of the present article.

If we change the measure dt in (1) to an arbitrary measure $d\mu(t)$ then the John equations are still valid but the invariance equation fails.

PROBLEM 1. (CLARKDOYLE). *Determine f and μ from F.*

The second problem has been studied by several participants including T. Quinto, B. Rubin and S. Gindinkin. Let Ω be a strictly convex smooth domain in $\mathbb{R}^n$. Let f be a "nice" function defined on the exterior of Ω. We form the Radon transform

(6) $$\mathbf{R}_\Omega f(s) = \int_S f$$

where S is a sphere in $\mathbb{R}^n$ tangent to Ω.

PROBLEM 2. *Is f uniquely determined by $\mathbf{R}_\Omega f$?*

There is an analogous problem for the interior of Ω.

The third problem was posed by L. Zalcman. Let f be a smooth function (no condition at infinity) defined on the strip $|y| \leq 1$ in the complex plane. Suppose that

(7) $$\int_{|z|=1} f(x_0 + z) z^n \, dz = 0$$

for all real x_0 and all $n \geq 0$. (7) can be written as

(7') $$z^n \delta_{|z|=1} *_x f = 0 \quad \text{on the } x\text{-axis}$$

where $*_x$ refers to convolution in x.

PROBLEM 3. *Is f holomorphic in the strip?*

We now present the essential ideas needed to solve these problems. Details will appear in [4].

I would like to thank Cristian Gurita for the help he has given me in preparing this manuscript.

Solution of Problem I

We are now studying the parametric Radon transform

(8) $$F_\mu(A, b) = \mathbf{R}_p^l(f, \mu)(A, b)$$
$$= \int f(At + b)\, d\mu(t).$$

For simplicity of notation we shall restrict our considerations to the case $l = 1$. There are no essential difficulties in passing to $l > 1$ as long as $n \geq l + 2$. Our methods apply to functions f in the Schwartz space $\mathcal{S}$. Moreover μ could be a distribution of a suitable type rather than a measure. Let $\widehat{F}_\mu$ be the Fourier transform of F_μ in the variables A, b.

(9) $$\widehat{F}_\mu(\widehat{A}, \widehat{b}) = \iint F_\mu(A, b) e^{iA \cdot \widehat{A} + ib \cdot \widehat{b}}\, dA\, db.$$

We work formally, as though $\widehat{F}_\mu$ is a function, though it is generally a distribution.

Since F_μ satisfies John's equations a formal argument suggests that

(10) $$\text{support } \widehat{F}_\mu \subset V$$

where V is the variety associated to (3), that is

(11) $$V = \{\widehat{A}_j \widehat{b}_k = \widehat{A}_k \widehat{b}_j\}$$
$$= \{\widehat{A} \| \widehat{b}\}$$
$$= \{\widehat{A} = \lambda \widehat{b}\}_\lambda \cup \{(\widehat{A}, 0)\}.$$

(This is an example of the Fundamental Principle of [3].)

We derive this result and, in fact, a more precise result by the following formal argument.

(12) $$\widehat{F}_\mu = \iiint e^{iA \cdot \widehat{A} + ib \cdot \widehat{b}} f(At + b)\, dA\, db\, d\mu(t)$$
$$= \iiint e^{iA \cdot \widehat{A} + i(y - At) \cdot \widehat{b}} f(y)\, dA\, dy\, d\mu(t)$$
$$= \int e^{iA \cdot \widehat{A}}\, dA \int e^{iy \cdot \widehat{b}} f(y)\, dy \int e^{-iAt \cdot \widehat{b}}\, d\mu(t)$$
$$= \widehat{f}(\widehat{b}) \int e^{iA \cdot \widehat{A}} \widehat{\mu}(-A \cdot \widehat{b})\, dA.$$

We decompose this A Fourier transform into a part perpendicular to $\widehat{b}$ and a part parallel to $\widehat{b}$. For each fixed $\widehat{b}$, $\widehat{\mu}(-A \cdot \widehat{b})$ is constant in the A direction orthogonal to $\widehat{b}$. Thus its Fourier transform in A is $\delta_{\widehat{b}^\perp = 0}$ (which is the δ function of the line consisting of those $\widehat{A}$ whose $\widehat{b}^\perp$ component is 0) times the (one dimensional) Fourier transform of $\widehat{\mu}(-A \cdot \widehat{b})$ on the line $A \| \widehat{b}$ which is $[\mu((\widehat{A} \cdot \widehat{b})/|\widehat{b}|]/|\widehat{b}|$. Finally

(13) $$\widehat{F}_\mu(\widehat{A} \cdot \widehat{b}) = [\widehat{f}(\widehat{b}) \mu((\widehat{A} \cdot \widehat{b})/|\widehat{b}|) \delta_{\widehat{b}^\perp}]/|\widehat{b}|,$$

when $\widehat{b} \neq 0$. It can be shown (see [4]) that for $f \in \mathcal{S}$ and suitable μ there is no essential contribution from $\widehat{b} = 0$. The factor $\delta_{\widehat{b}^\perp = 0}$ means that $\widehat{A} \| \widehat{b}$ in conformity with (11) and (12).

Since F_μ is given we can determine $\widehat{F}_\mu$. The problem is the determination of $\widehat{f}$ and μ from the products $\widehat{f}(\widehat{b})\mu(\widehat{A} \cdot \widehat{b}/|\widehat{b}|)$. (Since $|\widehat{b}| \neq 0$ we could multiply $\widehat{F}_\mu$ by $|\widehat{b}|$ and thereby remove the last factor $|\widehat{b}|^{-1}$ in (13).)

The solution of this problem reduces to the case when $\widehat{b}$ lies in a fixed line so we may as well assume at this point that $n = 1$. The case $\widehat{F}_\mu \equiv 0$ is uninteresting (and easily analyzed) so we assume $\widehat{F}_\mu \not\equiv 0$. Let $\widehat{A}_0, \widehat{b}_0$ be fixed with $\widehat{f}(\widehat{b}_0)\mu((\widehat{A}_0 \cdot \widehat{b}_0/|\widehat{b}_0|)) \neq 0$. In particular $\widehat{f}(\widehat{b}_0) = c \neq 0$. As $\widehat{A}$ varies we examine

(14) $$\widehat{F}_\mu(\widehat{A}, \widehat{b}_0) = c\mu((\widehat{A} \cdot \widehat{b}_0)/|\widehat{b}_0|).$$

The set $\{(\widehat{A} \cdot \widehat{b}_0)/|\widehat{b}_0|\}_A$ is the whole line. We have thus determined μ up to a constant multiple. Next, let $\widehat{b}$ vary. Since we are assuming that $n = 1$ we have

(15) $$\mu((\widehat{A} \cdot \widehat{b})/|\widehat{b}|) = \mu(\widehat{A} \cdot \widehat{b}_0/|\widehat{b}_0|) = \mu(\widehat{A}).$$

By (13)

(16) $$\widehat{F}_\mu(\widehat{A}, \widehat{b}) = \widehat{f}(\widehat{b})\mu(\widehat{A}).$$

Since $\widehat{F}_\mu$ is known and $\mu(\widehat{A})$ has been determined up to a constant multiple, $\widehat{f}$ is determined up to the reciprocal constant multiple.

This formal argument leads to

THEOREM 1. *Let $f \in \mathcal{S}$ and $\mu \in \mathcal{S}'$. Then f and μ can both be determined from F_μ, the only indeterminacy being that we can multiply μ by a constant $c \neq 0$ and f by c^{-1}.*

REMARK. *It is possible to put Problem 1 in a group theoretic setting. Call G the affine group of $\mathbb{R}^n$ (the $Ax + b$ group), which is the group of transformations of $\mathbb{R}^n$ of the form $x \mapsto Ax + b$, where $A \in GL(n)$ is an $n \times n$ invertible matrix and $b \in \mathbb{R}^n$. Equation (8) expresses $F(A, b)$ as a convolution of f and μ on G. (We "lift" f and μ to G in a convenient manner.) The possibility of determining f and μ up to a constant factor can be interpreted as the fact that G has only one irreducible representation which is significant for the classes of f, μ we consider.*

Solution of Problem II

For each point $p \in$ boundary Ω we construct rays $\rho_p^\pm \subset \mathbb{R}^{n+1}$ defined as the set of points (x_{pt}, t) where $x_{pt} \in \mathbb{R}^n$ lies on the exterior normal to Ω at distance $|t|$ from Ω. The rays $\rho_p^\pm$ have the property that the intersection of the retrograde (forward) light cone $\Gamma_{pt}^\pm$ with vertex at (x_{pt}, t) with the plane $\mathbb{R}^n = \{t = 0\}$ is the sphere with center x_{pt} which is tangent to Ω at p. All exterior tangent spheres to Ω are of this form.

Let f be a nice function, e.g., smooth and of compact support on $\mathbb{R}^n$. Let F be the solution of the wave equation $\square = \frac{\partial^2}{\partial t^2} - \frac{\partial^2}{\partial x_1^2} - \cdots - \frac{\partial^2}{\partial x_n^2}$ whose Cauchy Data on $\mathbb{R}^n$ is $(0, f)$, that is, $F(x, 0) = 0, F_t(x, 0) = f$. It is important to note that F is odd in t.

We now assume that n is odd and > 1. (The case $n = 1$ is uninteresting.) Then the solution of the wave equation at (x_{pt}, t) is given by the integral of f over $\Gamma_{pt}^\pm \cap \mathbb{R}^n$ (see e.g. [**2**] vol II, p. 201). Thus the integrals of f over the spheres externally tangent to Ω determine F on $\Gamma = \cup_{p,t}\Gamma_{pt}^\pm$.

We want to reverse things and determine f from the value of F on Γ. By Huygen's principle f is determined from the values of F and its normal derivatives on any space-like hypersurface S. In particular $f(x)$ is determined by the values of F and its normal derivatives on the intersection of S with the forward (retrograde) light cone $\Gamma_\pm^x$ with vertex at x. "Space-like" means, essentially, that $\Gamma_\pm^x \cap S$ is a compact cross section of the rays of $\Gamma_\pm^x$.

We want to choose $S = \Gamma$ (which is not space-like) since we have seen that the values of F on Γ are determined by $\mathbf{R}_\Omega f$. However there are two difficulties:

(a) $\Gamma_\pm^x \cap \Gamma$ is not compact.

(b) We only know F on Γ, not its normal derivatives.

To overcome (a) we study $F(x,t)$ for large t. As before the value $F(x,t)$ is the integral of f over the intersection of the retrograde (forward) cone $\Gamma_{xt}^\pm$ with vertex at x,t with the hyperplane $\{t = 0\}$. It is clear that for t large most of $\Gamma_{xt}^\pm \cap \{t = 0\}$ lies outside the support of f when x is on or outside Γ. In particular F is small at such points. The important geometric observation that we use to overcome (a) is that for any point x, which is exterior to Ω and near Ω, every *line* which is generator of $\Gamma_+^x \cup \Gamma_-^x$ meets Γ except for a lower dimensional set of generators. This is not true for *rays* in Γ_+^x or in Γ_-^x. However since F is even in t we can deal with lines instead of rays. Since F is small at infinity on Γ the lower dimensional set of lines can be seen to be insignificant.

The idea of overcoming difficulty (b) stems from a result of d'Adhémar [7]. He showed that solutions G of the wave equation at a point y inside the usual light cone $^0\Gamma$ can be written as an integral of G over the intersection with $^0\Gamma$ of the forward (retrograde) light cone with vertex at y; *there are no normal derivatives.* If one analyzes d'Adhémar's result one finds that its raison d'être is the fact that "normal derivative" means derivative in the Minkowski normal direction. The Minkowski normal to a point on $^0\Gamma$ is tangential hence can be computed in terms of the values of G on $^0\Gamma$. It is not difficult to show that the same is true if $^0\Gamma$ is replaced by Γ. In conclusion $F(x,0)$ can be computed in terms of $F|_\Gamma$. We have shown

THEOREM 2. *When n is odd, the Radon transform $\mathbf{R}_\Omega f$ defined by integrals of a C^∞ function f of compact support on spheres externally tangent to a smooth strictly convex domain $\Omega \subset \mathbb{R}^n$ detemine f near Ω.*

PROBLEM. *Is this result true when n is even?*

When Ω is a ball a positive answer is given by Quinto [6] and by Rubin [**this conference**].

REMARK. *Instead of using the space-like Cauchy Surface $\{t = 0\}$ for the wave equation we could apply our method to a time-like Cauchy surface such as $\{x_1 = 0\}$. Spheres are replaced by hyperboloids. There are numerous complications that arize because $\{x_1 = 0\}$ is time-like. We refer the reader to [4] for details.*

Solution of Problem III

We shall solve this problem (partially) using functional analysis. Let W be the space of functions which are C^∞ on the closed strip $|y| \leq 1$ in the x,y plane. W' is the dual of W. We cannot take the Fourier transform of functions in W because they may be arbitrarily large as $x \to \infty$. But the elements $T \in W'$ are distributions of compact support in the strip, so we can form the Fourier transform

$$\widehat{T}(\widehat{x},\widehat{y}) = T \cdot e^{ix\widehat{x}+iy\widehat{y}}. \tag{17}$$

We start with a function f satisfying (7). In functional analysis terms we can write (7) in the form

$$z^{n+1}\delta_{|z|=1} \cdot \tau_{x_0} f = 0 \tag{18}$$

for $n \geq 0$ and all x_0. Here τ_{x_0} refers to the translation by x_0 and $\delta_{|z|=1}$ is the measure $d\theta$ on the unit circle. (We normalize measures to suppress the factor 2π. There are also powers of i which we shall ignore; they play no serious role.)

By differentiating in x_0 we find

$$z^{n+1}\delta_{|z|=1} \cdot \frac{d^j}{dx^j}\tau_{x_0} f = 0 \tag{19}$$

for all j.

The Fourier transform of $T_{n+1} = z^{n+1}\delta_{|z|=1}$ is

$$\begin{aligned}\widehat{T}_{n+1}(\widehat{x},\widehat{y}) &= \widehat{T}_{n+1}(\widehat{r},\widehat{\theta}) \\ &= \int e^{i\widehat{r}\cos(\theta-\widehat{\theta})}e^{i(n+1)\theta}\,d\theta \\ &= e^{i(n+1)\widehat{\theta}}J_{n+1}(\widehat{r})\end{aligned} \tag{20}$$

where J represents the usual Bessel function (see [1] vol II).

The standard formula for the Bessel function is

$$J_{n+1}(\widehat{r}) = \sum_{m=0}^{\infty} \frac{(-1)^m \widehat{r}^{2m+n+1}}{2^{2m+n+1}m!(m+n+1)!} \tag{21}$$

so that

$$e^{i(n+1)\widehat{\theta}}J_{n+1}(\widehat{r}) = \sum \frac{(-1)^m}{2^{2m+n+1}m!(m+n+1)!}\widehat{z}^{n+1}\widehat{r}^{2m}. \tag{22}$$

The operator τ_{x_0} in (18) has the effect of multiplication by $e^{ix_0\widehat{x}}$ and d^j/dx^j becomes multiplication by $(i\widehat{x})^j$.

Our hypothesis is that f is orthogonal to all the distributions defined by (19). We want to conclude that f is holomorphic which, in function theoretic terms, means that f is orthogonal to $\overline{\partial}U$ for a dense set of U in W. Note that for each $n \geq 0$ the right side of (22) is divisible by $\widehat{z}$ which is the Fourier transform of $\partial/\partial\bar{z}$ in conformity with the fact that $z^{n+1}\delta_{|z|=1}$ is orthogonal to holomorphic functions.

Now suppose that f is real analytic in the strip $|y| \leq 1$. Then to show that f is holomorphic it suffices to show that all derivatives of $\overline{\partial}f$ vanish at the origin because $\overline{\partial}f$ is real analytic. By duality this is the same as saying that f is orthogonal to $\overline{\partial}P(D)\delta_0$ for any polynomial $P(\partial/\partial x, \partial/\partial y)$.

We translate this condition via Fourier transform: Since f is orthogonal to the inverse Fourier transforms of (22) as well as to the Fourier transforms of polynomials in $\widehat{x}$ times the right side of (22) [by (19)] we are left with the following

QUESTION. *Can we construct all polynomials as limits of linear combinations of the set of*

(23) $$\widehat{x}^l \widehat{z}^{-1} e^{i(n+1)\widehat{\theta}} J_{n+1}(\widehat{r})?$$

For example we want to write

(24) $$1 = \widehat{z}^{-1} \sum a_{ln}^1 x^l e^{i(n+1)\widehat{\theta}} J_{n+1}(\widehat{r}).$$

According to (22) the only term on the right side of (24) that contains the constant term in its power series corresponds to $l = 0, n = 0$. The next term in the power series for $e^{i\widehat{\theta}} J_1(\widehat{r}) \widehat{z}^{-1}$ is a (nonzero) multiple of r^2. The leading coefficients of $e^{3i\widehat{\theta}} J_3(\widehat{r}) \widehat{z}^{-1}$, of $\widehat{x} e^{2i\widehat{\theta}} J_2(\widehat{r}) \widehat{z}^{-1}$, and of $x^2 e^{i\widehat{\theta}} J_1(\widehat{r}) \widehat{z}^{-1}$ are nonzero multiples of $\widehat{z}^2$, $\widehat{x}\widehat{z}$, and of $\widehat{x}^2$ respectively.

Note that $\widehat{z}^2, \widehat{x}\widehat{z}, \widehat{x}^2$ form a basis for homogeneous polynomials of degree 2. Thus we can find a_{ln} for $l + n = 2$ so that (24) holds up to terms of degree 3.

To continue we use the fact that the monomial of lowest degree in the power series for (23) is $c\widehat{x}^l \widehat{z}^n$ where $c = c(l, n)$ is a nonzero constant. This means that for a given $N \geq 0$ we can find exactly $N + 1$ series of the form $\widehat{x}^l \widehat{z}^{-1} e^{i(n+1)\widehat{\theta}} J_{n+1}(\widehat{r})$ whose leading coefficients are $c\widehat{x}^l \widehat{z}^{N-l}$.

Observe that $\{\widehat{x}^l \widehat{z}^{N-l}\}$ is a basis for all polynomials in $\widehat{x}, \widehat{y}$. This follows, for example, from fact that the highest power of $\widehat{y}$ in $\widehat{x}^l \widehat{z}^{N-l}$ is $\widehat{y}^{N-l}$. For fixed N these powers are distinct from which it follows easily that when N is fixed $\{\widehat{x}^l \widehat{z}^{N-l}\}$ is a basis for all homogeneous polynomials of degree N.

In this way we can construct the a_{ln}^1 to satisfy (24). Similarly we can write any polynomial P as

(25) $$P = \widehat{z}^{-1} \sum a_{ln}^P x^l e^{i(n+1)\widehat{\theta}} J_{n+1}(\widehat{\theta}).$$

In particular, if we break up our procedure at the stage $N + 1$ then we have shown that every polynomial Q of degree $\leq N$ is of the form

(26) $$\widehat{z} Q = \sum_{l+n \leq N} b_{ln}^Q x^l e^{i(n+1)\widehat{\theta}} J_{n+1}(\widehat{\theta}) + \sum K_m(x, y).$$

The b_{ln}^Q are constants and the K_m are finite linear combinations of the

$$x^l e^{i(n+1)\widehat{\theta}} J_{n+1}(\widehat{\theta})$$

whose lowest power series terms are of degree $> N$. By duality and Fourier transform this means that if P is a polynomial of degree $\leq N$ which is orthogonal to

$$\left\{ \frac{\partial^l}{\partial x^l} z^n \delta_{|z|=1} \right\}_{l+n \leq N}$$

then P is holomorphic in z. To go beyond polynomials we must estimate the coefficients a_{ln}^P of (25) for homogeneous P. At present we cannot prove f is holomorphic unless f is real analytic in the strip $|y| \leq 1$ and the power series for f at the origin converges in $|x| \leq 1, |y| \leq 1$.

Mark Agranovsky has informed me (oral communication) that he has derived the same result by totally different means. His article will be published soon.

One might be tempted to use the wave equation method as developed for the solution of Problem 2 above. Suppose f is a smooth function, e.g. $f \in C^\infty$; we may suppose f is defined on all of $\mathbb{R}^2$. As before let F be the solution of the wave equation in (t, x, y) whose Cauchy data on $t = 0$ is $(0, f)$. If f were holomorphic in the strip $|y| \leq 1$ then the Cauchy Data of $\partial F/\partial \bar{z}$ would vanish in this strip so, by the finite speed of propagation for the wave equation

$$\text{(27)} \qquad \frac{\partial F}{\partial \bar{z}} = 0 \qquad |y| \leq 1 - |t|, \ |t| \leq 1.$$

It seems that a natural analog of Problem 2 is the question of whether f is holomorphic in $|y| \leq 1$ if

$$\text{(28)} \qquad \frac{\partial^n F}{\partial \bar{z}^n}(1, x, 0) = 0$$

for all x and n.

The difficulty in relating this question to Problem 2 is that there is no Huygens Principle for the wave equation in 2 space variables. If we trace the meaning of (28) in terms of f we find that the integral (7) on the circle on radius 1 is replaced by the weighted average of integrals on circles of radius ≤ 1 (same center) with the weight $\sqrt{1 - r^2}$ [**2, p. 205**].

Our method can be used (actually in a simpler form) to show that if f is real analytic then (28) implies it is complex analytic in the strip. But if f is only assumed to be C^∞ then (28) does not imply holomorphicity. For we can let the Cauchy data of F on $t = 1$ vanish in a neighborhood $\{y = 0\}$, which certainly implies (28), without F vanishing on $\{t = 0\}$. It follows easily that $F(0, x, y)$ cannot be real analytic.

All this leads me to conjecture that Problem 3 has a negative answer for C^∞ functions f.

We could define an analog $\widetilde{F}$ of F by

$$\text{(29)} \qquad \widetilde{F}(t, x, y) = \iint_{(x-x_1)^2 + (y-y_1)^2 = t^2} f(x - x_1, y - y_1)\, d\theta.$$

where $d\theta$ is the measure on the circle $(x - x_1)^2 + (y - y_1)^2 = t^2$ normalized to have total mass 1. $\widetilde{F}$ satisfies a pseudodifferential equation which we do not understand.

References

[1] N. Bateman, *Higher Transcendental Functions*, Mc Graw-Hill, 1953.
[2] R. Courant, D. Hilbert, *Methods of Mathematical Physics*, Interscience, 1966.
[3] Leon Ehrenpreis, *Fourier Analysis in Several Variables*, Interscience, 1970.
[4] Leon Ehrenpreis, *The Radon Transform*, Oxford, 2001, to appear.
[5] E. Grinberg, *Radon Transforms: Ranges and Restrictions. Integral Geometry.*, Contemp. Math. **63** (1987), 109-133.
[6] T. Quinto, *Oral communication*.
[7] M. Riesz, *L'integrale de Riemann-Liouville et le problème de Cauchy*, Acta Mathematica **81** (1949), 1-223.

DEPARTMENT OF MATHEMATICS TEMPLE UNIVERSITY PHILADELPHIA, PA 19122, USA

A PALEY-WIENER THEOREM FOR CENTRAL FUNCTIONS ON COMPACT LIE GROUPS

FULTON B. GONZALEZ

ABSTRACT. In this note we present a Paley-Wiener criterion for the Fourier coefficients of central functions on compact simply connected Lie groups, whose supports lie inside a ball of radius smaller than the injectivity radius of the group. The proof relies on an analogous result about standard Fourier series.

1. Introduction

In this paper we present a Paley-Wiener type theorem for functions on the n-torus, and, as an application, for central functions on compact semisimple Lie groups. Such theorems are useful in, for example, obtaining support theorems for integral transforms (see [2]), as well as in obtaining properties of solutions of various types of wave equations.

2. The Flat Case

Let $p : (x_1, \ldots, x_n) \mapsto (e^{ix_1}, \ldots, e^{ix_n})$ be the projection from $\mathbb{R}^n$ onto the torus $\mathbb{T}^n = \mathbb{R}^n/2\pi \mathbb{Z}^n = (S^1)^n$. For $I \in \mathbb{Z}^n$ and $t = (t_1, \ldots, t_n) \in \mathbb{T}^n$, we write $t^I = t_1^{i_1} \cdots t_n^{i_n}$. If $f \in C^\infty(\mathbb{T}^n)$, its Ith Fourier coefficient is given by $\widehat{f}(I) = \int_T f(t) t^{-I}\, dt$. Then the Fourier series $2\pi^{-n} \sum_{I \in \mathbb{Z}^n} \widehat{f}(I) t^I$ of f converges to f in (the topology of) $C^\infty(\mathbb{T}^n)$. Let $e = (1, \ldots, 1)$ denote the identity element of $\mathbb{T}^n$, and endow $\mathbb{T}^n$ with the flat (quotient) metric inherited from $\mathbb{R}^n$. The following "Paley-Wiener" result holds for Fourier series:

Proposition 1. *Suppose $R < \pi$. Then f is supported on the "ball" $B_R(e)$ if and only if $\widehat{f}$ extends to a holomorphic function on $\mathbb{C}^n$ of exponential type R.*

Proof. Suppose that $\widehat{f}$ extends to a holomorphic function on $\mathbb{C}^n$ of exponential type $R < \pi$. We denote this extension also by $\widehat{f}$. By the usual Paley-Wiener theorem (see [6], page 161), there exists a function $F \in C_c^\infty(\mathbb{R}^n)$, supported on the ball $B_R(0)$, whose Fourier transform coincides with $\widehat{f}$:

$$(2.1) \qquad \mathcal{F}F(z) = \int_{\mathbb{R}^n} F(x) e^{-ix \cdot z}\, dx = \widehat{f}(z). \qquad (z \in \mathbb{C}^n)$$

Define the periodic function F_0 on $\mathbb{R}^n$ by $F_0(x) = \sum_{I \in \mathbb{Z}^n} F(x + 2\pi I)$. (For each $x \in \mathbb{R}^n$, at most one of the summands on the right-hand side is nonzero.) Then of course, $F_0 = \varphi \circ p$ for some smooth function φ on $\mathbb{T}^n$, with support on $B_R(e)$. It is easy to see that $\mathcal{F}F(I) = \widehat{\varphi}(I)$ for all $I \in \mathbb{Z}^n$. Hence $\widehat{\varphi}(I) = \widehat{f}(I)$ so in fact $\varphi = f$.

Date: February 23, 2000.

2000 *Mathematics Subject Classification.* Primary: 43A77, Secondary: 43A90.

Key words and phrases. Paley-Wiener Theorem, Fourier Series.

Conversely, let $f \in C^\infty(\mathbb{T}^n)$ be supported in the ball $B_R(e)$. The function F on $\mathbb{R}^n$ defined by

$$(2.2) \qquad F(x) = \begin{cases} f \circ p(x), & \text{for } x_i \in (-\pi, \pi), i = 1, \ldots, n \\ 0, & \text{for all other } x \end{cases}$$

belongs to $C_c^\infty(\mathbb{R}^n)$, with support in $B_R(0)$. As with the above, $\mathcal{F}F(I) = \widehat{f}(I)$ for all $I \in \mathbb{Z}^n$, so by the easy part of the Paley-Wiener theorem $\widehat{f}$ extends to a holomorphic function of exponential type R on $\mathbb{C}^n$. □

3. The General Case

We use the proposition above to prove an analogous Paley-Wiener type result for central functions on a compact semisimple Lie group U. This is not entirely unexpected, since a central function on U is completely determined by its restriction to a maximal torus. To motivate the general theorem (Theorem 3 below) and its proof, we start with a low-dimensional example. This turns out to also allow us to obtain a variant of Proposition 1 for the Legendre series of an even function on the unit circle.

3.1. $SU(2)$ and Legendre Series on S^1.

The Lie group $U = SU(2)$ is compact, semisimple, and simply connected; in fact, it is isomorphic as a Lie group to S^3, the multiplicative group of unit quaternions, via the map $\begin{pmatrix} a & b \\ -\bar{b} & \bar{a} \end{pmatrix} \longrightarrow a + bj$ ($a, b \in \mathbb{C}, |a|^2 + |b|^2 = 1$). $SU(2)$ has a maximal torus T given by $\left\{ \begin{pmatrix} e^{i\theta} & 0 \\ 0 & e^{-i\theta} \end{pmatrix} \mid \theta \in \mathbb{R} \right\}$.

The usual representation of $SU(2)$ on $\mathbb{C}^2 = \{ \begin{pmatrix} z_1 \\ z_2 \end{pmatrix} \mid z_i \in \mathbb{C} \}$ gives rise to a representation π_n of $SU(2)$ on the (complex) vector space of degree n homogeneous polynomials in z_1, z_2. The representations π_n are easily seen to be irreducible, and are certainly inequivalent. Moreover, any irreducible representation of $SU(2)$ is equivalent to π_n for some n (see [5]). Thus the set of irreducible representations of $SU(2)$ is parametrized by $\mathbb{Z}^+$; the character χ_n of π_n is given on T by

$$(3.1) \qquad \chi_n \begin{pmatrix} e^{i\theta} & 0 \\ 0 & e^{-i\theta} \end{pmatrix} = \sum_{k=0}^{n} e^{i(n-2k)\theta} = \frac{e^{i(n+1)\theta} - e^{-i(n+1)\theta}}{e^{i\theta} - e^{-i\theta}}$$

The last expression above is the Weyl character formula for the representation π_n, and equals $\sin((n+1)\theta)/\sin\theta$, which is a polynomial $P_n(\cos\theta)$ of degree n in $\cos\theta$. ($P_n(\cos\theta)$ is the Gegenbauer polynomial $2C_n^{(3/2)}(\cos\theta)/(n+2)$, or the restriction to T of the zonal spherical harmonic of degree n on S^3.)

We now identify U with the unit quaternionic group S^3 in the manner above; the maximal torus T then corresponds to the great circle $\{e^{i\theta} \mid \theta \in \mathbb{R}\}$ through 1. The usual Riemannian structure on S^3 is invariant under left and right multiplication. The orbits $\{uxu^{-1} \mid u \in U\}$ correspond to two-spheres of latitude centered at 1. Each of these orbits intersects the great circle T at the points $e^{i\theta}$, $e^{-i\theta}$. A function φ on U is *central* iff φ is constant on the above orbits. Now using stereographic projections, for example, it is not too hard to see that a central function f is C^∞ on S^3 if and only if its restriction to the circle T is even and C^∞. This gives us a bijection between the C^∞ central functions on U and the C^∞ even fuctions on T.

Now let f be an even C^∞ function on T. By the above, we can extend f to be a C^∞ central function on U. On T, f can be expanded in a "Legendre" series

$$\begin{aligned} f(e^{i\theta}) &= \sum_{n=0}^\infty \widehat{f}(n)\, \chi_n(e^{i\theta}) \\ &= \sum_{n=0}^\infty \widehat{f}(n)\, P_n(\cos\theta) \end{aligned}$$
(3.2)

with rapidly decreasing coefficients $\widehat{f}(n)$ on $\mathbb{Z}^+$. The trigonometric polynomials $P_n(\cos\theta)$ form an orthonormal basis of the space of even L^2 functions on T, with respect to the measure $(\sin^2\theta/\pi)\,d\theta$ on T. In particular,

$$\begin{aligned} \widehat{f}(n) &= \frac{1}{\pi} \int_{-\pi}^{\pi} f(e^{i\theta})\, P_n(\cos\theta)\, \sin^2\theta\, d\theta \\ &= \frac{1}{\pi} \int_{-\pi}^{\pi} \left(f(e^{i\theta})\sin\theta\right) \sin(n+1)\theta\, d\theta \end{aligned}$$
(3.3)

Let $f_0(e^{i\theta}) = f(e^{i\theta})\sin\theta$; then $f_0(e^{i\theta})$ is an odd function on T, so

$$\begin{aligned} \widehat{f}(n) &= \frac{1}{2\pi i} \int_{-\pi}^{\pi} f_0(e^{i\theta}) \left(e^{i(n+1)\theta} - e^{-i(n+1)\theta}\right) d\theta \\ &= \frac{i}{\pi} \int_{-\pi}^{\pi} f_0(e^{i\theta})\, e^{-i(n+1)\theta}\, d\theta \end{aligned}$$
(3.4)

Let $W_{f_0}(n)$ denote the Fourier coefficient $1/(2\pi) \int_{-\pi}^{\pi} f_0(e^{i\theta})\, e^{-in\theta}\, d\theta$. From (3.4) we see that $\widehat{f}(n) = 2i\, W_{f_0}(n+1)$. Since f_0 is odd on T, we have $W_{f_0}(n) = -W_{f_0}(-n)$. Now suppose that $0 < R < \pi$. By Proposition 1 and its proof, f_0 has support in the arc $\{e^{i\theta}\mid -R < \theta < R\}$ if and only if the map $n \mapsto W_{f_0}(n)$ extends to an odd entire function $z \mapsto W_{f_0}(z)$ on $\mathbb{C}$, of exponential type R. Now f is supported on the arc $\{e^{i\theta}\mid -R < \theta < R\}$ if and only if f_0 is. Defining $\widehat{f}(z) = 2i\, W_{f_0}(z+1)$ we obtain the following proposition.

Proposition 2. *Let $f(e^{i\theta})$ be an even C^∞ function on the unit circle T, and let $0 < R < \pi$. Then f is supported in the arc $\{e^{i\theta}\mid -R < \theta < R\}$ iff the map $n \mapsto \widehat{f}(n)$ extends to an entire holomorphic function $z \mapsto \widehat{f}(z)$ on $\mathbb{C}$ of exponential type R, satisfying the skew condition $\widehat{f}(z) = -\widehat{f}(-z-2)$.*

3.2. Central Functions on Compact Semisimple Lie Groups. In this subsection, U will denote a compact semisimple Lie group. Fix a maximal torus $T \subset U$, and let $\mathfrak{u} = \mathrm{Lie}(U)$, $\mathfrak{t} = \mathrm{Lie}(T)$, $\mathfrak{g} = \mathfrak{u}^c$, $\mathfrak{h} = \mathfrak{t}^c$, $\Delta = \Delta(\mathfrak{g},\mathfrak{h})$, $W = \mathrm{Weyl}(\Delta)$. Fix a positive system $\Delta^+ \subset \Delta$, with corresponding Weyl chamber $C^+ \subset i\mathfrak{t}$. We identify $i\mathfrak{t}$ and its real dual $i\mathfrak{t}_*$ via the Killing form on $\mathfrak{g} = \mathfrak{u}^{\mathbb{C}}$. Finally, we endow U and T with normalized Haar measures du and dt, respectively.

A function φ on T is said to be *skew* if $\varphi(\exp(H)) = \det(s)\, \varphi(\exp(sH))$ for all $s \in W$.

Let f be a smooth central function on U (i.e. $f(uxu^{-1}) = f(x)$ for all $u, x \in U$.) If Λ denotes the highest weight lattice of U, we write

$$f = \sum_{\mu \in \Lambda} \widehat{f}(\mu)\chi_\mu$$
(3.5)

where χ_μ denotes the character of the irreducible representation π_μ of U with highest weight μ. Here
$$\widehat{f}(\mu) = \langle f, \chi_\mu \rangle = \int_U f(u)\chi_{\mu^*}(u)\,du$$
where $\mu^* = -s^*\mu$ is the highest weight of representation contragredient to π_μ. ($-s^*$ is the unique Weyl group element mapping C^+ to $-C^+$.) Note that the character χ_μ is a unit vector in $L^2(U)$.

Since both f and χ_{μ^*} are central, we have by [4], Chapter 1, Corollary 5.16,

(3.6) $$\widehat{f}(\mu) = \frac{1}{w}\int_T f(t)\chi_{\mu^*}(t)\delta(t)\,dt$$

where $\delta(t)$ is the density function $\delta(\exp H) = \left(\sum_{s\in W}(\det s)e^{s\rho(H)}\right)^2$. We take $\delta^{1/2}$ to be the sum inside the parenthesis. Then

(3.7) $$\widehat{f}(\mu) = \frac{1}{w}\int_T (f\cdot\delta^{1/2})(t)(\delta^{1/2}\cdot\chi_{\mu^*})(t)\,dt,$$

and by the Weyl character formula

(3.8) $$\left(\delta^{1/2}\cdot\chi_{\mu^*}\right)(t) = \left(\sum_{s\in W}(\det s)e^{s(\mu^*+\rho)}\right)(t).$$

Now the function $f_0(t) = (f\cdot\delta^{1/2})(t)$ is skew, so that

(3.9) $$\widehat{f}(\mu) = \frac{1}{w}\int_T f_0(t)\left(\sum_{s\in W}(\det s)e^{s(\mu^*+\rho)}\right)(t)\,dt$$
$$= \int_T f_0(t)e^{(\mu^*+\rho)}(t)\,dt.$$

Let $\widetilde{\Lambda}$ denote the weight lattice of T; for each L^2 function φ on T and $\lambda \in \widetilde{\Lambda}$, let $W_\varphi(\lambda)$ denote the toral Fourier coefficient $W_\varphi(\lambda) = \int_T \varphi(t)e^{-\lambda}(t)\,dt$. Since f_0 is skew and $s^*\rho = -\rho$,

(3.10) $$\widehat{f}(\mu) = (\det s^*)\int_T f_0((s^*)^{-1}\cdot t)e^{-\mu-\rho}((s^*)^{-1}\cdot t)\,dt$$
$$= (\det s^*)\int_T f_0(t)e^{-\mu-\rho}(t)\,dt$$
$$= (\det s^*)W_{f_0}(\mu+\rho).$$

Let δ be a root of maximum length and fix $0 < R < \frac{2\pi}{|\delta|}$ (the injectivity radius of U). Suppose that f is supported in the ball $B_R(e) \subset U$. Then f_0 has support in $B_R(e) \subset T$, so by Proposition 1, $\widehat{f}(\mu) = (\det s^*)W_{f_0}(\mu+\rho)$ extends to a holomorphic function on $(\mathfrak{t}^*)^c = \mathfrak{h}^*$ of exponential type R in $\mu + \rho$ (hence in μ). Since f_0 is skew, this extension (which we also denote by $\widehat{f}$) is skew in $\mu + \rho$: $W_{f_0}(\mu+\rho) = (\det s)W_{f_0}(s(\mu+\rho))$, so that $\widehat{f}(\mu) = (\det s)\widehat{f}(s(\mu+\rho)-\rho)$.

Conversely, suppose that $\widehat{f}$ extends to a holomorphic function on $(\mathfrak{t}^*)^c$ of exponential type R, satisfying the condition $\widehat{f}(\mu) = (\det s)\widehat{f}(s(\mu+\rho)-\rho)$ for all $s \in W$, $\mu \in (\mathfrak{t}^*)^c$. Then the function $W(\mu) = (\det s^*)\widehat{f}(\mu-\rho)$ satisfies $W(\mu+\rho) = (\det s)W(s(\mu+\rho))$ for all $s \in W, \mu \in (\mathfrak{t}^*)^c$, and is holomorphic of exponential type R in $\mu - \rho$, hence in μ.

Since $\mu \mapsto \mu + \rho$ is a bijection of $\widetilde{\Lambda}$, the function

(3.11)
$$f_0(t) = (\det s^*) \sum_{\mu \in \widetilde{\Lambda}} W(\mu) e^{\mu}(t)$$
$$= (\det s^*) \sum_{\mu \in \widetilde{\Lambda}} W(\mu + \rho) e^{\mu+\rho}(t)$$

is a skew C^∞ function on T, with support in $B_R(e) \cap T$ by Proposition 1. Now C^+ is a transversal set for the W-orbits in $i t_*$, so we rewrite the above as

(3.12) $$f_0(t) = (\det s^*) \sum_{\mu+\rho \in \widetilde{\Lambda} \cap \overline{C^+}} W(\mu+\rho) \sum_{s \in W} (\det s) e^{s(\mu+\rho)}(t).$$

In the above sum, we note that for $\mu + \rho \in \widetilde{\Lambda} \cap \overline{C^+}$, $W(\mu+\rho) \neq 0$ only if $\mu \in \Lambda$. For, if $\mu \notin \Lambda$, then $\langle \mu, \alpha \rangle < 0$ for some simple root α; since $2\langle \rho, \alpha \rangle / \langle \alpha, \alpha \rangle = 1$ and $\langle \mu + \rho, \alpha \rangle \geq 0$ we must have $\langle \mu + \rho, \alpha \rangle = 0$. If s_α denotes the reflection on the hyperplane $\alpha = 0$, we obtain $W(\mu+\rho) = -W(s_\alpha(\mu+\rho)) = -W(\mu+\rho)$, so $W(\mu+\rho) = 0$.

Thus,

(3.13)
$$f_0(t) = (\det s^*) \sum_{\mu \in \Lambda} W(\mu+\rho) \sum_{s \in W} (\det s) e^{s(\mu+\rho)}(t)$$
$$= \det(s^*) \delta^{1/2}(t) \sum_{\mu \in \Lambda} W(\mu+\rho) \xi_\mu(t).$$

Since f_0 is skew, the function $\psi(t) = \delta^{-1/2}(t) f_0(t)$ belongs to $C^\infty(T)$. (See [4], page 504.) Moreover, ψ is W-invariant and has support in $B_R(e) \cap T$, since f_0 does. By (3.13),

(3.14) $$\psi(t) = \sum_{\mu \in \Lambda} \widehat{f}(\mu) \chi_\mu(t)$$

has support in $B_R(e) \cap T$. But the right hand side represents $f|_T$; since f is central, it follows that f is supported in $B_R(e)$. We have thus obtained the following theorem.

Theorem 3. *A central function $f \in C^\infty(U)$ has support in the ball $B_R(e)$ if and only if $\widehat{f}(\mu)$ extends to a holomorphic function on $(\mathfrak{t}^*)^c$ of exponential type R satisfying the skew condition $\widehat{f}(\mu) = (\det s)\widehat{f}(s(\mu+\rho) - \rho)$.*

We note that this result is the compact analogue (for groups) of the spherical Paley-Wiener theorem ([1, 2]). Although Proposition I surely exists in the literature, the author could not find a reference for it.

References

[1] S. Helgason, *An analogue of the Paley-Wiener theorem for the Fourier transform on certain symmetric spaces*, Math. Ann. **165** (1966), pp297–308.

[2] S. Helgason, *The surjectivity of invariant differential operators on symmetric spaces*, Ann. of Math. **98** (1973), pp451–480.

[3] S. Helgason, *Differential Geometry, Lie Groups, and Symmetric Spaces*, Academic Press, 1978.

[4] S. Helgason, *Groups and Geometric Analysis*, Academic Press, 1984.
[5] M. Sugiura *Unitary Representations and Harmonic Analysis*, 2nd ed. North Holland and Kodansha, Amsterdam and Tokyo, 1990.
[6] K. Yosida, *Functional Analysis*, Springer-Verlag , 1968.

DEPARTMENT OF MATHEMATICS, TUFTS UNIVERSITY, MEDFORD, MA 02155
E-mail address: `fulton.gonzalez@tufts.edu`

Inversion of the spherical Radon transform by a Poisson type formula

Isaac Pesenson and Eric L. Grinberg

ABSTRACT. We present an analog of the Poisson summation formula for approximate reconstruction of an even smooth function on the unit sphere using a discrete set of values of its integrals along great sub-spheres.

1. Introduction and main results

The spherical Radon transform, also known as the Funk transform, associates to a function $\varphi(\theta)$ its integral over great subspheres: $R\varphi(\theta^\perp \cap S^d) = \int_{\theta^\perp \cap S^d} d\mu$, where $\theta^\perp \cap S^d$ is the great (equatorial) subsphere of S^d whose plane has normal θ. This transform is of interest in pure and applied mathematics. A good deal of the analysis of the spherical Radon transform can be done using spherical harmonics and the representation theory of the orthogonal group. See, [3] and also, e.g., [1] for a discussion of projective spaces that applies equally well to spheres. Here we will deal with even functions on the sphere. The entire discussion could be phrased in terms of real projective spaces. But this usage is not common in applications and, in any case, does not add much to the present context. Analysis of the spherical Radon transform using spherical wavelets, the spherical Calderon reproducing formula and their applications to inversion of the Radon transform on symmetric spaces is considered in [**7, 8, 9, 10**].

To outline our approach to the inversion of the spherical Radon transform let us begin by recalling the classical Poisson formula for the Fourier transform. One of the forms of the classical Poisson summation formula for a function φ belonging to $L^2(\mathbb{R})$ with support in $[-\omega, \omega]$ is

(1.1) $$\varphi(t) = \sum_{n \in Z} \hat{\varphi}(n\Omega) e^{int}, \quad \Omega = \pi/\omega.$$

This shows that the Fourier coefficients of a function with compact support are regularly spaced samples of its Fourier transform. Taking the limit of the right

---

1991 *Mathematics Subject Classification.* 44A12; Secondary 94A12, 94A20.

Key words and phrases. Poisson summation formula, spherical Radon transform Laplace-Beltrami operator, fundamental solutions, polyharmonic splines.

The author ELG was supported in part by NSF Grant #DMS9971828.

hand side when ω goes to infinity we obtain a formula that makes sense for any function belonging to $L^2(\mathbb{R})$. This formula can be treated as an inversion formula for the Fourier transform.

A simple way to obtain formula (1.1) is by using the famous Shannon-Whittaker sampling theorem. This result states that if a function $f \in L^2(\mathbb{R})$ has band width ω, i.e., its Fourier transform $\hat{f}$ has support in $[-\omega, \omega]$, then f is completely determined by its values at points $n\Omega$, where $\Omega = \pi/\omega$ and, in the L^2-sense,

$$f(t) = \sum f(n\Omega) \frac{\sin(\pi(t - n\Omega))}{\pi(t - n\Omega)}.$$

It was shown in [4] that a band limited function belonging to $L^2(\mathbb{R}^d)$ can be reconstructed from an appropriate irregular set of points as a limit of polyharmonic splines. To give a precise formulation of this statement we will need the following notations.

$B_\sigma(\mathbb{R}^d)$:
The set of all band limited functions in $L^2(\mathbb{R}^d)$ with band width σ is denoted by $B_\sigma(\mathbb{R}^d)$.

$X(\lambda)$:
The symbol $X(\lambda) = \{x_\nu\}$ is used for a set of sample points, or a *knot set* in $\mathbb{R}^d$. Mention of a knot set $X(\lambda)$ will carry tacitly the assumption such that there is a triangulation of $\mathbb{R}^d$ with the following properties:

1) Each element of the triangulation contains just one of the points x_ν;
2) The diameter of every element of the triangulation is not greater than λ.

We consider the operator $\Delta + \varepsilon$ in the space $L^2(\mathbb{R}^d)$, where Δ is the Laplace operator in $L^2(\mathbb{R}^d)$. A *polyharmonic Lagrangian spline* $L_\nu^k \in L^2(\mathbb{R}^d)$ is a special linear combination of the fundamental solutions of the shifted Laplace operator $(\Delta + \varepsilon)^k$. (See more details in [4].) Note that Lagrangian splines have exponential decay at infinity. The following result can be found in [4], [2].

THEOREM 1.1 (Approximation by polyharmonic splines with bounds).
There exists a constant $c = c(d, \epsilon)$ that depends only on the dimension d and the parameter ε such that for any $\sigma > 0$, for every knot set $X(\lambda)$ with $\lambda < (c(\sigma + \varepsilon))^{-1}$, for every integer $r \geq [d/2] + 1$, and for every band limited function $f \in B_\sigma(\mathbb{R}^d)$, we have the following interpolation by polyharmonic Lagrangian splines:

$$(1.2) \qquad f = \lim_{l \to \infty} \sum_{x_\nu \in X(\lambda)} f(x_\nu) L_\nu^{2^l r}, \; l \in \mathbb{N},$$

whose approximations enjoy the error estimate below:

$$(1.3) \qquad \|f - \sum_\nu f(x_\nu) L_\nu^{2^l r}\| \leq 2(c\lambda(\sigma + \varepsilon))^{2^{l+1} r} \|f\|.$$

Allowing λ to go to zero we obtain (1.3) for any function belonging to $L^2(\mathbb{R}^d)$ that is sufficiently smooth. This sampling theorem can be used to obtain the following irregular analog of the Poisson summation formula [2]

THEOREM 1.2. *If the band-limited function $\varphi \in L^2(\mathbb{R}^d)$ has support in the ball $B(\sigma, 0)$ and Λ_ν^k is the inverse Fourier transform of the spline function L_ν^k then*

(1.4) $$\varphi = \lim_{l \to \infty} \sum_{\xi_\nu \in \Xi(\lambda)} \hat{\varphi}(\xi_\nu) \Lambda_\nu^{2^l r},$$

assuming $l \in \mathbb{N}$, $r \geq [d/2] + 1$, $\lambda < (c(\sigma + \varepsilon))^{-1}$, $c = c(d, \varepsilon)$ is the same as in the previous theorem and the $\Xi(\lambda)$ is an appropriate discrete set in the space of the dual variable ξ.

An error estimate for this approximation is

$$\left\| \varphi - \sum_{\xi_\nu \in \Xi(\lambda)} \hat{\varphi}(\xi_\nu) \Lambda_\nu^{2^l r} \right\| \leq 2(c\lambda(\sigma + \varepsilon))^{2^{l+1} r} \|\varphi\|.$$

The above result can also be considered as an inverse Fourier transform. The main objects of study in the present paper are: the unit sphere S^d, the corresponding space $L^2(S^d)$, the Laplace-Beltrami operator in the space $L^2(S^d)$, and the spherical Radon transform (also called the Funk transform). In this situation we obtain a very close analog of Theorem 1.2 with the spherical Radon transform in place of the Fourier transform. The method used to obtain such a Theorem is similar to that of $\mathbb{R}^d$. Namely, we go to the Radon Transform side and use an appropriate sampling theorem on the sphere [4]. For example, if φ is an even harmonic polynomial on the unit sphere S^d and $R\varphi(\xi_\nu)$ are the values of its Spherical Radon transform on an appropriate grid $\{\xi_\nu\}$ we obtain the formula

(1.5) $$R\varphi = \lim_{k \to \infty} \sum_\nu R\varphi(\xi_\nu) L_\nu^k,$$

where L_ν^k is a Lagrange spline on the dual sphere i.e. L_ν^k is a specific linear combination of the fundamental solutions of the shifted Laplace-Beltrami operator on the sphere.

Taking the inverse Radon Transform of both sides of (1.5) we obtain our main formula:

(1.6) $$\varphi = \lim_{k \to \infty} \sum_\nu R\varphi(\xi_\nu) \Lambda_\nu^k,$$

where Λ_ν^k is the inverse Radon transform of L_ν^k.

Using a nested sequence of knots whose union is dense in the sphere S^d we can extend (1.6) to any smooth function on S^d by means of the formula

(1.7) $$\varphi = \lim \sum_\nu R\varphi(\xi_\nu) \Lambda_\nu^k,$$

where the limit is taken over a nested sequence of knots and λ is fixed.

Formulas (1.6) and (1.7) are inversion formulas for the spherical Radon Transform. Their advantage is that they suggest a way to approximate a function using just a finite number of samples of its Radon transform. Note that according to the uncertainty principle we have the usual trade-off between smoothness and localization: the lower the index k the better localization we obtain for the functions Λ_ν^k. At the end of the paper we give a constructive way to determine the Fourier-Laplace coefficients of the functions Λ_ν^k. We hope to conduct numerical experiments to test the reconstruction schemes presented here. The results of such experiments will be published in a future paper.

2. Splines on Spheres and a Poisson summation formula for the Spherical Radon Transform

In this section we recall the notion of splines on manifolds and their construction. We prove that a sequence of splines that interpolates a given smooth function converges to it in the appropriate Sobolev norms. Then we prove the inversion formula for the spherical Radon transform (1.6).

Let $\{r_\nu\}, \nu = 1, \ldots N$ be a simplicial cover of S^d such that every simplex r_ν has a diameter not greater than λ. Let $X(\lambda) = \{x_\nu\}$ be a set of points of S^d such that x_ν belongs to the interior of r_ν. In this paper we will be interested in even functions on S^d and as a result in even spline functions. Because of this it is natural to assume that the set of knots $X(\lambda)$ is symmetric or *even*, i.e., $X(\lambda) = -X(\lambda)$.

Given a knot set $X(\lambda)$ and a sequence of numbers $\{t_\gamma\} \in l^2$ we will be interested to find a function $t_k \in H^{2k}(S^d)$, for k large enough, such that the following conditions hold.

a) *Interpolation*: $t_k(x_\gamma) = t_\gamma$, $x_\gamma \in X(\lambda)$;

b) *Minimality*:   the function t_k minimizes the functional $u \to \|(1+\Delta)^k u\|$.

Since $\|(1+\Delta)^k u\|$ is a norm this minimization problem has a unique solution $t_k \in H^{2k}(S^d)$ which will be called a *spline* function (see [**4**]). We will often refer to this *minimization problem* below.

Notation. If f is a smooth function on S^d then the (unique) spline function belonging to $H^{2k}(S^d)$ that interpolates f on $X(\lambda)$ is denoted by $t_k(f)$. Note that if f is even and the set $X(\lambda)$ is symmetric, then the spline $t_k(f)$ is an even function.

It was shown in [**4**] that every smooth function f can be interpolated and approximated by splines $t_k(f)$ in the L^2-norm. In order to realize our approach to the formula (1.6). We have to prove that the above convergence takes place not only in the L^2 norm but in fact in any Sobolev norm and even in the uniform norm. We begin with a lemma that will be useful in obtaining Sobolev norm estimates.

LEMMA 2.1. *Let Q be any self-adjoint operator in a Hilbert space E. If f in the domain of Q satisfies*

(2.1) $$\|f\| \leq A + a\|Qf\|$$

for some $a > 0$ then for the same f and all integers $m = 2^{l_1}, l_1 = 0, 1, 2, \ldots$

(2.2) $$\|f\| \leq mA + 8^{m-1}a^m\|Q^m f\|.$$

Moreover, if $A = 0$ then for any nonnegative r and every $m = 2^{l_1} \geq r, l_1 = 0, 1, \ldots$ there exists a positive constant $b(r, m)$ such that for all $n = 2^{l_2}, l_2 = 0, 1, 2, \ldots,$

(2.3) $$\|Q^r f\| \leq \left(b(r,m)a^{r-m}\right)^n \|Q^{n(m-r)+r} f\|$$

as long as f belongs to the domain of $Q^{n(m-r)+r}$.

PROOF. The assertions here are quite similar to those of Lemmas 3.1 and 3.2 of [6], so we merely sketch the proof. If Q is a self-adjoint operator then iQ generates a strongly continuous group of unitary operators e^{itQ}. Because of this we have the following Laplace transform representations for the resolvents of the operators iQ and $-iQ$

$$(\lambda I - iQ)^{-1}f = \int_0^\infty e^{-\lambda t} e^{itQ} f \, dt,$$

$$(\lambda I + iQ)^{-1}f = \int_0^\infty e^{-\lambda t} e^{-itQ} f \, dt.$$

for any $\lambda > 0$. This gives us the following inequalities

$$\|(\lambda I - iQ)^{-1}\| \leq \lambda^{-1}, \qquad \|(\lambda I + iQ)^{-1}\| \leq \lambda^{-1}.$$

In other words for any positive ε we have

$$\|(I - i\varepsilon Q)^{-1}\| \leq 1, \qquad \|(I + i\varepsilon Q)^{-1}\| \leq 1.$$

This gives

$$\varepsilon\|Qf\| \leq \|(I - i\varepsilon Q)f\| + \|f\| \leq \|(I + \varepsilon^2 Q^2)f\| + \|f\| \leq \varepsilon^2\|Q^2 f\| + 2\|f\|.$$

So, for any f from the domain of Q^2 we have the inequality

(2.4) $$\|Qf\| \leq \varepsilon\|Q^2 f\| + 2/\varepsilon\|f\|, \varepsilon > 0.$$

This inequality holds true for any self-adjoint operator and *any* $\varepsilon > 0$ with the *same* constant 2. Since any power of a self-adjoint operator is self-adjoint, we can formulate the following inductive proof.

The claim is evident for $m = 1$ and combining this case with the last equation above verifies the case $m = 2$. Continuing in this way, or, equivalently, using the induction hypothesis for a given m we conclude that

$$\|f\| \leq mA + 2^{3m-3} a^m (\varepsilon \|Q^{2m} f\| + 2/\varepsilon \|f\|).$$

Setting $\varepsilon = 2^{3m-1}(a)^m$, we obtain

(2.5) $$\|f\| \leq 2mA + 2^{6m-3} a^{2m} \|Q^{2m} f\|.$$

Thus the induction continues and the first part of the lemma is proved. In particular the inequality (2.5) implies, for $A = 0$, that

(2.6) $$\|f\| \leq (8a)^m \|Q^m f\|, m = 2^l, l = 0, 1, \ldots$$

Next, since

$$\|Q^r f\| \leq c(m,r) \|Q^m f\|^{r/m} \|f\|^{r-m}, 0 \leq r \leq m$$

we have

$$\|Q^r f\| \leq c(m,r)(8a)^{r-m} \|Q^m f\|, m = 2^l, l = 0, 1, \ldots, 0 \leq r \leq m.$$

For $g = Q^r f$ this gives

$$\|g\| \leq c(m,r)(8a)^{r-m} \|Q^{m-r} g\|$$

and then by (2.5)

$$\|g\| \leq \left(b(m,r)a^{r-m}\right)^n \|Q^{n(m-r)}g\|, m = 2^{l_1}, n = 2^{l_2}, \tag{2.7}$$

where the constant b is of the form

$$b(m,r) = c(m,r)8^{r-m}. \tag{2.8}$$

In other words with the same b as above we have for $l_1, l_2 = 0, 1, \ldots$

$$\|Q^r f\| \leq (ba^{r-m})^n \|Q^{n(m-r)+r} f\| \quad , \quad m = 2^{l_1} \quad , \quad n = 2^{l_2}. \tag{2.9}$$

$\square$

The following lemma is a consequence of a result from [4].

LEMMA 2.2. *There exist constants λ_0 and C such that for any smooth function $f \in C_0^\infty(S^d)$ and any λ with $0 < \lambda < \lambda_0$*

$$\|f\| \leq C\left(\lambda^{d/2}(\sum_\gamma |f(x_\gamma)|^2)^{1/2} + \lambda^d \|\Delta^{d/2} f\|\right).$$

The Lemma 2.2 implies that if λ is small enough then for a suitable C

$$\|f\| \leq C\left(\lambda^{d/2}(\sum_\gamma |f(x_\gamma)|^2)^{1/2} + \lambda^d \|(1+\Delta)^{d/2} f\|\right). \tag{2.10}$$

Now we use Lemma 2.1 with $Q = (1+\Delta)^{d/2}$ which gives us

$$\|f\| \leq C\left(m\lambda^{d/2}(\sum |f(x_\gamma)|^2)^{1/2} + 8^{m-1}(\lambda^d)^m \|(1+\Delta)^{md/2} f\|\right),$$

where $m = 2^l, l = 0, 1, 2, \ldots$. In particular if the restriction of f to the set $X(\lambda)$ is zero then the second part of the Lemma 2.1 gives for $m = 2r, 2k = rd, n = 2^l, l = 0, 1, \ldots$

$$\|(1+\Delta)^k f\| \leq (C_1 \lambda^{2k})^n \|(1+\Delta)^{k(n+1)} f\|. \tag{2.11}$$

Putting all this together we obtain the following estimate for convergence in the Sobolev norm

$$\|f - t_{k(n+1)}(f)\|_{2k} \leq (C_2 \lambda^{2k})^n \|(1+\Delta)^{k(n+1)} f\|. \tag{2.12}$$

Here the norm $\|\cdot\|_{2k}$ is the *Sobolev Norm* $\|\cdot\|_{L^2} + \|\Delta^k(\cdot)\|_{L^2}$. The right hand side above goes to zero if n is fixed and λ goes to zero.

The formula (2.12) can be applied to interpolate and approximate any smooth function in $L^2(S^d)$.

In particular, if f is a spherical harmonic polynomial of degree ω then we also have another way to approximate f by keeping λ fixed. Indeed for a polynomial of degree ω we have the estimate

$$\|f - t_{k(n+1)}(f)\|_{2k} \leq (C_2 \lambda^{2k})^n (1+\omega)^{k(n+1)} \|f\| \tag{2.13}$$

and the right hand side goes to zero given that n goes to infinity and, for a fixed λ, that the following condition holds true:

$$C_2\lambda(1+\omega) < 1.$$

In what follows we will apply the estimates (2.12) and (2.13) to obtain our main result. Let us recall the basic properties of the spherical Radon transform. If a function φ belonging to $L^2(S^d)$ has Fourier coefficients φ_j^m then its Radon Transform is given by the formula

$$(2.14) \qquad R\varphi = \pi^{-1/2}\Gamma((d+1)/2)\sum_{j,i} c_j \varphi_j^m Y_j^i$$

where Y_j^i is the standard notation for spherical harmonic polynomials and where

$$c_j = (-1)^{j/2}\Gamma((j+1)/2))/\Gamma((j+d)/2)$$

if j is even and $c_j = 0$ otherwise.

Because the coefficients c_j have asymptotics $(-1)^{j/2}(j/2)^{(1-d)/2}$ as j goes to infinity we have that R is a continuous operator from the Sobolev space of even functions $H_{even}^\alpha(S^d)$ onto the space $H_{even}^{\alpha+(d-1)/2}(S^d)$. Its inverse is a continuous operator from the space $H_{even}^{\alpha+(d-1)/2}(S^d)$ onto the space $H_{even}^\alpha(S^d)$.

Let $\{s_\nu\}$ be a set of equatorial subspheres on S^d of codimension one. We assume that the corresponding set of points ξ_ν on the dual sphere form a set of type $X(\lambda)$. According to the estimate (2.12) we have that for the spherical Radon transform $R\varphi$ of the smooth function φ,

$$(2.15) \qquad \|R\varphi - t_{k(n+1)}(R\varphi)\|_{2k} \leq (C_2\lambda^{2k})^n \|(1+\Delta)^{k(n+1)} R\varphi\|$$

where k is large enough. Taking the inverse Radon transform we obtain

$$(2.16) \qquad \|\varphi - R^{-1}t_{k(n+1)}(R\varphi)\|_{2k-(d-1)/2} \leq (C_3\lambda^{2k})^n \|(1+\Delta)^{k(n+1)} R\varphi\|$$

where the right side goes to zero for any smooth φ as long as λ goes to zero. If φ is a polynomial of degree ω than we have the estimate

$$(2.17) \qquad \|\varphi - R^{-1}t_{k(n+1)}(R\varphi)\|_{2k-(d-1)/2} \leq (C_3\lambda^{2k})^n (1+\omega)^{k(n+1)} \|R\varphi\|$$

and the right hand side goes to zero also when λ is fixed and the following condition holds:

$$C_3\lambda(1+\omega) < 1.$$

The goal of the next section is to give an effective procedure to determine Fourier coefficients of the terms $R^{-1}t_{k(n+1)}(R\varphi)$.

3. Fourier coefficients of $R^{-1}t_{k(n+1)}(R\varphi)$

A fundamental solution E_x^k of the operator $(1+\Delta)^k, k > d/2$, is a solution of the equation

$$(3.1) \qquad (1+\Delta)^k E_x^k = \delta(x),$$

where $\delta(x)$ is the Dirac measure at x. Since $(1+\Delta)^k$ is an isomorphism between Sobolev spaces H^α and $H^{\alpha-k}$ and since $\delta(x)$ belongs to $H^{-d/2-\varepsilon}$ for any positive ε, the fundamental solution E_x^k for $k > d/2$ belongs to a Sobolev space with a positive index.

Now let us consider the notion of even Lagrangian splines on the sphere. Even Lagrangian splines L_ν^{2k} are the even functions that enjoy the same minimality condition as in section 2 above and also satisfy the following conditions:

1) if $-x_\nu = x_\mu$ then $L_\nu^{2k}(x_\nu) = L_\nu^{2k}(x_\mu) = 1$,
2) $L_\nu^{2k}(x_\rho) = 0$, if ρ is different from ν and μ.

The following theorem is consequence of a result from [4].

THEOREM 3.1. *For any smooth function f on the dual sphere the following representations hold true:*

$$(3.2) \qquad t_k(f) = \sum_\nu f(\xi_\nu) L_\nu^{2k},$$

$$(3.3) \qquad t_k(f) = \sum_\nu \alpha_\nu E_\nu^k,$$

where $E_\nu^k = E_{x_\nu}^k$ and $\{\alpha_\nu\}$ are suitable constants.

From this we obtain that

$$(3.4) \qquad R^{-1} t_{k(n+1)}(R\varphi) = \sum_\nu (R\varphi)(\xi_\nu) \Lambda_\nu^{2k(n+1)}$$

where $\Lambda_\nu^{2k(n+1)} = R^{-1} L_\nu^{2k(n+1)}$.

Since every function $t_k(f)$ is a linear combination of Lagrangian splines it is enough to find the Fourier coefficients of L_ν^{2k}. Suppose that

$$(3.5) \qquad L_\nu^{2k} = \sum_{j=0}^\infty \sum_i \beta_{\nu,j}^{k,i} Y_j^i.$$

Then we have the formula

$$(3.6) \qquad (1+\Delta)^{2k} L_\nu^{2k} = \sum_{j=0}^\infty \sum_i [1+(j(j+1))]^k \beta_{\nu,j}^{k,i} Y_j^i.$$

Using the following representation of the delta measure δ_{x_γ}

$$(3.7) \qquad \delta_{x_\gamma} = \sum_j \sum_i Y_j^i(x_\gamma) Y_j^i$$

and the equations (3.3) and (3.6) we obtain

$$(3.8) \qquad \sum_{\gamma=1}^N \alpha_{\gamma,\nu} Y_j^i(x_\gamma) = [1+(j(j+1))]^k \beta_{\nu,j}^{k,i}$$

where

$$(1+\Delta)^{2k} L_\nu^{2k} = \sum_{\gamma=1}^N \alpha_{\gamma,\nu} \delta(x_\gamma).$$

So for a fixed ν we can express unknowns $\beta_{\nu,j}^{k,i}$ in terms of N unknowns $\alpha_{\gamma,\nu}, \gamma = 1, 2, \ldots, N$. Using conditions 1) and 2) from the definition of the splines L_ν^{2k} we obtain another N equations which help us to determine $\alpha_{\gamma,\nu}, \gamma = 1, 2, \ldots, N$. Note that such a system of N equations in N unknowns has a unique solution because the corresponding minimization problem (alluded to earlier) for the functions L_ν^{2k} does have a unique solution [4].

All together this gives an algorithm for finding Fourier coefficients $\beta_{\nu,n}^{j,i}$ of L_ν^{2k} and then for the functions Λ_ν^{2k} we have

$$(3.9) \qquad \Lambda_\nu^{2k} = \sum_{j=even}^{\infty} \sum_i c_j^{-1} \beta_{\nu,j}^{k,i} Y_j^i.$$

where

$$c_j = (-1)^{j/2} \Gamma((j+1)/2))/\Gamma((j+d)/2)$$

if j is even.

We can now give a precise formulation of our main result.

THEOREM 3.2. *If ξ_ν is a discrete set of the type $X(\lambda)$ on the dual sphere S^{d*} then, for any natural number k, there are special functions $\Lambda_\nu^{2k(n+1)}, n = 1, \ldots,$ such that for any smooth even function φ on the sphere S^d we have the following estimate for the approximation of φ by discrete samples of its spherical Radon transform:*

$$\|\varphi - \sum_{\nu=1}^{N} R\varphi(\xi_\nu) \Lambda_\nu^{2k(n+1)}\|_{2k-(d-1)/2} \leq (C\lambda^{2k})^n \|(1+\Delta)^{k(n+1)} R\varphi\|.$$

If φ is an even harmonic spherical polynomial of (even) degree ω then

$$\|\varphi - \sum_{\nu=1}^{N} R\varphi(\xi_\nu) \Lambda_\nu^{2k(n+1)}\|_{2k-(d-1)/2} \leq (C\lambda(1+\omega))^{2kn} \|R\varphi\|.$$

The functions $\Lambda_\nu^{2k(n+1)}, n = 1, \ldots$ are inverse spherical Radon transforms of Lagrangian splines on the sphere S^d. Moreover, by solving N linear systems of size $N \times N$ one can determine N^2 constants $\alpha_{\gamma,\nu}$ so that the induced constants

$$\beta_{\nu,j}^{k,i} = [1 + (j(j+1))]^{-k} \sum_{\gamma=1}^{N} \alpha_{\gamma,\nu} Y_j^i(x_\gamma)$$

give the following presentation of the special functions $\Lambda_\nu^{2k(n+1)}$:

$$(3.10) \qquad \Lambda_\nu^{2k(n+1)} = \sum_{j=even}^{\infty} \sum_i \beta_{\nu,j}^{k,i} R^{-1} Y_j^i.$$

Note that the last linear combinations can be calculated in advance as long as the discrete set of points $X(\lambda)$ is chosen. This could lead to a table lookup scheme for tomography on the sphere.

References

1. Grinberg, E., Spherical harmonics and integral geometry on projective spaces. Trans. Amer. Math. Soc. 279 (1983), no. 1, 187–203.
2. Grinberg, E. and Pesenson, I., Irregular sampling and the Radon transform, Contemp. Math. Vol. 251, AMS (2000), 255-268.
3. Helgason, S., The Radon transform Birkhäuser, Boston, Second edition, (1999)

4. I. Pesenson, *A sampling theorem on homogeneous manifolds*, Trans. AMS Vol. 352 (9), (2000), 4257-4270.
5. Pesenson, I., *Reconstruction of band limited functions in $L^2(R^d)$*, Proceed. AMS, Vol.127(12), (1999), 3593- 3600.
6. Pesenson, I., Sampling of band-limited vectors Jour. Fourier Anal. Applic. 2001, 7(1), (2001) 93-100.
7. Rubin, B., Spherical Radon Transforms and Intertwining Fractional Integrals, to appear.
8. _____, Fractional integrals and wavelet transforms associated with Blaschke-Levy representations on the sphere, Israel J. Math. 114 (1999), 1–27.
9. _____, Inversion of Radon transforms using wavelet transforms generated by wavelet measures. Math. Scand. 85 (1999), no. 2, 285–300.
10. _____ Continuous wavelet transforms on a sphere. Signal and image representation in combined spaces, Wavelet Anal. Appl., 7, (1998), 457–476.

DEPARTMENT OF MATHEMATICS, TEMPLE UNIVERSITY, PHILADELPHIA, PA 19122
E-mail address: **pesenson@math.temple.edu**

DEPARTMENT OF MATHEMATICS, TEMPLE UNIVERSITY, PHILADELPHIA, PA 19122
E-mail address: **grinberg@math.temple.edu**

Application of the Radon Transform to Calibration of the NASA-Glenn Icing Research Wind Tunnel

Steven H. Izen and Timothy J. Bencic

ABSTRACT. In order to measure the distribution of super-cooled water in the Icing Research Wind Tunnel, an optical sensor system has been proposed by researchers at the NASA-Glenn Research Center. This system will generate data which must be analyzed using tomographic algorithms.

A numerical feasibility study was performed and determined that the data which can be provided by the optical sensors appears to be robust enough to provide meaningful information about the distribution of super-cooled water in the Icing Research Wind Tunnel when sources and detectors are located on all four walls.

In the second phase, a laboratory model of the proposed sensor system was constructed by NASA-Glenn personnel. Tomographic algorithms developed in the first phase are applied to data acquired from the laboratory model.

1. Introduction

Ice formation on wings has been implicated in a number of aviation accidents, so understanding the physical processes in ice formation as well as testing the performance of component designs under icing conditions is of significant interest to the aeronautics industry. The NASA-Glenn Icing Research Tunnel (IRT) is a wind tunnel which has the capability of providing a controlled environment for icing experiments. The wind tunnel (see Figure 1) differs from conventional wind tunnels in the air is cooled to freezing temperatures and that there is a system of nozzles that inject supercooled water into the airflow. This forms a cloud of water droplets which is carried past an object which has been placed in the 9'×6' test section. In order to make sense of the ice growth which occurs on the model in the test section, it is necessary to know the liquid water content as a function of position of the cloud as it arrives at the test section. Ideally, the particle distribution in the cloud would be as uniform as possible. The measurement of this cloud distribution, followed by adjustment of the nozzles to produce a more uniform water distribution, is known as the calibration of the wind tunnel.

An optical sensor system to help in the calibration of the IRT was proposed by Tim Bencic of the Glenn Research Center's Optical Instrumentation Technology Branch of the Instrumentation and Controls Division. This system to measure the

---

1991 *Mathematics Subject Classification.* Primary 44A12, 44-04, 92F05; Secondary 65F22, 65-04 .

This paper work was supported by NASA Grant NAG 3-2210.

FIGURE 1. Looking downstream through the spray bars towards the test section of the NASA Glenn Icing Research Tunnel. The spray bars inject supercooled water into the air flow.

cloud distribution consists of an array of optical sources and detectors placed on the walls of a section of wind tunnel proximate to the test section. The optical density of the water cloud provides a measurement of the liquid water content. The advantages of this sensor system are that it could operate in almost real-time, would have no impact on the flow, and could obtain distribution profiles at a higher resolution than is currently available.

As a first step in the development of this system, a feasibility study was performed to investigate to what extent liquid water content could be reliably reconstructed from the data to be measured. Included in this study was the consideration of different acquisition geometries. Since the measurements can be modeled as the Radon transform of liquid water content, an investigation of the practical aspects of performing a Radon inversion in various rectangular geometries was performed.

This paper reports on this feasibility study and presents preliminary results of the reconstruction of water content from a laboratory scale model of the wind tunnel and the sensor system.

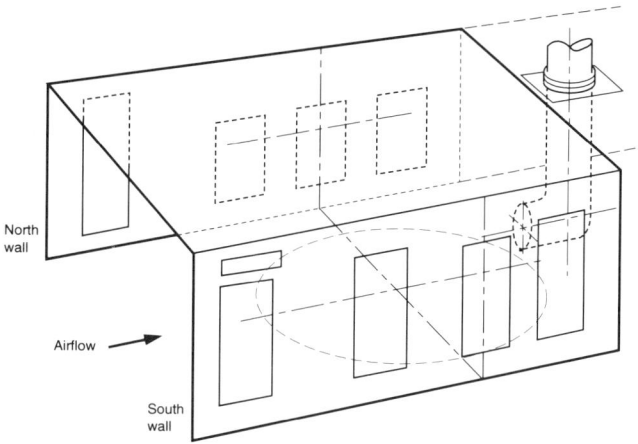

FIGURE 2. Diagram of the IRT test section showing the turntable and the windows for visual access

2. The Experimental Model

Light sources and detectors are to be placed along the walls of the wind tunnel. The sources are illuminated rapidly in sequence, and the attenuation from each source to each detector not on the same wall is recorded.

The liquid water content is measured by ρa^2, where ρ is the particle density and a is the particle radius. For the expected particle density and size distributions[1] and the optical path lengths from the sources to the detectors, a physical model predicts that multiple scattering interactions are not significant.[2] Accordingly, the probability of a photon being scattered at a particular location along a line from the source to the detector is proportional to the liquid water content at the point. This is the same model used to describe x-ray attenuation in medical CT. Given the intensity I_0 of a source, and the intensity I measured at a detector, the log of the ratio is proportional to the Radon transform of the liquid content on the line connecting the source to the detector. That is, with the attenuation coefficient $\mu = c\rho a^2$, c being the proportionality constant,

$$-\ln\left(\frac{I}{I_0}\right) = \int_{l_i} \mu(x)ds. \tag{1}$$

Here, s denotes length along the line l_i from source to detector.

An important issue from both a hardware and a reconstruction viewpoint is the placement of the optical sensors. The simplest geometry from the hardware perspective is to take advantage of the test section windows. See Figure 2. The windows provide optical access through three of the walls of the wind tunnel. This would allow the sensor system to be built with minimal wind tunnel modification, as the sensor system would be physically outside the wind tunnel. While sensors cannot be placed along the entire length of the three walls, for the purposes of this study, such placement was allowed. This will be called the *three wall geometry*.

If the sensor system is to be placed physically inside the tunnel (which raises a host of engineering issues not discussed here), other geometries are possible. One arises from the desire to take measurements as close as possible to the model or aircraft part in the test section. The part is mounted on a turntable which allows the experimenter to change the angle of attack. This turntable occupies a portion of the floor. If the cross section of the wind tunnel on which the liquid water content is to be determined intersects the turntable, it will not be possible to place sources and detectors on part of the floor covered by the turntable. The geometry corresponding to this scenario is to completely cover three walls with sensors, and to partially cover the floor, omitting the region in the middle of the floor corresponding to the turntable. This will be referred to as the 3.5 *wall geometry*.

The last geometry considered was full coverage along all four walls in the wind tunnel. This would require the detection apparatus to be placed inside the wind tunnel upstream from the turntable. By placing the sensors on a movable collar it might be possible to investigate the cloud distribution at different distances from the test section. This is the *four wall geometry*.

It was expected that the three and 3.5 wall geometries would lead to severely ill-conditioned reconstructions, especially for the recovery of liquid water content in the region of interest to icing researchers, the region in the middle of the cross section where the part being tested resides. However, because of the engineering difficulties and added cost of the four wall geometry, it was necessary to conclusively demonstrate that a sensor system with the first two geometries could not be used to recover the cloud distribution.

3. The Mathematical Model

The sensor geometry determines the set of lines on which the Radon transform of the unknown function μ is sampled. For each sampling geometry, let L denote the set of lines on which the sample of the Radon transform

$$(2) \qquad R\mu(l_i) = \int_{l_i} \mu \, ds$$

is known.

Each line $l_i \in L$ connects a source on one wall to a detector on another wall, and each detector sees all sources not on the same wall. The equation of the line l_i is

$$(3) \qquad x \cdot \omega_i = p_i,$$

where ω_i is the outward normal and p_i is the distance of the line from the origin.

The rectangular placement of the the sensors leads to an unavoidable irregular sampling in p and ω. As a result, an application of direct reconstruction algorithms such as convolution-backprojection would require interpolation. Moreover, direct methods perform poorly when large sets of samples are missing, as is the case for the first two sampling geometries. Direct limited angle algorithms would be difficult to apply because of the irregular sampling and the relative sparsity of measurements.

In order to handle the irregular sampling and to more easily work with the sampling geometries that do not well cover the space of lines in $\mathbb{R}^2$, the problem was treated as a discrete inversion. That is, instead of discretizing the inverse Radon transform R^{-1}, the Radon transform R is discretized before inversion.

To accomplish the pre-inversion discretization, the reconstruction region is divided into J pixels. The discretized Radon transform operator becomes a matrix R.

$$R = \{r_{ij}\}, \tag{4}$$

$$r_{ij} = \int_{l_i} \Phi_j ds, \tag{5}$$

where the ij^{th} entry is the Radon transform of Φ_j, the characteristic function for the jth pixel, evaluated on the ith line.

The attenuation is reconstructed from the image vector $x \in \mathbb{R}^J$ by

$$\mu = \sum_{j \in J} x_j \Phi_j, \tag{6}$$

and the image vector is recovered from the vector $b \in \mathbb{R}^I$ of data points by solving the linear system

$$Rx = b. \tag{7}$$

The reconstruction region was placed on the rectangle $[0, 9] \times [0, 6]$, each unit corresponding to a foot in the wind tunnel test section. For each of three geometries, the sources and detectors were distributed evenly along the permitted locations on the walls. This corresponds to evenly spaced endpoints for the lines on the sides of the rectangles. For the three-sided geometry, endpoints were excluded from the bottom side. That is, no endpoints were allowed with coordinates $(x, 0), 0 < x < 9$. For the 3.5 sided geometry, endpoints were excluded from the middle third of the rectangle's bottom. That is, endpoints on the segment $\{(x, 0) \mid 3 < x < 6\}$ were excluded. The pixel size on the rectangle was matched to the spacing between endpoints. Figures 3,4,5 illustrate the lines for which data are available for each of the three geometries.

3.1. Solving the Linear System. The linear system (7) will, in general, be a fairly large ill-conditioned system. In order to obtain a meaningful image vector x, some sort of regularizing solution method must be applied. A good discussion of solution methods for problems of this type appears in [6].

Applying the discretization method leading to equation (7) to medical tomography sampling geometries generates extremely large linear systems. Due to the size of such systems, they must be solved by iterative techniques. For the wind tunnel geometries, the sampling densities, while still large, are significantly smaller than those arising in medical tomography. The sizes of the largest wind tunnel systems under consideration in this study are just small enough to allow the non-iterative methods to be tractable on workstation class computers. For the repeated solution of equation (7) in a production environment, iterative techniques may be preferable.

From among the regularizing solution methods available, we have chosen to use the Truncated Singular Value Decomposition (TSVD). The TSVD is not a particular computationally efficient non-iterative algorithm, and does not necessarily give the "best" reconstructions. However, it is able give a qualitative and quantitative sense of which features in the image can be recovered and which features will be sensitive to the presence of noise in the data. The TSVD has previously been applied to industrial tomography problems, in particular those with very small sampling densities.[3].

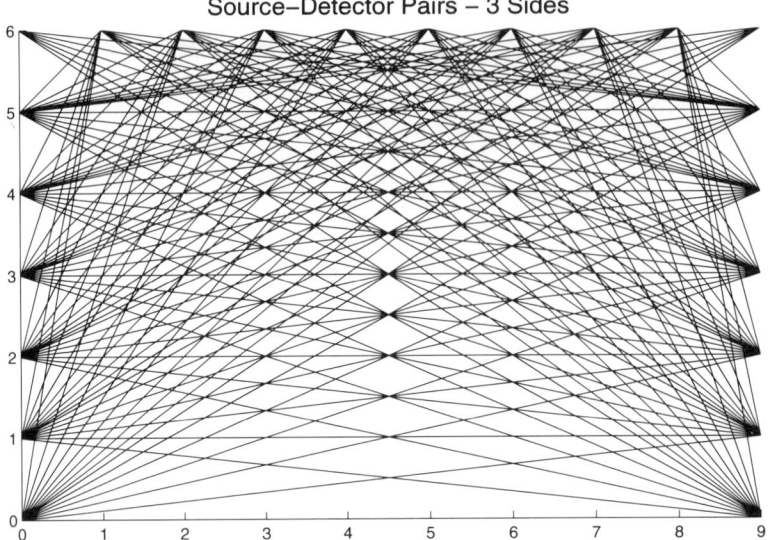

FIGURE 3. The lines on which data are available for the three side geometry when sensors are located one foot apart. The pixel size for a reconstruction with the lines shown would be 1'×1'.

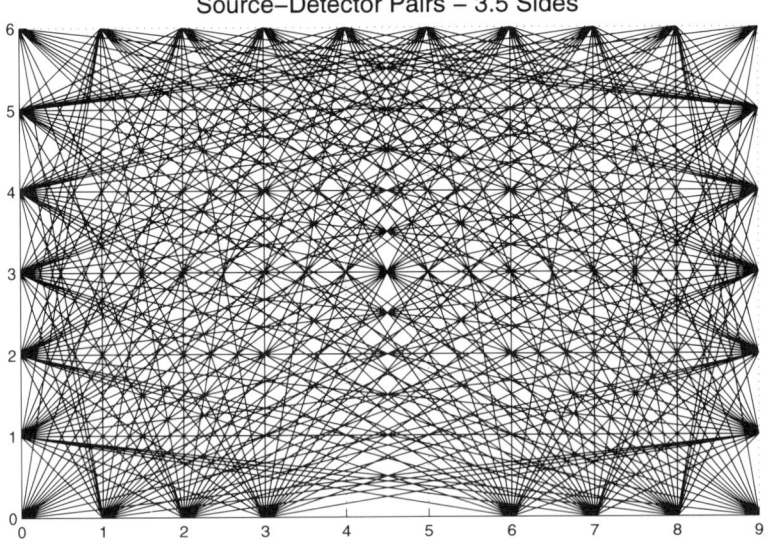

FIGURE 4. The lines on which data are available for the three and a half sided geometry when sensors are located one foot apart. The pixel size for a reconstruction with the lines shown would be 1'×1'.

The first step in solving equation (7) with the TSVD is the computation of a Singular Value Decomposition (SVD) of the matrix R. The SVD computation is expensive and scales poorly in both memory requirement and execution time. This poor scaling limited the largest density of sensors which could be examined for this

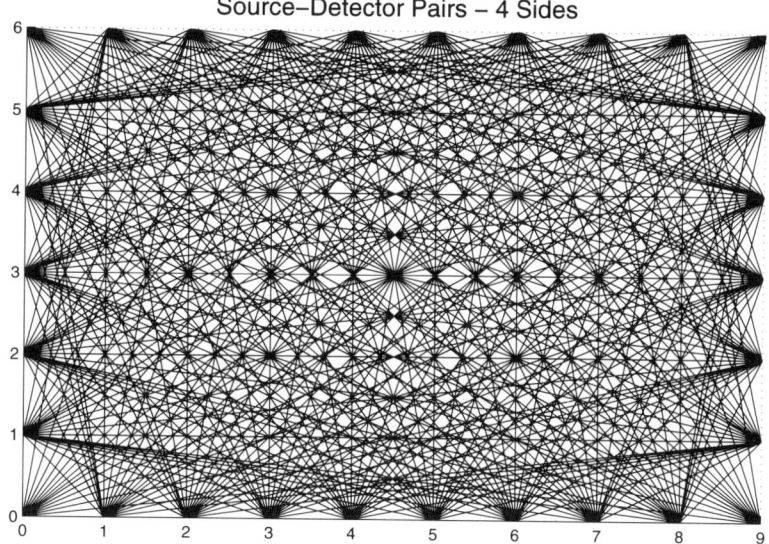

FIGURE 5. The lines on which data are available for the four sided geometry when sensors are located one foot apart. The pixel size for a reconstruction with the lines shown would be 1'×1'.

study. One practical advantage of the use of the TSVD method is that off the shelf computer codes to perform the SVD are readily available.[5]

3.2. The Truncated Singular Value Decomposition. The SVD for a linear map $R : \mathbb{R}^J \to \mathbb{R}^I$ is a representation

$$(8) \qquad Rx = \sum_{i=1}^{K} \sigma_i \langle x, v_i \rangle u_i,$$

where $\{u_i \mid i = 1, \cdots, K\}$ and $\{v_i \mid i = 1, \cdots, K\}$ are orthonormal bases for Range(R) and Null(R)$^\perp$ respectively, and $\{\sigma_i\}$ is a decreasing sequence of positive numbers. The bases can be extended to $\mathbb{R}^I$ and $\mathbb{R}^J$, respectively, by choosing an orthonormal basis $\{u_i \mid i = K+1, \cdots, I\}$ for Range(R)$^\perp$ and $\{v_i \mid i = K+1, \cdots, J\}$, for Null($R$). $\{u_i\}$ are called the *left singular vectors*, and $\{v_i\}$ are called the *right singular vectors*.

The Moore-Penrose pseudo-inverse $R^+ : \mathbb{R}^I \to \mathbb{R}^J$ selects the unique element $x \in \mathbb{R}^J$ which minimizes the residual $\|Rx - b\|$ and also has minimum norm. This is, in an L^2 sense, the best solution which can be obtained for an overconstrained or underconstrained linear system. In terms of the SVD, it has the representation

$$(9) \qquad R^+ b = \sum_{k=1}^{K} \sigma_k^{-1} \langle b, u_k \rangle v_k.$$

To interpret equation (9), observe that the data b can be expanded in terms of the left singular vectors u_i,

$$(10) \qquad b = \sum_{i=1}^{I} \langle b, u_i \rangle u_i.$$

Only the components of b which are in the range of R have influence on R^+b. Each such component $\langle b, u_k \rangle u_k$ is translated to the component $\sigma_k^{-1}\langle b, u_k \rangle v_k$ in the reconstructed image. The size of the kth component in the data b is amplified by a factor of σ_k^{-1} in R^+b.

When the data b suffers from white noise, that noise is evenly distributed in all the directions u_i. The noise is reconstructed, along with the data by equation (9). In particular, the resulting noise along direction v_k in the image vector gets amplified by a factor of σ_1/σ_k compared to the noise in the direction v_1. When σ_k/σ_1 is comparable to the noise level in the data, the noise level in the reconstruction is comparable to the signal level. When this ratio is smaller than the noise level, the reconstructed image is swamped by noise.

In order to attenuate the effects of this noise amplification, a regularization of the Moore-Penrose inverse must be used. That is, instead of equation (9), a regularization parameter λ is chosen, and the regularized solution $x_\lambda = R_\lambda^+ b$ is used.

$$(11) \qquad R_\lambda^+ b = \sum_{k=1}^{k} f_\lambda(\sigma_k) \langle b, u_k \rangle v_k,$$

where $f_\lambda(\sigma_k)$ is a function which approximates σ_k^{-1} for small k, but decays to 0 for large k. The parameter λ governs where the transition between the two behaviors occurs. Each choice of f_λ leads to a different regularization scheme.

The Truncated Singular Value Distribution is obtained by using

$$(12) \qquad f_\lambda(\sigma_k) = \begin{cases} \sigma_k^{-1} & k \leq \lambda \\ 0 & k > \lambda \end{cases}.$$

That is, the sum in equation (9) is truncated to

$$(13) \qquad R_\lambda^+ b = \sum_{k=1}^{\lambda} \sigma_k^{-1} \langle b, u_k \rangle v_k.$$

The components u_k, with $k > \lambda$ are being treated as if they are effectively part of Null(R). The rate of decay of the singular values combined with the data noise level determines where the sum is to be truncated. If the truncation parameter is set too high, the resulting image will be degraded by noise. Setting the parameter too low will further reduce the effects of the noise in the data, but useful data will be excluded as well. More systematic error will be incurred than is necessary. When the singular values σ_k exhibit a sharp decay, it is easy to select the choice of the truncation parameter λ. For a more gradual decay, the best choice for λ may not be so clear. Developing selection criteria for the parameter λ is an active area of research.[6] When the parameter λ has been properly selected, the reconstruction $R_\lambda^+ b$ arguably contains all the information that is available from the data b. It includes all the components v_k which are reconstructible from the data and it excludes all the components for which it is not realistic to attempt reconstruction.

4. Numerical Experiments

The Truncated Singular Value Decomposition solution for equation (7) was implemented for each of the three geometries described in Section (2).

All the computations were performed on an SGI Octane with dual 225MHZ R10K processors and 384 MB of RAM. For each geometry, the matrix R was generated by tracing the line from each source to each detector as it progresses through the pixel array. Although R is a sparse array, for simplicity (at the expense of computing resources), the sparseness was not exploited. A dense Singular Value Decomposition was computed using the SGI optimized parallel versions of LAPACK[5] with a C++ front end. Unfortunately, only the first phase of the SVD computation, the bidiagonalization, was parallelized so the full power of the dual CPU computer was not utilized. The largest sized reconstructions repeatedly attempted were with an image vector of size $J = 1944$. This was for an image size of a 54×36 pixel array which corresponds to a sensor separation of 2" in the wind tunnel. The data vector sizes were 5255, 10007, and 12170 for the three, three and a half, and four sided geometries, respectively. R, and also the matrix U of left singular vectors required 78, 150, and 181 megabytes of storage. The SVD computation required about 100 minutes for the four sided geometry. Halving the reconstruction resolution (to the equivalent of 1") and doubling the sampling density increased the memory requirements by a factor of 16, and increased execution time to approximately 36 hours. That size of reconstruction was not practical with the implementation used.

For each geometry, phantom images were created, and projection data for these images were created. Also, sample cloud distributions obtained by other experimental techniques were projected.

Full sets of the right singular vectors can be viewed at
http://zubenelgenubi.math.cwru.edu/~shi/windtunnel/.

4.1. Three Sides and Three and a Half Sides.
An analysis of the three sided geometry was performed on a 54×36 pixel array, corresponding to a sensor separation of 2". R was 5255×1944.

Figure 6 is a log plot of the singular values compared to the largest singular value. The plot shows the decay over ten orders of magnitude. The fall off of the singular value ratio is rapid. Even in the presence of data with only 1% noise, only a bit more than half of the right singular vectors can be used in the reconstruction. The ones corresponding to the largest singular values have the lowest spatial frequencies. The spatial frequencies increase as the singular values decrease. As the higher spatial frequency right singular vectors are needed to reconstruct details in the image, the details in the reconstructed image are more susceptible to noise than the gross features. But more importantly for this feasibility study, the right singular vectors which have significant variation in the horizontal direction in the region near the bottom of the rectangle by the missing sensors are all in the range of singular values which cannot be used in the reconstruction. This can be explained by noting that horizontal variation must be measured by vertical or nearly vertical line integrals. These lines are precisely those for which no data are available. In other words, if the boundary of a feature is tangent to a line from which measurements are missing, the boundary cannot be recovered. Conversely, the right singular vectors which can be used are flat and provide little information near the bottom edge. As an illustration of these points, Figure 7 shows the right singular vectors $u_1, u_{50}, u_{1000}, u_{1944}$ corresponding to the singular value ratios $\sigma_1/\sigma_1 = 1$, $\sigma_{50}/\sigma_1 = 0.2982$, $\sigma_{1000}/\sigma_1 = 0.058725$, and $\sigma_{1944}/\sigma_1 = 1.2815 \times 10^{-19}$.

A similar analysis was done for the three and a half wall geometry. Figure 8 is a log plot of the singular values compared to the largest singular value. The plot shows the decay over four orders of magnitude. The fall off of the singular value

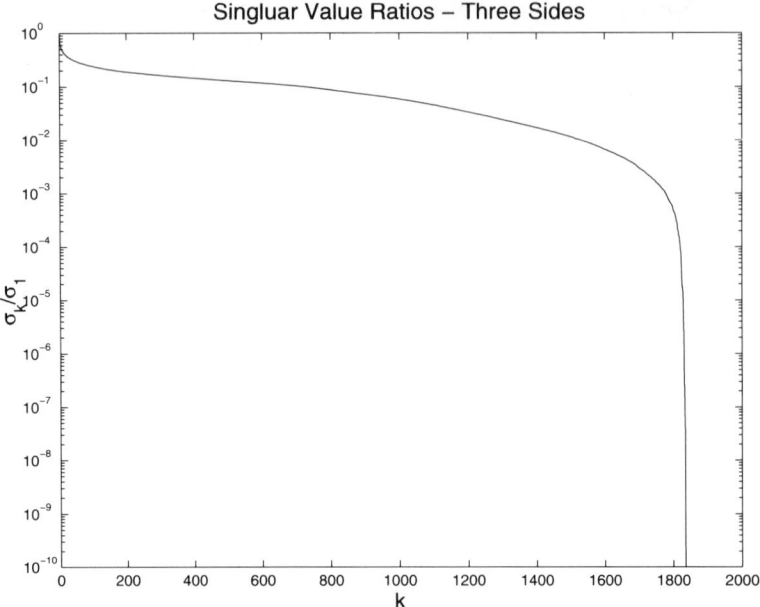

FIGURE 6. A log plot showing 10 orders of magnitude of the singular value ratio σ_k/σ_1 for the three wall geometry with sources and detectors spaced every two inches. There are 1944 singular values. The rapid decay of σ_k/σ_1 (10 orders of magnitude) indicates that the inversion of R is severely ill-conditioned.

ratio is more gradual, but even so, a dramatic drop off starts soon after the singular value number 1600. As expected, more of the image can be recovered than in the three wall case, but there is still a fundamental problem in the recovery of the liquid water content in the region which is of most interest, the area in the center, just above the missing sensors. The right singular vectors in Figure 9 illustrate that the unusable right singular vectors have in the central region the horizontal variation necessary for horizontal resolution in that region. Again note that these oscillations are transverse to the lines for which no data are available.

After reviewing the results of the three-sided and 3.5-sided reconstructions, all plans to proceed with those geometries were dropped.

4.2. Four sides. When sensors are placed on all four walls, the reconstruction becomes only mildly ill-posed. That is, there is only a relatively gentle singular value decay, as can be observed in Figure 10. The ratio of the largest singular value to the smallest is about 17, so a robust reconstruction is possible without making unreasonable demands on the experimental hardware noise levels. Even allowing for some noise suppression by truncating at the slight drop off near the singular value number 1800 doesn't significantly affect the reconstruction. As shown in Figure 11, the last right singular vectors primarily provide the highest horizontal resolution, and information used to reconstruct in the corners, neither of which are that significant.

Since this geometry appears to be feasible, further tests were performed. Data was simulated for the projection of a single point located slightly off center. The

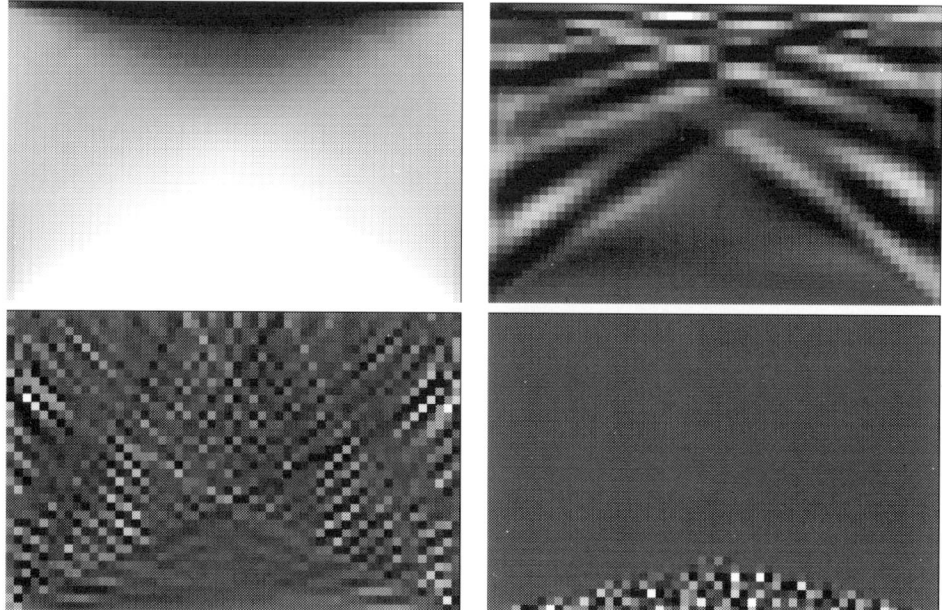

FIGURE 7. The right singular vectors u_1, u_{50}, u_{1000}, and u_{1944} for the three wall geometry with sources and detectors spaced every two inches. The corresponding singular value ratios are $\sigma_1/\sigma_1 = 1$, $\sigma_{50}/\sigma_1 = 0.2982$, $\sigma_{1000}/\sigma_1 = 0.058725$, and $\sigma_{1944}/\sigma_1 = 1.2815 \times 10^{-19}$.

noise-free reconstruction from this data was essentially perfect. A reconstruction truncated at 1916 out of 1944 singular values reconstructs the point to within 1%. The reconstruction of the point is not sensitive to added noise, but the systematic error induced by an early truncation is significant. See Figure 12.

To test the reconstruction on the types of distributions that are likely to be seen in the wind tunnel, cloud distributions obtained using different technology were projected. Noise was added and the data were reconstructed with varying degrees of truncation. Figure 13 shows a typical liquid water distribution. In the absence of added noise, the reconstruction is perfect. Adding 5% noise produces a degraded image. See Figure 14. The noise can be suppressed by truncating the reconstruction. The degradation of the image due to the omission of many right singular vector components was much less than that of the point phantom, and was less than was originally anticipated. See Figure 15. The explanation is that the actual liquid water content distributions have very little high spatial frequency content, and therefore only the right singular vectors with low spatial frequency content, the ones corresponding to the largest singular values, are needed to get a good reconstruction. In other words, the *a priori* smoothness of the attenuation μ makes the reconstruction inherently less ill-conditioned. A quantitative version of this argument in an infinite-dimensional setting using Sobolev norms to measure smoothness appears in Section V.2 of [**4**].

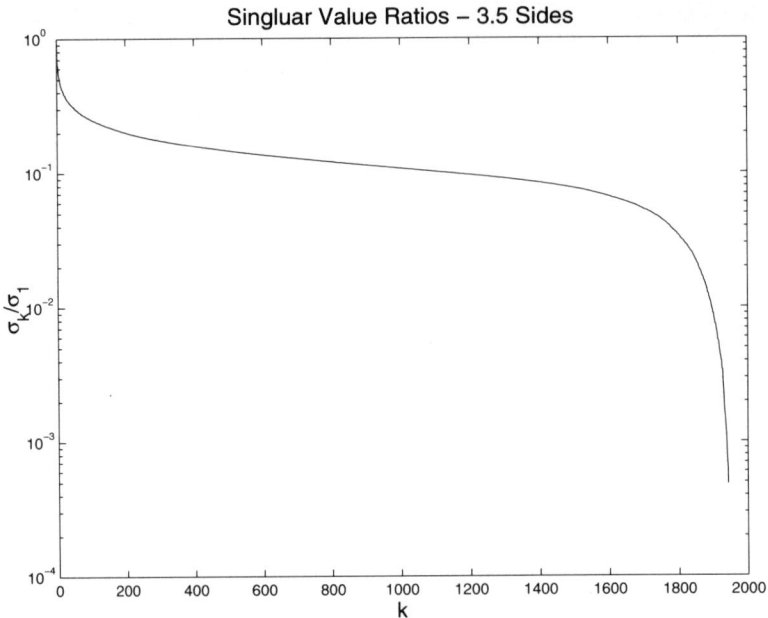

FIGURE 8. A log plot over four orders of magnitude of the singular value ratio σ_k/σ_1 for the three and a half wall geometry with sources and detectors spaced every two inches. There are 1944 singular values. The rapid decay of σ_k/σ_1 indicates that the inversion of R is severely ill-conditioned.

5. Preliminary Laboratory Results

Based on the encouraging results of the feasibility study, it was decided to build a laboratory mock up to demonstrate the hardware proof of concept. A scale model (18"×12") of the wind tunnel acquisition system is being constructed in a laboratory. In this section preliminary results of reconstructions from this apparatus will be given. The detailed description of the hardware acquisition system will appear elsewhere.

Due to hardware implementation considerations, the model was constructed with only 24 sources and 86 detectors. In addition, the collection angle for the detectors was not a full 180 degrees, but is currently only 25 degrees. Moreover, the sources were only visible to detectors on the opposite wall. The set of lines for which data is available is given in Figure 16. The reduced number of detectors combined with the lack of data from diagonal lines makes this setup unlikely to produce high quality reconstructions. Fortunately, at this preliminary stage, high quality reconstructions are not required. A reconstruction was performed on a 12 × 9 grid. The corresponding discretized Radon transform R was a 218 × 216 matrix. In Figure 17 it can be seen that the singular values display a very sharp drop off, starting at σ_{117}.

To create an image to be reconstructed, a jet of water was directed through the test section. Figure 18 shows the laboratory apparatus. The small dots visible on the frame are the detectors. Each is connected to a fiber optic cable which

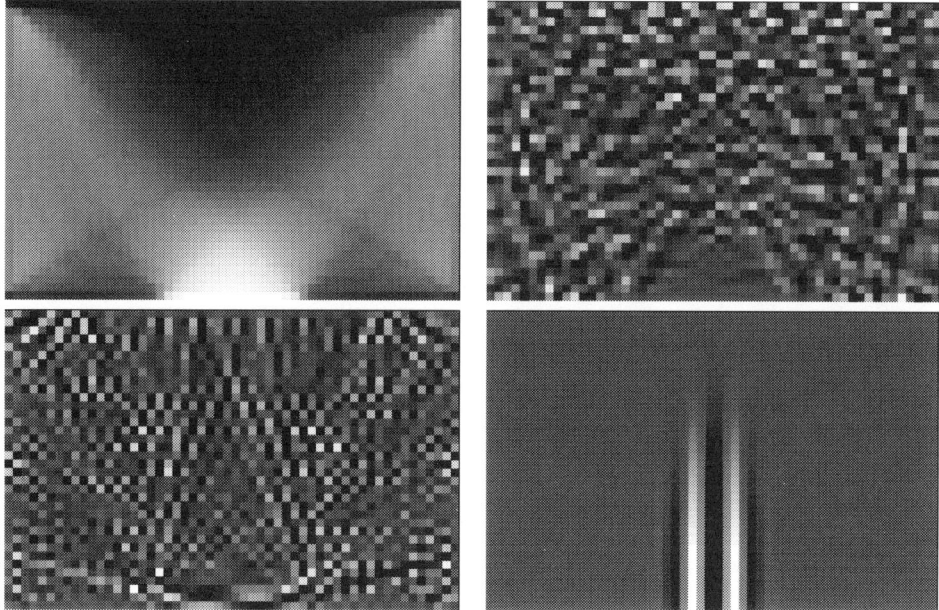

FIGURE 9. The right singular vectors u_1, u_{569}, u_{1542}, and u_{1944} for the three wall geometry with sources and detectors spaced every two inches. The corresponding singular value ratios are $\sigma_1/\sigma_1 = 1$, $\sigma_{569}/\sigma_1 = 0.13952$, $\sigma_{1542}/\sigma_1 = 0.07279$, and $\sigma_{1944}/\sigma_1 = 4.811 \times 10^{-4}$.

channels the light to a location on a CCD camera. Barely visible, just adjacent to the detector array is a sparser array of LED sources. A jet shoots a spray of water through the center of the laboratory scale model.

Figure 19 shows a reconstruction from data acquired by this system. The reconstruction was truncated at $k = 55$, corresponding to a singular value ratio of $\sigma_{55}/\sigma_1 = 0.1754$. There is clearly a bright spot in the reconstruction corresponding to the water jet. The absence of data from diagonal lines leads to the boxiness in the image. This effect should be attenuated once the collection angle for the detectors is increased from 25 degrees to 90 degrees. This enhancement is currently being implemented.

6. Future Directions and Conclusions

The first phase of this feasibility study demonstrated that the proposed optical tomographic system to measure cloud uniformity was feasible, provided sources and detectors can be located along all four walls of the wind tunnel. Geometries with less than full sensor coverage were shown to be unsatisfactory.

The second phase is a laboratory study to test how well the reconstruction algorithms will work with real data acquired from sensor hardware. Preliminary results are encouraging. Work is continuing on expanding the number of sources. Also, efforts are underway to expand the collection angle of the fiber detectors.

The ultimate cloud resolution objective is 1"×1", the discrete matrix R will be very large. Accordingly, the regularization method used for the feasibility study

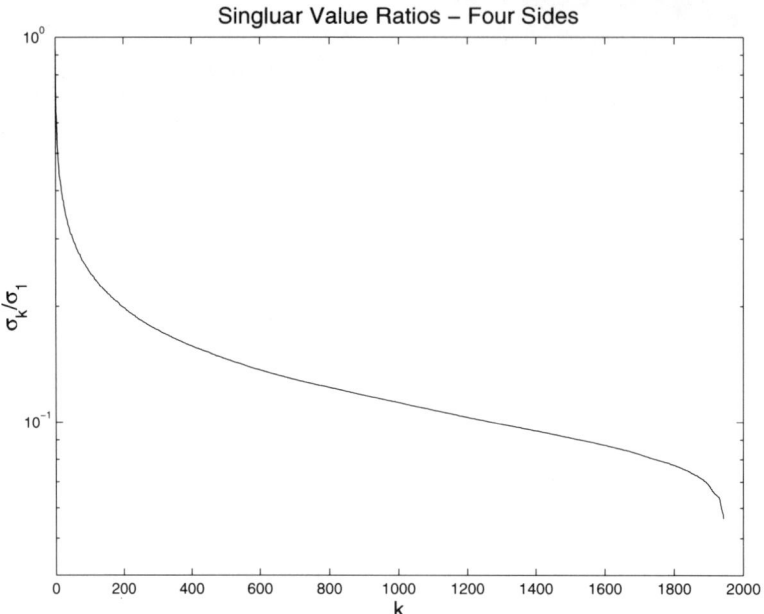

FIGURE 10. A log plot over two orders of magnitude of the singular value ratio σ_k/σ_1 for the four wall geometry with sources and detectors spaced every two inches. There are 1944 singular values. The slow drop indicates that the inversion of R is mildly ill-conditioned.

is probably not the optimal choice for a production system. Investigations are underway to select a more suitable algorithm for a production system. Under consideration are iterative methods and the selection of alternate representations for the image vector to exploit the expected smoothness of the cloud distribution.

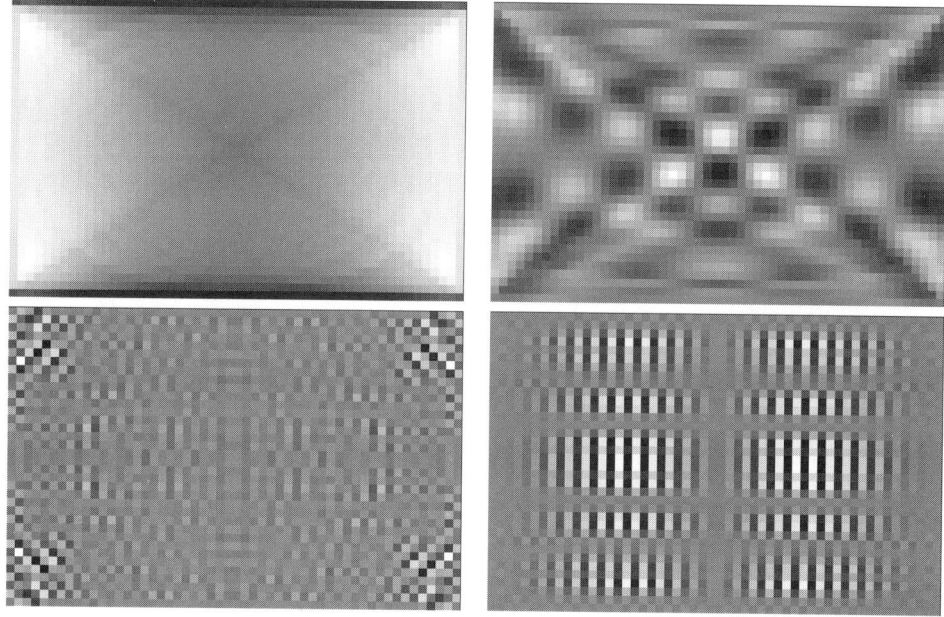

FIGURE 11. The right singular vectors u_1, u_{85}, u_{1810}, and u_{1944} for the four wall geometry with sources and detectors spaced every two inches. The corresponding singular value ratios are $\sigma_1/\sigma_1 = 1$, $\sigma_{85}/\sigma_1 = 0.2587$, $\sigma_{1810}/\sigma_1 = 0.0768$, and $\sigma_{1944}/\sigma_1 = 0.0563$.

FIGURE 12. The left figure shows the four wall reconstruction using all singular vectors of a single pixel phantom. The right figures shows the truncated reconstruction when only 550 out of 1944 singular vectors are used.

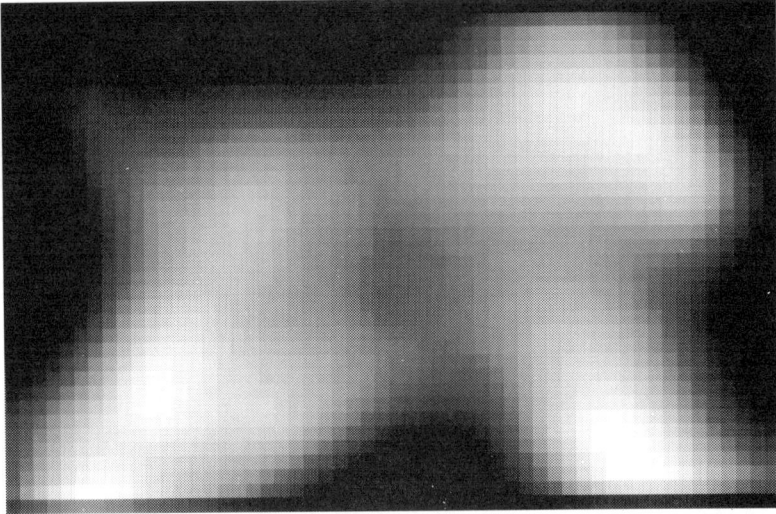

FIGURE 13. A sample liquid water cloud in the Icing Research Tunnel. This distribution is the central region of a cloud measured by other experimental techniques.

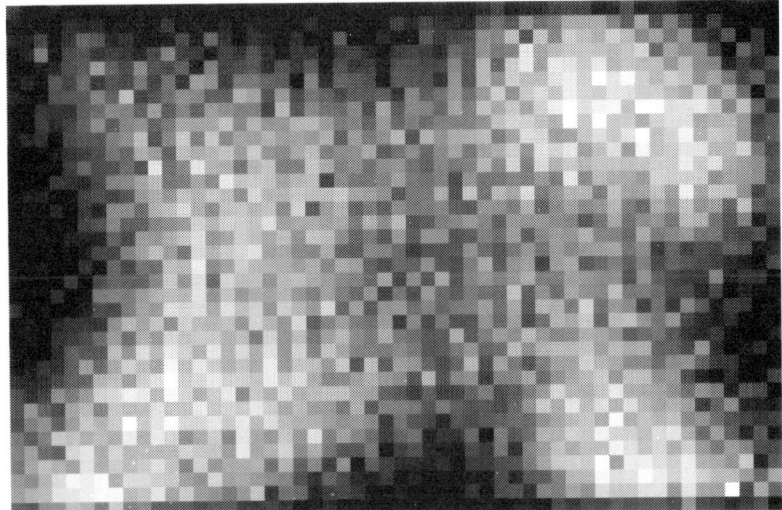

FIGURE 14. Reconstruction of the sample liquid water cloud after 5% noise was added to the projection data. All 1944 right singular vectors were used.

CALIBRATION OF THE ICING RESEARCH WIND TUNNEL 163

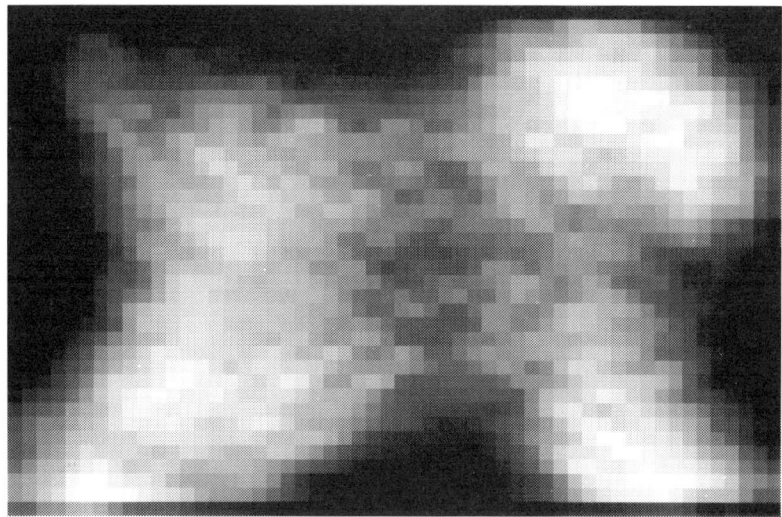

FIGURE 15. The sample cloud reconstructed from projection data with 5% added noise using 550 of 1944 right singular vectors.

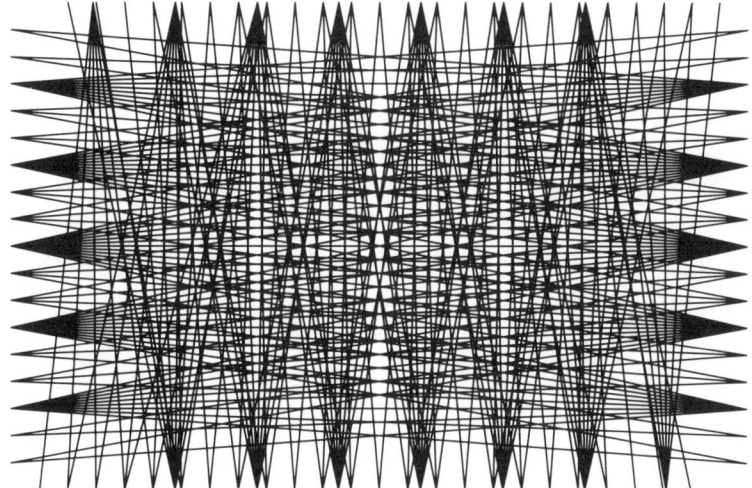

FIGURE 16. Diagram of lines from which data is measured in the laboratory scale model. Sources are only visible to detectors on the opposite wall and the collection angle is only 25 degrees.

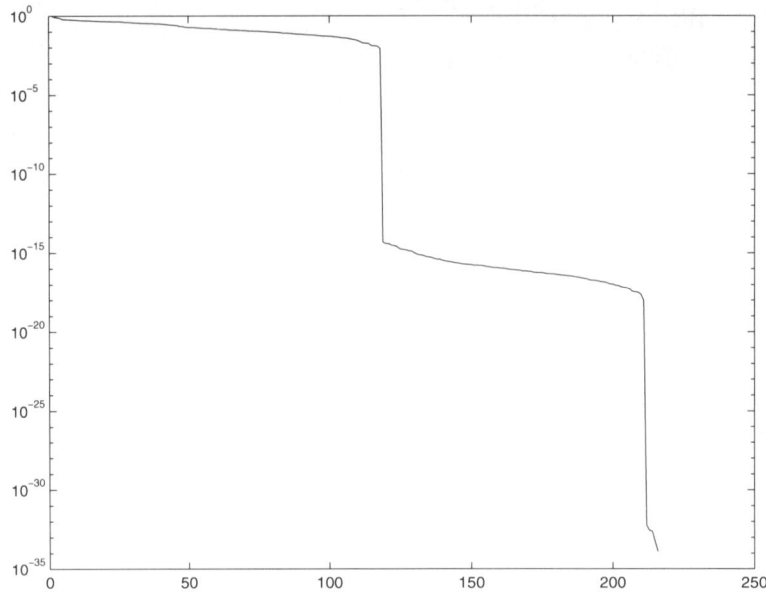

FIGURE 17. Log plot over 35 orders of magnitude of the singular value ratios σ_k/σ_1. Note the dramatic drop at $k = 118$.

FIGURE 18. Photograph of the lab scale model. The small disks appearing around the edge of the frame are the detectors, which are connected by fiber-optic cables to a CCD camera. The water stream which is imaged by the apparatus is emitted from the jet which appears in the middle, just left of center.

FIGURE 19. Reconstruction of data acquired from laboratory scale model. The reconstruction used 55 of 216 singular vectors, truncating at a singular value ratio of $\sigma_{55}/\sigma_1 = 0.1754$.

References

[1] Ronald H. Soeder, David W. Sheldon, Charles R. Andracchio, Robert F. Ide, David A. Spera, Nick M. Lalli, *NASA Lewis Icing Research Tunnel User Manual*, NASA Technical Memorandum 107159, Cleveland, OH, 1996.

[2] James Lock, *Mapping the Position-Dependence of the Liquid Water Content in the NASA Glenn Icing Research Tunnel Using Light Scattering Tomography*, NASA Grant NAG3-2232.

[3] Helmut Sielschott, Willy Derichs, *Use of collocation methods under inclusion of a priori information in acoustic pyrometry*, in *Process Tomography—95: Implementation for Industrial Processes*, UMIST, Manchester, 1995.

[4] Frank Natterer, *The Mathematics of Computerized Tomography*, Wiley, New York, NY, 1986.

[5] E. Anderson, Z. Bai, C. Bischof, J. Demmel, J. Dongarra, J. Du Croz, A. Greenbaum, S. Hammerling, A. McKenney, S. Ostrouchov, D. Sorenson, *LAPACK User's Guide*, SIAM, Philadelphia, PA, 1992.

[6] Per Christian Hansen, *Rank-Deficient and Discrete Ill-Posed Problems: Numerical Aspects of Linear Inversion*, SIAM, Philadelphia, PA, 1998.

CASE WESTERN RESERVE UNIVERSITY, CLEVELAND, OHIO.
E-mail address: shi@po.cwru.edu
URL: http://zubenelgenubi.math.cwru.edu/~shi/

NASA-GLENN RESEARCH CENTER, CLEVELAND, OHIO
E-mail address: Timothy.J.Bencic@grc.nasa.gov

Range theorems for the Radon transform and its dual

Alexander Katsevich

ABSTRACT. Established is a series of range theorems and continuity properties for the Radon transform R and its dual R^* on natural classes of functions and distributions with bounded rate of decay at infinity.

1. Introduction

A number of range theorems for the Radon transform R is known (see papers [**GGV66, Hel99, Hel82, Her83, Her84, Kat97, LP70, Lou84, Lud60, Ram95, SSW77**]). Most of them deal with either compactly supported or rapidly decreasing functions and distributions. A preimage under the Radon transform of $\mathcal{S}(Z_n)$ even functions which satisfy none or finitely many moment conditions is studied in [**Sol87**]. Here $Z_n = S^{n-1} \times \mathbb{R}$ and S^{n-1} is the unit sphere in $\mathbb{R}^n$. Throughout the paper we will be using the standard notations $C_0^\infty, \mathcal{S}, \mathcal{D}'$, and $\mathcal{S}'$ for spaces of compactly supported smooth functions, rapidly decreasing Schwartz functions, distributions, and tempered distributions, respectively. Among other results, it is proved in [**Sol87**] that the more moment conditions $\mu \in \mathcal{S}(Z_n)$ satisfies, the faster $f := R^{-1}\mu$ decays at infinity. In [**SM88**] a partial range theorem for R on functions which decay only polynomially is derived. In [**Kat97**] an exact range theorem is obtained in the case when R acts on a class of C^∞-functions which admit an asymptotic expansion at infinity:

$$(1.1) \qquad f(r\beta) \sim \sum_{k=0}^\infty \frac{f_k(\beta)}{r^{n+k}}, \ f_k \in C^\infty(S^{n-1}).$$

However, this is not a very general class, because it does not contain, for example, functions which are $\sim f_0(\beta) r^{-(s+n)}, r \to \infty, f_0 \not\equiv 0$, where s is not an integer. In [**Her83**] the Radon transform of integrable distributions is defined and it is shown that they constitute, essentially, the "maximal" space beyond which all extensions of R are nonunique.

Less is known about the dual Radon transform R^* and how it acts on different spaces of functions. In particular, it is shown in [**Her84**] that $R^* : \mathcal{D}'(Z_n) \to \mathcal{D}'(\mathbb{R}^n)$ is not surjective. An asymptotic expansion of $f(r\beta) = R^*\mu, \mu \in \mathcal{S}(Z_n)$, as

---

2000 *Mathematics Subject Classification.* Primary 44A12, 46F12.
Key words and phrases. Radon transform, dual Radon transform, range theorems, continuity.
This research was supported in part by NSF grant DMS-9704285.

$r \to \infty$, and an inversion formula for R^* have been obtained in [**Sol87, SM88**] (see also [**Hel65, Gon84, Gon87**] for earlier inversion formulas and [**Ram96**] for a generalization to nonsmooth functions). A relationship between singular supports of $f = R^*\mu$ and μ is established in [**Ram96**]. Most of the known explicit characterizations of ranges of R^* are based on the identity $R^*KRf = f, f \in \mathcal{S}(\mathbb{R}^n)$, that is the range of R^* acting on the space $X = KR\mathcal{S}$ is $\mathcal{S}$ (see e.g. [**Lud60, Ram96, RK96**]). Here K is certain singular operator. In [**Kat01**] the author established three range theorems for R^*: on $C_0^\infty(Z_n), \mathcal{S}(Z_n)$, and on C^∞ even functions admitting an asymptotic expansion at infinity:

$$(1.2) \qquad \mu(\alpha, p) \sim \sum_{k=0}^{\infty} \sum_{m=0}^{M_k} \mu_{k,m}(\alpha) \frac{\ln^m p}{p^{k+1}}, \ p \to +\infty, \ \mu_{k,m} \in C^\infty(S^{n-1}).$$

Again, any $\mu(\alpha, p) \sim \mu_0(\alpha) p^{-(s+1)}, p \to +\infty, \mu_0 \not\equiv 0$, where $s > 0$ is not an integer, cannot be written in the form (1.2).

Moreover, it appears that no theorem describing ranges of either R or R^* on spaces of distributions with bounded rate of decay at infinity existed in the literature.

In this paper we establish a series of range theorems for R and R^* on functions and distributions with bounded rate of decay at infinity. Fix arbitrarily $q, 0 < q < 1$, and an integer m (not necessarily positive). In Section 2 we establish various range theorems for R^* on C^∞ functions satisfying:

$$(1.3) \qquad \partial_\alpha^\vartheta \partial_p^j \mu(\alpha, p) = o(|p|^{-(q+m+j)}), p \to \infty,$$

for any multiindex $\vartheta \in \mathbb{N}^{n-1}$ and integer $j \geq 0$. Here $\partial_\alpha^\vartheta$ denotes the derivative on the unit sphere. Assume now $m \geq 0$. In Section 3 we establish a range theorem for R on C^∞ functions satisfying:

$$(1.4) \qquad \partial_x^\gamma f(x) = o(|x|^{-(q+m+|\gamma|+(n-1))}), |x| \to \infty,$$

for any multiindex $\gamma \in \mathbb{N}^n$. We also define R and R^* on distributions and obtain corresponding range theorems. In particular, we find the range of R on distributions, which are, roughly speaking, of order $O(|x|^{-(q+m+(n-1))}), |x| \to \infty$. Just as in the classical range theorems of [**GGV66, Hel99, Lud60**], moment conditions here play a central role. Both functions and distributions from the range of R are supposed to satisfy a finite number of moment conditions. If m is increased by one, the number of moment conditions is also increased by one. Since q is not an integer, the theorems mentioned above do not describe "transition" cases where $f(r\beta) \sim f_0(\beta) r^{-(n+m)}, r \to \infty, f_0 \not\equiv 0$, either as a function or in the sense of distributions. In the case of functions such a theorem was already mentioned above (see [**Kat97**]). It turns out that in the case of distributions a similar transition theorem can be obtained. It is stated and proved in Section 4. Auxiliary lemmas are collected in Section 5. Results of this paper generalize previously known results regarding maximal extension of R to distributions, inverse continuity of R, as well as extend inversion formulas for R and R^* to wider classes of functions and distributions (see papers [**Her83, Her84, Sol87, SM88**] and Remarks 2.2–3.5 below).

2. Range theorems for the dual Radon transform

Let $\dot{S}_t(\mathbb{R}^n), t \in \mathbb{R}$, be the space of all functions $f \in C^\infty(\mathbb{R}^n)$ such that

(2.1) $$(1+|x|)^{t+|\gamma|}\partial^\gamma f(x) \to 0$$

uniformly as $|x| \to \infty$ for any multiindex $\gamma \in \mathbb{N}^n$. Similarly, $\dot{S}_t(Z_n)$ is the space of all even functions $\mu \in C^\infty(Z_n)$, $\mu(\alpha, p) = \mu(-\alpha, -p)$, such that

(2.2) $$(1+|p|)^{t+j}\partial_\alpha^\vartheta \partial_p^j \mu(\alpha, p) \to 0$$

uniformly with respect to $\alpha \in S^{n-1}$ as $|p| \to \infty$ for any multiindex $\vartheta \in \mathbb{N}^{n-1}$ and integer $j \geq 0$. Here $\partial_\alpha^\vartheta$ denotes the derivative on the unit sphere:

(2.3) $$\partial_\alpha^\vartheta \mu(\alpha, p) := \left(\frac{\partial}{\partial \omega}\right)^\vartheta \mu(\sqrt{1-|\omega|^2}\alpha + \omega, p)\bigg|_{\substack{\alpha \cdot \omega = 0 \\ |\omega| = 0}}.$$

Endowing spaces $\dot{S}_t(\mathbb{R}^n)$ and $\dot{S}_t(Z_n)$ with seminorms

(2.4)
$$\|f\|_\gamma := \sup_{x \in \mathbb{R}^n}(1+|x|)^{t+|\gamma|}|\partial^\gamma f(x)|,$$
$$\|\mu\|_{\vartheta,j} := \sup_{(\alpha,p) \in Z_n}(1+|p|)^{t+j}|\partial_\alpha^\vartheta \partial_p^j \mu(\alpha, p)|,$$

respectively, converts them into Frechet spaces. It is easy to see that the canonical injections $C_0^\infty(\mathbb{R}^n) \hookrightarrow \dot{S}_t(\mathbb{R}^n)$ and $C_0^\infty(Z_n) \hookrightarrow \dot{S}_t(Z_n)$ have dense images.

The following important distributions will be used in the paper. One is the Riesz potential (see [**Hel99**], pp. 164–168):

(2.5)
$$(\mathcal{I}^s f)(x) = \mathcal{F}^{-1}(|\xi|^{-s}\tilde{f}(\xi)) = \frac{1}{H_n(s)}\int_{\mathbb{R}^n}|x-y|^{s-n}f(y)dy, \ s-n \notin 2\mathbb{Z}^+,$$
$$H_n(s) = 2^s \pi^{n/2}\frac{\Gamma(s/2)}{\Gamma((n-s)/2)},$$

where $f \in \mathcal{S}(\mathbb{R}^n)$ and $\mathcal{F}$ is the Fourier transform:

(2.6) $$\tilde{f}(\xi) := (\mathcal{F}f)(\xi) = \int_{\mathbb{R}^n}f(x)e^{i\xi \cdot x}dx.$$

In particular, when $s = -1$ we get the Calderon-Zygmund operator Λ, which can also be defined in terms of the Riesz transforms $\mathcal{H}_j$ (see [**Sol87**] for a brief discussion and references):

(2.7)
$$\Lambda f = \mathcal{F}^{-1}(|\xi|\tilde{f}(\xi)) = \sum_{j=1}^n \partial_{x_j}\mathcal{H}_j f,$$
$$\mathcal{H}_j f(x) = \mathcal{F}^{-1}(i\xi_j |\xi|^{-1}\tilde{f}(\xi)) = \frac{\Gamma((n+1)/2)}{\pi^{(n+1)/2}}\int_{\mathbb{R}^n}\frac{x_j - y_j}{|x-y|^{n+1}}f(y)dy, \ 1 \leq j \leq n.$$

If $n = 1$, there is only one $\mathcal{H}_j$, which coincides with the Hilbert transform $\mathcal{H}$.

Recall that a linear map $A : X \to Y$ between topological vector spaces is an isomorphism if and only if A and A^{-1} are both continuous. Denote $\gamma_n := 1/[2(2\pi)^{n-1}]$.

THEOREM 2.1. *Let $0 < q < 1$. Then R^* is an isomorphism of $\dot{S}_q(Z_n)$ onto $\dot{S}_q(\mathbb{R}^n)$ and $(R^*)^{-1} = \gamma_n R \Lambda^{n-1}$.*

REMARK 2.2. Since $\mathcal{S}(Z_n) \subsetneq \dot{S}_q(Z_n)$, the inversion formula of Theorem 2.1 generalizes that of Theorem 8.1 in [**Sol87**].

PROOF. First, let us prove that $R^*(\dot{S}_q(Z_n)) \subset \dot{S}_q(\mathbb{R}^n)$. Consider

$$f(x) = \int_{S^{n-1}} \mu(\alpha, \alpha \cdot x) d\alpha, \tag{2.8}$$

where $\mu \in \dot{S}_q(Z_n)$. Obviously, $f \in C^\infty(\mathbb{R}^n)$. Let $x = r\beta, \beta \in S^{n-1}, r > 0$. Then

$$f(r\beta) = \int_{-1}^{1} \nu_\beta(t, rt)(1-t^2)^{(n-3)/2} dt,$$
$$\nu_\beta(t, p) = \int_{\substack{|\omega|=1 \\ \beta \cdot \omega = 0}} \mu(t\beta + \sqrt{1-t^2}\,\omega, p) d\omega. \tag{2.9}$$

Pick $\chi \in C_0^\infty((-1,1)), \chi(t) = 1$ if $|t| \leq 1/2$. Then

$$f(r\beta) = \int_{-1}^{1} \nu_\beta(t, rt)(1-t^2)^{(n-3)/2} \chi(t) dt$$
$$+ \int_{0.5 \leq |t| \leq 1} \nu_\beta(t, rt)(1-t^2)^{(n-3)/2}(1-\chi(t)) dt. \tag{2.10}$$

Clearly, the function $\nu_\beta(t, s)(1-t^2)^{(n-3)/2}\chi(t)$ and its asymptotic properties as $|s| \to \infty$ depend smoothly and uniformly on the parameter $\beta \in S^{n-1}$. Therefore, Lemma 5.1 implies that the first integral in (2.10) is $o(r^{-q}), r \to \infty$, uniformly with respect to $\beta \in S^{n-1}$. Since t is bounded away from zero in the second integral in (2.10), it is also $o(r^{-q}), r \to \infty$. In a similar fashion,

$$\partial_x^\gamma f(x) = \int_{S^{n-1}} \alpha^\gamma \partial_p^{|\gamma|} \mu(\alpha, \alpha \cdot x) d\alpha$$
$$= \int_{-1}^{1} \nu_\beta(t, rt)(1-t^2)^{(n-3)/2} \chi(t) dt$$
$$+ \int_{0.5 \leq |t| \leq 1} \nu_\beta(t, rt)(1-t^2)^{(n-3)/2}(1-\chi(t)) dt, \tag{2.11}$$

where $x = r\beta$ and

$$\nu_\beta(t, p) = \int_{\substack{|\omega|=1 \\ \beta \cdot \omega = 0}} (t\beta + \sqrt{1-t^2}\,\omega)^\gamma \partial_p^{|\gamma|} \mu(t\beta + \sqrt{1-t^2}\,\omega, p) d\omega. \tag{2.12}$$

By the assumptions about μ, $\partial_t^l \nu_\beta(t, p) = o(|p|^{-(q+|\gamma|)}), p \to \infty, l \geq 0$, uniformly with respect to $\beta \in S^{n-1}$ and $t \in \operatorname{supp} \chi$, and

$$\int_{-\infty}^{\infty} \nu_\beta(t, p) p^j dp \equiv 0, \ 0 \leq j \leq |\gamma| - 1. \tag{2.13}$$

Thus, by Lemma 5.2, the first integral on the right in (2.11) is $o(r^{-(q+|\gamma|)}), r \to \infty$, uniformly with respect to $\beta \in S^{n-1}$. Since t is bounded away from zero in the second integral, we conclude $\partial_x^\gamma f(x) = o(|x|^{-(q+|\gamma|)}), |x| \to \infty$, that is $f \in \dot{S}_q(\mathbb{R}^n)$.

To prove the converse statement, fix $f \in \dot{S}_q(\mathbb{R}^n)$ and find $\mu \in \dot{S}_q(Z_n)$ which satisfies $f = R^*\mu$. Define

$$\mu = Rg, \ g = \gamma_n \Lambda^{n-1} f, \ \gamma_n = 1/(2(2\pi)^{n-1}). \tag{2.14}$$

It is well known that

(2.15) $\quad \Lambda^{n-1} = (-\Delta)^{(n-1)/2}$, n odd; $\Lambda^{n-1} = (-\Delta)^{(n-2)/2} \sum_{j=1}^{n} \partial_{x_j} \mathcal{H}_j$, n even,

where Δ is the Laplacian. By assumption (and Lemma 5.5 if n is even), $g \in \dot{S}_{q+(n-1)}(\mathbb{R}^n)$. Further, by Lemma 5.4, $\mu \in \dot{S}_q(Z_n)$.

Fix now any $\psi \in C_0^\infty(\mathbb{R}^n)$. From (2.14),

(2.16) $\quad (R^*\mu, \psi)_{\mathbb{R}^n} = (\gamma_n R^* R \Lambda^{n-1} f, \psi)_{\mathbb{R}^n} \stackrel{(1)}{=} (R\Lambda^{n-1} f, \gamma_n R\psi)_{Z_n}$
$\stackrel{(2)}{=} (\Lambda^{n-1} f, \gamma_n R^* R\psi)_{\mathbb{R}^n} \stackrel{(3)}{=} (f, \gamma_n \Lambda^{n-1} R^* R\psi)_{\mathbb{R}^n} = (f, \psi)_{\mathbb{R}^n},$

that is $R^*\mu = f$ and the inversion formula is established. Equality (1) in (2.16) holds because $(R\Lambda^{n-1} f)(\alpha, \alpha \cdot x)\psi(x) \in L^1(\mathbb{R}^n \times S^{n-1})$ and, by the Fubini theorem, one can change the order of integration. As is well-known,

(2.17) $\quad \int_{S^{n-1}} |(R\psi)(\alpha, \alpha \cdot x)| d\alpha = O(1/|x|), \, |x| \to \infty.$

Hence the map

(2.18) $\quad x \to |(\Lambda^{n-1} f)(x)| \int_{S^{n-1}} |(R\psi)(\alpha, \alpha \cdot x)| d\alpha \in L^1(\mathbb{R}^n),$

and by the second part of the Fubini theorem ([**Lan93**], p. 165), one can change the order of integration. This establishes equality (2). To justify (3) notice that $R\psi$ is compactly supported and, in particular, $R\psi \in \dot{S}_{q_1}(Z_n)$ for all $0 < q_1 < 1$. By what we already proved, $R^* R\psi \in \dot{S}_{q_1}(\mathbb{R}^n)$ for any $0 < q_1 < 1$. By assumption, $f \in \dot{S}_q(\mathbb{R}^n)$. Choosing $q_1, 0 < q_1 < 1$, so that $q + q_1 > 1$ and using Lemma 5.8 we prove (3).

To prove injectivity of R^*, take $\mu \in \dot{S}_q(Z_n)$ and suppose $R^*\mu \equiv 0$. This implies

(2.19) $\quad 0 = \mathcal{P}_1(\partial_x) R^* \mu = R^*(\mathcal{P}_1(\alpha) \partial_p \mu) = \frac{1}{\pi} \int_0^\infty \int_{S^{n-1}} (\tilde{\nu}(\alpha, \lambda) \lambda^{1-n}) e^{-i\lambda \alpha \cdot x} \lambda^{n-1} d\lambda d\alpha,$

where $\nu(\alpha, p) = \mathcal{P}_1(\alpha) \partial_p \mu = o(|p|^{-(q+1)}), p \to \infty$, and $\mathcal{P}_1$ is any homogeneous polynomial of degree one. Since $\tilde{\nu}(\alpha, \lambda) \lambda^{1-n}$, regarded as a function of $\xi \in \mathbb{R}^n, \lambda = |\xi|, \alpha = \xi/\lambda$, is integrable at the origin, $\tilde{\nu}(\alpha, \lambda)$ is continuous and bounded (in fact, $\tilde{\nu}$ decays faster than any negative power of λ as $\lambda \to \infty$), we conclude from (2.19) that $\nu \equiv 0$ and $\mu = \text{const}(p)$. Since $\mu \to 0$ as $p \to \infty$, this implies $\mu \equiv 0$.

Lemma 5.3 combined with (2.10), (2.11), and (2.13) implies that the graph of R^* is closed. By the closed graph theorem, R^* is continuous. Since R is one-to-one and onto, appealing to Corollary 8 in [**Hus65**], p. 42, we conclude that R^* is an isomorphism. $\square$

THEOREM 2.3. *Fix $0 < q < 1$ and an integer $m \geq 1$. Consider the following subspaces*

$$\dot{X}_{q,m} := \{\mu \in \dot{S}_{q+m}(Z_n) : \int \mu(\alpha, p) p^j dp = 0, \, 0 \leq j \leq m-1\},$$

(2.20)
$$\dot{Y}_{q,m} := \begin{cases} \dot{S}_{q+m}(\mathbb{R}^n), & m \leq n-1, \\ \{f \in \dot{S}_{q+m}(\mathbb{R}^n) : \int f(x) x^\gamma dx = 0, \, |\gamma| \leq m-n\}, & m \geq n, \end{cases}$$

endowed with the topologies that are induced by $\dot{S}_{q+m}(Z_n)$ and $\dot{S}_{q+m}(\mathbb{R}^n)$, respectively. Then R^* is an isomorphism of $\dot{X}_{q,m}$ onto $\dot{Y}_{q,m}$.

PROOF. Using an argument analogous to (2.9)–(2.13), taking into account that the resulting function $\nu_\beta(t,p)$ will have $m + |\gamma|$ vanishing moments, and appealing to Lemma 5.2, we prove the inclusion $R^*\dot{X}_{q,m} \subset \dot{S}_{q+m}(\mathbb{R}^n)$. If $m \geq n$, the Fourier transform of $\mu(\alpha,p)$ with respect to p can be written as $\tilde{\mu}(\alpha,\lambda) = \lambda^{m-1}\nu(\alpha,\lambda)$, where ν is continuous, $\nu(\alpha,0) = 0$, and decays rapidly as $\lambda \to \infty$. Thus

$$(2.21) \quad R^*\mu(x) = \frac{1}{\pi}\int_0^\infty \int_{S^{n-1}} \lambda^{m-n}\nu(\alpha,\lambda)e^{-i\lambda\alpha\cdot x}\lambda^{n-1}d\alpha d\lambda, \quad m \geq n,$$

and the zero moment conditions follow. Hence, $R^*(\dot{X}_{q,m}) \subset \dot{Y}_{q,m}$.

To prove the reverse inclusion assume first $m \leq n-1$ and use the inversion formula of Theorem 2.1. Define

$$(2.22) \quad \mu := \gamma_n R\Lambda^{n-1}f, \quad f \in \dot{Y}_{q,m} = \dot{S}_{q+m}(\mathbb{R}^n).$$

We establish now an auxiliary identity. Let $g \in C_0^\infty(\mathbb{R}^n)$. Arguing in the Fourier transform domain it is easy to derive that the following holds:

$$(2.23) \quad R\Lambda^l g = (\mathcal{H}\partial_p)^l Rg = \partial_p^l \mathcal{H}^l Rg, \quad l \geq 0.$$

First, we derive (2.23) in the sense of distributions. By Lemmas 5.4 and 5.5 the right-hand side of (2.23) is C^∞. Clearly,

$$(2.24) \quad \Lambda^l g = R^*\nu, \quad \nu(\alpha,p) = \frac{1}{2(2\pi)^n}\int |\lambda|^{l+n-1}\tilde{g}(\lambda\alpha)e^{-i\lambda p}d\lambda.$$

This implies $\nu(\alpha,p) = O(p^{-(l+n)}), p \to \infty$, $\int \nu(\alpha,p)p^j dp = 0$, $0 \leq j \leq l+n-2$, and $\nu \in \dot{X}_{q,l+n-1}$ for any $q, 0 < q < 1$. By what we have already proved,

$$(2.25) \quad \Lambda^l g = R^*\nu \in \dot{Y}_{q,l+n-1} \subset \dot{S}_{q+l+n-1}(\mathbb{R}^n) \subset \dot{S}_{q+n-1}(\mathbb{R}^n).$$

By Lemma 5.4, $R\Lambda^l g \in C^\infty$, and (2.23) holds pointwise. Since $C_0^\infty(\mathbb{R}^n)$ is dense in $\dot{S}_t(\mathbb{R}^n)$, by Lemmas 5.4 and 5.5 the right-hand side of (2.23) extends by continuity to all of $\dot{S}_{q+n-1}(\mathbb{R}^n)$.

Returning to (2.22) and using (2.23),

$$(2.26) \quad \mu = \gamma_n(\mathcal{H}\partial_p)^m R\Lambda^{n-1-m}f, \quad f \in \dot{Y}_{q,m} = \dot{S}_{q+m}(\mathbb{R}^n).$$

As was noted in [**Sol87**],

$$(2.27) \quad \Lambda^m \mathcal{I}^m g = g, \quad g \in L^p(\mathbb{R}^n), \quad 1 < p < n/m.$$

Since $\dot{S}_{q+m}(\mathbb{R}^n) \subset L^p(\mathbb{R}^n), n/(q+m) < p < n/m$, we have $\Lambda\mathcal{I}g = g$ on $\dot{S}_{q+1}(\mathbb{R}^n)$. From this and Lemma 5.6, $\Lambda(\dot{S}_q(\mathbb{R}^n)) = \dot{S}_{q+1}(\mathbb{R}^n)$. Therefore, using (2.27) with $m = 2$ and Lemma 5.6 with $t = q+1, q+2 < n$, we conclude $\dot{S}_{q+2}(\mathbb{R}^n) = \Lambda^2(\dot{S}_q(\mathbb{R}^n)) = \Lambda(\dot{S}_{q+1}(\mathbb{R}^n))$. Continuing in a similar fashion, $\Lambda^m(\dot{S}_{q+j}(\mathbb{R}^n)) = \dot{S}_{q+j+m}(\mathbb{R}^n)$, $0 \leq j \leq j+m < n$. Applying this result to (2.26), $\Lambda^{n-1-m}f \in \dot{S}_{q+n-1}(\mathbb{R}^n)$. By Lemma 5.4, $R: \dot{S}_{q+n-1}(\mathbb{R}^n) \to \dot{S}_q(Z_n)$ is continuous. Using that $\mathcal{H}$ and ∂_p commute and applying $\mathcal{H}$ first we see that $\mu \in \dot{S}_{q+m}(Z_n)$ and its first m moments vanish. Thus, $\mu \in \dot{X}_{q,m}$.

If $m \geq n$, define

$$(2.28) \quad \mu := \gamma_n(\mathcal{H}\partial_p)^{n-1}Rf, \quad f \in \dot{Y}_{q,m}.$$

By Lemma 5.4, $Rf \in \dot{S}_{q+m-(n-1)}(Z_n)$. By assumption, $\int f(x)x^\gamma dx = 0, |\gamma| \leq m - n$. This implies that Rf has $m - n + 1$ first vanishing moments, that is $Rf = \partial_p^{m-n+1}\nu$ for some $\nu \in \dot{S}_q(Z_n)$. Since $\mathcal{H}$ and ∂_p commute, $\mu \in \dot{S}_{q+m}(Z_n)$ and its first m moments vanish. Thus, $\mu \in \dot{X}_{q,m}$.

By Theorem 2.1, R^* is injective. As in the proof of Theorem 2.1, appealing to Lemma 5.3 we establish that the graph of R^* is closed. Since $\dot{X}_{q,m}$ is a closed subspace of the Frechet space $\dot{S}_{q+m}(Z_n)$, it is also a Frechet space. By the closed graph theorem, R^* is continuous. Since R^* is one-to-one, onto, and $\dot{Y}_{q,m}$ is Frechet, R^* is an isomorphism. □

Let Y_j denote a spherical harmonic of degree j.

COROLLARY 2.4. *Fix $0 < q < 1$ and an integer $m \geq 1$. The range of R^* on $\dot{S}_{q+m}(Z_n)$ consists of C^∞ functions that can be represented in the form:*

$$(2.29) \quad f(r\beta) = \sum_{k=0}^{m-1} \frac{f_k(\beta)}{r^{k+1}} + g(r\beta), \; r \geq 1, g \in \dot{Y}_{q,m},$$

$$f_k \in C^\infty(S^{n-1}), \; f_k(-\beta) = (-1)^k f_k(\beta).$$

and, if $m \geq n$, satisfy additional conditions:

$$(2.30)$$
$$\int_{S^{n-1}} f_k(\beta)Y_j(\beta)d\beta = 0, \; 0 \leq j \leq k - (n-1), n-1 \leq k \leq m-1, \; \text{if } n \text{ is odd,}$$

or, for any polynomial P of degree not exceeding $m - n$, the finite part of the limit

$$(2.31) \quad F.p.(f,P) := F.p.\lim_{\epsilon \to 0^+} \int_{\mathbb{R}^n} f(x)P(x)e^{-\epsilon|x|}dx = 0, \; \text{if } n \text{ is even.}$$

First, let us state a theorem which was proved in [**Kat97**]:

THEOREM 2.5. *Let Y be the space of $C^\infty(\mathbb{R}^n)$ functions that admit an expansion*

$$(2.32) \quad f(r\beta) \sim \sum_{k=0}^{\infty} \frac{f_k(\beta)}{r^{k+1}}, \; r \to +\infty, \; f_k \in C^\infty(S^{n-1}), \; f_k(-\beta) = (-1)^k f_k(\beta),$$

which is uniform with respect to $\beta \in S^{n-1}$ and can be differentiated with respect to $x = r\beta$ any number of times. Suppose, in addition, that

$$(2.33) \quad \int_{S^{n-1}} f_k(\beta)Y_m(\beta)d\beta = 0, \; m \leq k - (n-1), \; \text{if } n \text{ is odd,}$$

and, for any polynomial P, the finite part of the limit

$$(2.34) \quad F.p.\lim_{\epsilon \to 0^+} \int_{\mathbb{R}^n} f(x)P(x)e^{-\epsilon|x|}dx = 0, \; \text{if } n \text{ is even.}$$

Then $R^*(\mathcal{S}(Z_n)) = Y$.

PROOF OF COROLLARY 2.4. Clearly, $\dot{S}_{q+m}(Z_n) = \mathcal{S}(Z_n) + \dot{X}_{q,m}$. Therefore, by Theorems 2.3 and 2.5,

$$(2.35) \quad R^*(\dot{S}_{q+m}(Z_n)) = R^*(\mathcal{S}(Z_n)) + R^*(\dot{X}_{q,m}) = Y + \dot{Y}_{q,m}.$$

First, let n be odd. It is obvious that the space of functions described in Corollary 2.4 is a subset of $Y + \dot{Y}_{q,m}$. Thus, it suffices to show that any $f \in Y + \dot{Y}_{q,m}$

can be represented in the form (2.29), (2.30). Let $f = g_1 + g_2, g_1 \in Y, g_2 \in \dot{Y}_{q,m}$. By assumption, we can write

(2.36) $$g_1 = g_{11} + g_{12}, \ g_{11}(r\beta) = \sum_{k=0}^{m-1} \frac{f_k(\beta)}{r^{k+1}}, \ r \geq 1, g_{12} \in \dot{S}_{q+m}(\mathbb{R}^n),$$

and $f_k, 0 \leq k \leq m-1$, have all the required properties. If $m \geq n$, find $w \in C_0^\infty$, $w(x) = 0$ for $|x| \geq 1$, such that

(2.37) $$\int w(x) x^\gamma dx = \int g_{12}(x) x^\gamma dx, \ 0 \leq |\gamma| \leq m - n.$$

By construction, $g_{13} := g_{12} - w \in \dot{Y}_{q,m}$, and the desired decomposition is $f = (g_{11} + w) + (g_{13} + g_2)$.

Now let n be even. Take f of the form (2.29), (2.31):

(2.38) $$f = g_1 + g_2, \ g_1(r\beta) = \sum_{k=0}^{m-1} \frac{f_k(\beta)}{r^{k+1}}, \ r \geq 1, \ g_2 \in \dot{Y}_{q,m}.$$

By (2.31), $F.p.(g_1, P) = 0$ if $\deg(P) \leq m - n$ and $m \geq n$. Denote $c_\gamma = F.p.(g_1, x^\gamma)$. By construction,

(2.39) $$c_\gamma = 0, \ |\gamma| \leq m - n, \ \text{if } m \geq n.$$

Using Borel's theorem in the Fourier transform domain we can find $g_3 \in \mathcal{S}(\mathbb{R}^n)$ such that $(g_3, x^\gamma) = c_\gamma$ for all γ. By (2.39), $g_3 \in \dot{Y}_{q,m}$. Thus,

(2.40) $$f = (g_1 - g_3) + (g_2 + g_3) \in Y + \dot{Y}_{q,m}.$$

The reverse inclusion is immediate. Indeed, we showed above that any $f = g_1 + g_2, g_1 \in Y, g_2 \in \dot{Y}_{q,m}$, can be represented in the form (2.29). By construction, all the generalized moments of g_1 and all the moments of g_2 up to the order $m - n$ (if $m \geq n$) vanish. Thus, if $m \geq n$, f satisfies (2.31) and the corollary is proved. □

Recall that a continuous linear map between topological vector spaces $A : X \to Y$ is a strict morphism if the associated injection $A : X/\mathrm{Ker}A \to Y$ is an isomorphism onto $\mathrm{Ran}A$, the range of A ([**Hor66**], Definition 2, p. 106).

THEOREM 2.6. *Fix $0 < q < 1$ and an integer $m \geq 1$. Then $R^* : \dot{S}_{q-m}(Z_n) \to \dot{S}_{q-m}(\mathbb{R}^n)$ is continuous, the kernel of R^* is given by*

(2.41)
$$\mathrm{Ker}R^* = \{\mu \in \dot{S}_{q-m}(Z_n) : \mu(\alpha, p) = \sum_{j=0}^{m-1} A_j(\alpha) p^j, A_j(\alpha) \in C^\infty(S^{n-1}),$$
$$A_j(-\alpha) = (-1)^j A_j(\alpha), \int_{S^{n-1}} A_j(\alpha) \alpha^\gamma d\alpha = 0, 0 \leq j = |\gamma| \leq m - 1\},$$

and R^* is a strict morphism onto $\mathrm{Ran}R^* = \dot{S}_{q-m}(\mathbb{R}^n)$.

PROOF. Just as in the proof of Theorem 2.1 (cf. (2.8)–(2.13)), by appealing to Lemmas 5.1 and 5.2 (with q replaced by $q - m$) we prove that $R^*(\dot{S}_{q-m}(Z_n)) \subset \dot{S}_{q-m}(\mathbb{R}^n)$. Let $\mathcal{P}_m$ be any homogeneous polynomial of degree m. Pick $\mu \in \dot{S}_{q-m}(Z_n)$ and let $f = R^*\mu$. Then

(2.42) $$\mathcal{P}_m(\partial_x)f = R^*(\mathcal{P}_m(\alpha)\partial_p^m \mu(\alpha, p)) \in \dot{S}_q(\mathbb{R}^n).$$

This implies that solutions $\mu \in \dot{S}_{q-m}(Z_n)$ of $R^*\mu = f$ are defined up to a polynomial of degree $m - 1$ in p with coefficients depending on α. By Theorem 2.1, this also implies that if $\mu \in \text{Ker}R^*$, then μ is such a polynomial. To find all polynomials which are in $\text{Ker}R^*$ we have to solve

$$(2.43) \qquad R^*\left(\sum_{j=0}^{m-1} A_j(\alpha)p^j\right) = \int_{S^{n-1}} \sum_{j=0}^{m-1} A_j(\alpha)(\alpha \cdot x)^j d\alpha \equiv 0.$$

Considering individual terms (they are of different degrees in x and, therefore, linearly independent), we get

$$(2.44) \qquad \int_{S^{n-1}} A_j(\alpha)(\alpha \cdot x)^j d\alpha \equiv 0, \ 0 \le j \le m-1,$$

which immediately implies (2.41). The symmetry property of A_j follows from the requirement that $\mu(\alpha, p)$ be even.

To prove the inclusion $\dot{S}_{q-m}(\mathbb{R}^n) \subset R^*(\dot{S}_{q-m}(Z_n))$, we first need to establish the following identity (compare with (2.23)):

$$(2.45) \qquad R^*(\mathcal{H}\partial_p)^m\mu = \Lambda^m R^*\mu, \ \mu \in \dot{S}_{q-j}(Z_n), \ -\infty < j \le m.$$

By Lemma 5.7 (with the dimension of the space $n = 1$), $(\mathcal{H}\partial_p)^m\mu \in \dot{S}_{q-j+m}(Z_n)$. By Theorem 2.1, $R^*(\mathcal{H}\partial_p)^m\mu \in \dot{S}_q(\mathbb{R}^n) \subset \mathcal{S}'(\mathbb{R}^n)$. By what we already proved, $R^*\mu \in \dot{S}_{q-j}(\mathbb{R}^n)$. By Lemma 5.7, $\Lambda^m R^*\mu \in \dot{S}_{q-j+m}(\mathbb{R}^n) \subset \mathcal{S}'(\mathbb{R}^n)$. Thus, it makes sense to take the Fourier transform on both sides of (2.45). By choosing $\phi \in \mathcal{S}(\mathbb{R}^n)$ such that ϕ vanishes identically near $x = 0$ and applying both sides of (2.45) to $\mathcal{F}\phi$, we immediately conclude that the Fourier transform of $g = R^*(\mathcal{H}\partial_p)^m\mu - \Lambda^m R^*\mu$ vanishes identically outside the origin. Hence, g is a polynomial. By what we already know, $g \to 0$, $|x| \to \infty$, hence $g \equiv 0$.

Now pick $f \in \dot{S}_{q-m}(\mathbb{R}^n)$ and find $\nu \in \dot{S}_q(Z_n)$ such that $R^*\nu = \Lambda^m f \in \dot{S}_q(\mathbb{R}^n)$. Using (2.45) with $j = m$ and the fact that $\mathcal{H}$ and ∂_p commute on $\dot{S}_q(Z_n)$, denote

$$(2.46) \qquad \mu_1(\alpha, p) := \int_0^p \frac{(p-t)^{m-1}}{(m-1)!}((-\mathcal{H})^m\nu)(\alpha, t)dt, \ f_1 := R^*\mu_1.$$

From (2.46) and Lemma 5.5 (if m is odd), $\mu_1 \in \dot{S}_{q-m}(Z_n)$. By what we already proved, $f_1 \in \dot{S}_{q-m}(\mathbb{R}^n)$. By construction, $\Lambda^m(f - f_1) \equiv 0$, $f - f_1 \in \dot{S}_{q-m}(\mathbb{R}^n)$ and, in particular, $|f(x) - f_1(x)| \le c(1 + |x|)^{m-q}, x \in \mathbb{R}^n$, for some $c > 0$. Thus, in the sense of distributions, the Fourier transform of $f - f_1$ is supported at the origin. Together with the restriction on the growth of $f - f_1$ at infinity this implies that $f - f_1$ is a polynomial of degree at most $m - 1$. Given $P_{m-1} = f - f_1$, one can find a polynomial in p: $\mu_2(\alpha, p) = \sum_{j=0}^{m-1} A_j(\alpha)p^j \in \dot{S}_{q-m}(Z_n)$, $A_j(-\alpha) = (-1)^j A_j(\alpha)$, such that $R^*\mu_2 = P_{m-1}$. Indeed, equating coefficients in front of the identical powers of x we arrive at the equations

$$(2.47) \qquad \int_{S^{n-1}} A_j(\alpha)\alpha^\gamma d\alpha = c_\gamma, \ |\gamma| = j, \ 0 \le j \le m-1,$$

where c_γ is the coefficient of x^γ in $P_{m-1}(x)$. Since the set of functions $\alpha^\gamma, |\gamma| = j$, is linearly independent on S^{n-1}, solutions A_j can be found. Thus, by construction, $\mu = \mu_1 + \mu_2 \in \dot{S}_{q-m}(Z_n)$ and $R^*\mu = f$.

To prove sequential continuity of R^* consider a sequence $\mu_j \to 0$ in $\dot{S}_{q-m}(Z_n)$. For $|\gamma| < m$,

$$\sup_{x \in \mathbb{R}^n} (1+|x|)^{q-m+|\gamma|} \left| \partial_x^\gamma \int_{S^{n-1}} \mu_j(\alpha, \alpha \cdot x) d\alpha \right| \tag{2.48}$$

$$\leq \sup_{x \in \mathbb{R}^n} \int_{S^{n-1}} \frac{\left|\partial_p^{|\gamma|} \mu_j(\alpha, \alpha \cdot x)\right|}{(1+|\alpha \cdot x|)^{m-(q+|\gamma|)}} d\alpha$$

$$\leq |S^{n-1}| \sup_{(\alpha,p) \in Z_n} (1+|p|)^{q-m+|\gamma|} \left|\partial_p^{|\gamma|} \mu_j(\alpha, p)\right| \to 0, \ j \to \infty.$$

If $|\gamma| \geq m$, then

$$\begin{aligned} & \alpha^\gamma \partial_p^{|\gamma|} \mu_j \in \dot{S}_t(Z_n), \ \alpha^\gamma \partial_p^{|\gamma|} \mu_j \to 0 \text{ in } \dot{S}_t(Z_n), \\ & \partial_x^\gamma(R^*\mu_j) = R^*(\alpha^\gamma \partial_p^{|\gamma|} \mu_j) \in \dot{S}_t(\mathbb{R}^n), \ t = q - m + |\gamma| > 0, \end{aligned} \tag{2.49}$$

and the convergence follows from Theorem 2.1 ($|\gamma| = m$) or Theorem 2.3 ($|\gamma| > m$). By the closed graph theorem, R^* is continuous on $\dot{S}_{q-m}(Z_n)$ and, therefore, on $\dot{S}_{q-m}(Z_n)/\text{Ker}R^*$ as well (cf. [**Tre67**], Proposition 4.6). Since R^* is continuous, $\text{Ker}R^*$ is closed, and the quotient space $\dot{S}_{q-m}(Z_n)/\text{Ker}R^*$ is Frechet. Since $R^* : \dot{S}_{q-m}(Z_n)/\text{Ker}R^* \to \dot{S}_{q-m}(\mathbb{R}^n)$ is continuous, one-to-one, and onto, R^* is an isomorphism of $\dot{S}_{q-m}(Z_n)/\text{Ker}R^*$ onto $\dot{S}_{q-m}(\mathbb{R}^n)$. The last assertion of the theorem is proved. $\square$

3. Range theorems for the Radon transform

THEOREM 3.1. *Fix $0 < q < 1$ and an integer $m \geq 0$. Introduce the space*

$$\dot{H}_{q,m} = \{\mu \in \dot{S}_{q+m}(Z_n) : \int \mu(\alpha,p) p^j dp = \mathcal{P}_j(\alpha), 0 \leq j \leq m-1\}, \tag{3.1}$$

where $\mathcal{P}_j$ is a homogeneous polynomial of degree j, that will be endowed with the topology induced by $\dot{S}_{q+m}(Z_n)$. Then R is an isomorphism of $\dot{S}_{q+m+(n-1)}(\mathbb{R}^n)$ onto $\dot{H}_{q,m}$, and inversion formulas $R^{-1} = \gamma_n \Lambda^{n-1} R^* = \gamma_n R^*(\mathcal{H}\partial_p)^{n-1}$ hold on $\dot{H}_{q,m}$. In particular, if $m = 0$, R is an isomorphism of $\dot{S}_{q+(n-1)}(\mathbb{R}^n)$ onto $\dot{S}_q(Z_n)$.

REMARK 3.2. Since $\mathcal{S}(Z_n) \subsetneq \dot{S}_q(Z_n)$, the results of Theorem 3.1 generalize the inversion formula of Theorem 7.7 in [**Sol87**].

PROOF. By Lemma 5.4, $R(\dot{S}_{q+m+(n-1)}(\mathbb{R}^n)) \to \dot{S}_{q+m}(Z_n)$ is continuous, and the moment conditions in (3.1) are verified easily.

Fix now $\mu \in \dot{H}_{q,m}$ and define

$$f = \gamma_n \Lambda^{n-1} R^* \mu. \tag{3.2}$$

We need to show that $f \in \dot{S}_{q+m+(n-1)}(\mathbb{R}^n)$ and $Rf = \mu$. Since we can always find a compactly supported C^∞ function whose Radon transform has any finite number of a priori prescribed moments, we can assume without loss of generality that the first m moments of μ vanish:

$$\int \mu(\alpha,p) p^j dp \equiv 0, \ 0 \leq j \leq m-1, \tag{3.3}$$

if $m > 0$. By Lemma 5.5, $\mu = (\mathcal{H}\partial_p)^m \eta$ for some $\eta \in \dot{S}_q(Z_n)$. From (2.45) with $j = 0$:

(3.4) $$R^*(\mathcal{H}\partial_p)^m \mu = \Lambda^m R^* \mu, \ \mu \in \dot{S}_q(Z_n).$$

Hence $f = \gamma_n \Lambda^{n-1} R^* (\mathcal{H}\partial_p)^m \eta = \gamma_n \Lambda^{m+n-1} R^* \eta$. By Theorem 2.1 and (2.15), $f \in \dot{S}_{q+m+(n-1)}(\mathbb{R}^n)$. Fix any $\nu \in C_0^\infty(Z_n)$. We have

(3.5)
$$(Rf, \nu)_{Z_n} \stackrel{(1)}{=} (f, R^*\nu)_{\mathbb{R}^n} = (\gamma_n \Lambda^{n-1} R^* \mu, R^* \nu)_{\mathbb{R}^n}$$
$$\stackrel{(2)}{=} (R^*\mu, \gamma_n \Lambda^{n-1} R^* \nu)_{\mathbb{R}^n} \stackrel{(3)}{=} (\mu, \gamma_n R \Lambda^{n-1} R^* \nu)_{Z_n} \stackrel{(4)}{=} (\mu, \nu)_{Z_n}.$$

The steps in (3.5) can be justified similarly to (2.16). Equality (1) follows from the second part of the Fubini theorem analogously to (2.17), (2.18). Equality (2) follows from Lemma 5.8. To prove (3), find a sequence $\mu_k \in C_0^\infty(Z_n)$, $\mu_k \to \mu$ in $\dot{S}_{q+m}(Z_n) \subset \dot{S}_q(Z_n)$. Using Theorem 2.1, the second part of the Fubini theorem, Lemma 5.5, and the Lebesgue dominated convergence theorem analogously to the proof of Lemma 5.8, we have

(3.6)
$$(R^*\mu, \gamma_n \Lambda^{n-1} R^* \nu)_{\mathbb{R}^n} = \lim_{k \to \infty} (R^* \mu_k, \gamma_n \Lambda^{n-1} R^* \nu)_{\mathbb{R}^n}$$
$$= \lim_{k \to \infty} (\mu_k, \gamma_n R \Lambda^{n-1} R^* \nu)_{Z_n} = (\mu, \gamma_n R \Lambda^{n-1} R^* \nu)_{Z_n}.$$

Finally, (4) follows from Theorem 8.1 in [**Sol87**]. Since Rf and μ are C^∞, $Rf = \mu$ and the first inversion formula is proved. The second formula follows from (3.4) with $m = n - 1$. Uniqueness of f follows from the results of [**Sol87**], pp. 325, 326 ($f \in L^p(\mathbb{R}^n), n/(q+n-1) < p < n/(n-1)$).

Since the topology of $\dot{H}_{q,m}$ is inherited from $\dot{S}_{q+m}(Z_n)$, $R: \dot{S}_{q+m+n-1}(\mathbb{R}^n) \to \dot{H}_{q,m}$ is continuous. Using the same argument as in [**Her83**], p. 187, it is easy to see that $\dot{H}_{q,m}$ is closed and, therefore, Frechet. Since R is continuous, one-to-one, and onto, R is an isomorphism. $\square$

Let $\dot{S}'_t(\mathbb{R}^n)$ and $\dot{S}'_t(Z_n)$ denote the spaces of continuous linear functionals on $\dot{S}_t(\mathbb{R}^n)$ and $\dot{S}_t(Z_n)$, respectively. As was already mentioned, the image of the canonical injection $\imath: C_0^\infty(\mathbb{R}^n) \hookrightarrow \dot{S}_t(\mathbb{R}^n)$ is dense. Obviously, $\imath$ is sequentially continuous. Hence, by the closed graph theorem, $\imath$ is continuous and the spaces $\dot{S}'_t(\mathbb{R}^n), t \in \mathbb{R}$, can be regarded as spaces of distributions (cf. [**Tre67**], pp. 243, 244). Further, let $\delta_1(x) \in C_0^\infty(\mathbb{R}^n)$ by a nonnegative function supported in the unit ball and $\int \delta_1(x) dx = 1$. Define $\delta_\epsilon(x) := \epsilon^{-n} \delta_1(x/\epsilon), \epsilon > 0$. As is easily seen, $\delta_\epsilon * f \to f, \epsilon \to 0$, in $\dot{S}_t(\mathbb{R}^n)$ for any $f \in \dot{S}_t(\mathbb{R}^n), t \in \mathbb{R}$. Also, $\chi(\epsilon x) f(x) \to f(x), \epsilon \to 0$, in $\dot{S}_t(\mathbb{R}^n)$, where $\chi \in C_0^\infty(\mathbb{R}^n), \chi \equiv 1$ near the origin. Using this, the standard argument implies that $C_0^\infty(\mathbb{R}^n)$ is dense in $\dot{S}'_t(\mathbb{R}^n), t \in \mathbb{R}$, in the sense of weak topology. The same, of course, applies to the distribution spaces $\dot{S}'_t(Z_n), t \in \mathbb{R}$. When we consider the continuity properties of R, R^*, and their inverses, it is assumed that all the dual spaces $\dot{S}'_t$ carry an identical topology, which can be arbitrarily selected from the four standard ones: the weak dual topology, the topology of convex compact convergence, the topology of compact convergence, or the strong dual topology (cf. [**Tre67**], pp. 197, 198).

Using Theorem 2.1, define now the Radon transform on distributions $R: \dot{S}'_q(\mathbb{R}^n) \to \dot{S}'_q(Z_n)$ via duality

(3.7) $$(Rf, \mu) = (f, R^*\mu), \ f \in \dot{S}'_q(\mathbb{R}^n), \ \mu \in \dot{S}_q(Z_n), \ 0 < q < 1.$$

In a similar fashion, using Theorem 3.1, we can define the dual Radon transform on distributions $R^* : \dot{S}'_q(Z_n) \to \dot{S}'_{q+(n-1)}(\mathbb{R}^n)$:

(3.8) $\quad (R^*\mu, f) = (\mu, Rf), \ \mu \in \dot{S}'_q(Z_n), \ f \in \dot{S}_{q+(n-1)}(\mathbb{R}^n), \ 0 < q < 1.$

Since $R^*(\dot{S}_q(Z_n)) = \dot{S}_q(\mathbb{R}^n)$, it is clear that (3.7) gives a direct, unique, and injective extension of the classical Radon transform. Analogously, $R(\dot{S}_{q+n-1}(\mathbb{R}^n)) = \dot{S}_q(Z_n)$, and (3.8) gives a direct, unique, and injective extension of the classical dual Radon transform. We will also need to extend Λ^m to distributions. Since $\Lambda^m : \dot{S}_q(\mathbb{R}^n) \to \dot{S}_{q+m}(\mathbb{R}^n)$ is continuous, we can define $\Lambda^m : \dot{S}'_{q+m}(\mathbb{R}^n) \to \dot{S}'_q(\mathbb{R}^n)$ by the formula

(3.9) $\quad (\Lambda^m f, g) = (f, \Lambda^m g), \ f \in \dot{S}'_{q+m}(\mathbb{R}^n), \ g \in \dot{S}_q(\mathbb{R}^n), \ 0 < q < 1.$

Since $C_0^\infty(\mathbb{R}^n)$ is dense in $\dot{S}'_t(\mathbb{R}^n)$ in the sense of weak topology, (3.9) gives a direct and unique extension of Λ^m (see also Lemma 5.8 and note that $\dot{S}_{q_1}(\mathbb{R}^n) \subset \dot{S}'_{q+m}(\mathbb{R}^n)$ if $q_1 + q + m > n$. If $m = n - 1$, we get the usual $q + q' > 1$).

Suppose $m \geq 1$. In view of Corollary 2.4, $R^*(\dot{S}_{q+m}(Z_n)) \not\subset \dot{S}_{q+m}(\mathbb{R}^n)$. Thus, attempts to extend R to classes of distributions wider than $\dot{S}'_q(\mathbb{R}^n)$ will lead to theoretical difficulties. Given $f \in \dot{S}'_{q+m}(\mathbb{R}^n)$, we can regard Rf as a continuous linear functional on the test function space $(R^*)^{-1}(\dot{S}_{q+m}(\mathbb{R}^n))$. If n is odd, Λ^{n-1} is a differential operator, and the inversion formula $(R^*)^{-1} = \gamma_n R\Lambda^{n-1}$ of Theorem 2.1 together with Theorem 3.1 imply that $(R^*)^{-1}(\dot{S}_{q+m}(\mathbb{R}^n)) \subset \dot{H}_{q,m}$, which is not dense in $\dot{S}_{q+m}(Z_n)$. Thus, this extension is not unique. If n is even, we can write the inversion formula as follows: $\mu(\alpha, p) = \gamma_n/(2\pi) \int |\lambda|^{n-1} \tilde{f}(\lambda\alpha) \exp(-i\lambda p) d\lambda$. Since $|\lambda|^{n-1}$ is not smooth at $\lambda = 0$, $(R^*)^{-1}(\dot{S}_{q+m}(\mathbb{R}^n))$, in general, is not a subset of $\dot{S}_{q+m}(Z_n)$ and, therefore, Rf is not a distribution from $\dot{S}'_{q+m}(Z_n)$. On the other hand, restricting the set of test functions μ in (3.7) to the subspace $\{\mu \in \dot{S}_{q+m}(Z_n) : R^*\mu \in \dot{S}_{q+m}(\mathbb{R}^n)\}$ for even n will also lead to a difficulty. Indeed, using Proposition 1 in [**Kat01**], which relates the coefficients $f_k(\beta)$ in expansion (2.32) of $f = R^*\mu$ with properties of μ, it is easy to establish that this subspace is not dense in $\dot{S}_{q+m}(Z_n)$, and the resulting extension of R is not unique. These problems arise because R^* converts rapidly decreasing functions into slowly decreasing functions. On the other hand, R converts rapidly decreasing (compactly supported) functions into rapidly decreasing (compactly supported) functions. Thus, R^* can be defined even on distributions $\mathcal{D}'(Z_n)$ ([**Her83, Her84**]). Theorem 3.1 implies, however, that R^* loses its injectivity already on spaces $\dot{S}'_{q+m}(Z_n), m \geq 1$. Another extension of R^* to tempered distributions and distributions was proposed in [**Lud60**]. One assumes that its domain is either $(R\mathcal{S})'$ or $(R\mathcal{D})'$, and in this case $R^* : (R\mathcal{S})' \to \mathcal{S}'$ and $R^* : (R\mathcal{D})' \to \mathcal{S}'$ are onto and bicontinuous [**Lud60**]. However, neither $(R\mathcal{S})'$ nor $(R\mathcal{D})'$ are, actually, spaces of distributions.

Theorems 2.1, 3.1, and Corollary on p. 199 of [**Tre67**] immediately imply the following result.

COROLLARY 3.3. *Fix q, $0 < q < 1$. Then R^* is an isomorphism of $\dot{S}'_q(Z_n)$ onto $\dot{S}'_{q+(n-1)}(\mathbb{R}^n)$ and $(R^*)^{-1} = \gamma_n R\Lambda^{n-1}$. Also, R is an isomorphism of $\dot{S}'_q(\mathbb{R}^n)$ onto $\dot{S}'_q(Z_n)$ and $R^{-1} = \gamma_n \Lambda^{n-1} R^*$.*

REMARK 3.4. Since for every $q_1, 0 < q_1 < 1$, there is $q, 0 < q < 1$, such that $\dot{S}_{q_1}(\mathbb{R}^n) \subset \dot{S}'_{q+(n-1)}(\mathbb{R}^n)$ and $\dot{S}_{q_1}(Z_n) \subset \dot{S}'_q(Z_n)$, the inversion formulas of Corollary 3.3 generalize those of Theorems 2.1 and 3.1.

PROOF. The assertion that the maps $R^* : \dot{S}'_q(Z_n) \to \dot{S}'_{q+(n-1)}(\mathbb{R}^n)$ and $R : \dot{S}'_q(\mathbb{R}^n) \to \dot{S}'_q(Z_n)$ are isomorphisms follows immediately from duality (see [**Tre67**], Proposition 19.5 and Corollary on p. 199) and Theorems 3.1 and 2.1, respectively. The inversion formula for R^* can be proved if we understand the equalities below in the sense of distributions and use the inversion formula of Theorems 2.1:
(3.10)
$$(\gamma_n R \Lambda^{n-1} R^* \mu, \nu)_{Z_n} = (\Lambda^{n-1} R^* \mu, \gamma_n R^* \nu)_{\mathbb{R}^n} = (R^* \mu, \gamma_n \Lambda^{n-1} R^* \nu)_{\mathbb{R}^n}$$
$$= (\mu, \gamma_n R \Lambda^{n-1} R^* \nu)_{Z_n} = (\mu, \nu)_{Z_n}, \ \mu \in \dot{S}'_q(Z_n), \ \nu \in \dot{S}_q(Z_n).$$

As is easily checked, the properties of operators R, Λ^{n-1}, and R^* ensure that all the expressions in (3.10) make sense. The inversion formula for R on distributions can be proved in a similar fashion. □

REMARK 3.5. The results of Corollary 3.3 push the limit of the "maximal extension" of the Radon transform derived in [**Her83**] a little further. Indeed, let $\rho(x)$ be a positive continuous function and define
(3.11) $\quad \dot{C}^\infty(\mathbb{R}^n, \rho) = \{f \in C^\infty(\mathbb{R}^n) : \rho(x)\partial_x^\gamma f(x) \to 0 \text{ as } |x| \to \infty, \forall \gamma \in \mathbb{N}^n\},$

which will be equipped with its canonical seminorms $\|f\|_\gamma = \max_x \rho(x)|\partial_x^\gamma f(x)|$. In [**Her83**] it was shown how to extend R to distributions on $\dot{C}^\infty(\mathbb{R}^n, \rho_\epsilon)$, where $\rho_\epsilon(x) = (1 + |x|)^{1-\epsilon}, 1/2 < \epsilon < 1$. Thus, the spaces $\dot{C}^\infty(\mathbb{R}^n, \rho_\epsilon), 1/2 < \epsilon < 1$, always contain functions that decrease at infinity slower than $O(|x|^{-1/2})$. This implies $\dot{S}_q(\mathbb{R}^n) \subsetneq \dot{C}^\infty(\mathbb{R}^n, \rho_\epsilon)$ and $\mathcal{D}'_{L_1}(\mathbb{R}^n, \rho_\epsilon) \subsetneq \dot{S}'_q(\mathbb{R}^n)$ when $1/2 \leq q < 1, 1/2 < \epsilon < 1$. Here $\mathcal{D}'_{L_1}(\mathbb{R}^n, \rho_\epsilon)$ is the space of weighted integrable distributions, which is the strong dual of $\dot{C}^\infty(\mathbb{R}^n, \rho_\epsilon)$. In a similar fashion, $\mathcal{D}'_{L_1}(\mathbb{R}^n) \subsetneq \dot{S}'_q(\mathbb{R}^n)$, where $\mathcal{D}'_{L_1}(\mathbb{R}^n) := \mathcal{D}'_{L_1}(\mathbb{R}^n, \rho \equiv 1)$ is the space of integrable distributions. Thus, inversion formula of Theorem 3.6 is an extension of the inversion formula in Proposition 4.3 of [**Her83**]. Note also that Corollary 3.3 extends the continuity property of R^{-1} to a wider class of distributions than in [**Her83**], Corollary 4.8, and [**Her84**]. It was shown in [**Her83**] and [**Her84**] that $R : \mathcal{E}' \to \mathcal{E}'$ is continuous in both directions and $R : \mathcal{O}'_c \to \mathcal{O}'_c$ is sequentially continuous in both direction. Here $\mathcal{E}'$ and $\mathcal{O}'_c$ are the spaces of compactly supported and rapidly decreasing distributions, respectively.

Let $m \geq 1$. When one considers the transpose of a linear map, the orthogonal of the kernel of the map plays an important role. By definition, $\eta \in (\text{Ker} R^*)^\perp \subset \dot{S}'_{q-m}(Z_n)$ if and only if $(\eta, \mu) = 0$ for all $\mu \in \text{Ker} R^*$. By (2.41), this means
(3.12) $$\int_{S^{n-1}} \int_{-\infty}^\infty \eta(\alpha, p) A_j(\alpha) p^j \, dp \, d\alpha = 0, \ 0 \leq j \leq m-1,$$

for all $A_j \in C^\infty(S^{n-1})$ which satisfy $A_j(-\alpha) = (-1)^j A_j(\alpha)$ and $(A_j, \alpha^\gamma)_{S^{n-1}} = 0$, $0 \leq j = |\gamma|$. This implies that, in the sense of distributions, $\eta(\alpha, p)$ satisfies the first m moment conditions.

THEOREM 3.6. *Fix $0 < q < 1$ and an integer $m \geq 1$. Then $R : \dot{S}'_{q-m}(\mathbb{R}^n) \to \dot{S}'_{q-m}(Z_n)$ is continuous, injective, and the range of R consists of all $\eta \in \dot{S}'_{q-m}(Z_n)$ such that, in the sense of distributions, $\int_{-\infty}^\infty \eta(\alpha, p) p^j \, dp$ is a homogeneous polynomial of degree j in α for $0 \leq j \leq m-1$.*

PROOF. Continuity of R follows immediately from Theorem 2.6 and the duality theory. Injectivity of R follows from Corollary 3.3 because $\dot{S}'_{q-m}(\mathbb{R}^n) \subset \dot{S}'_q(\mathbb{R}^n)$. By [**Tre67**], Proposition 35.4, $(\mathrm{Ker}\,R^*)^\perp$ coincides with the weak closure of $\mathrm{Ran}\,R$, the range of R. Since $R^* : \dot{S}_{q-m}(Z_n) \to \dot{S}_{q-m}(\mathbb{R}^n)$ is continuous, Proposition 35.8 in [**Tre67**] implies that R^* is continuous from $\dot{S}_{q-m}(Z_n)$, equipped with the weak topology $\sigma(\dot{S}_{q-m}(Z_n), \dot{S}'_{q-m}(Z_n))$, into $\dot{S}_{q-m}(\mathbb{R}^n)$, equipped with $\sigma(\dot{S}_{q-m}(\mathbb{R}^n), \dot{S}'_{q-m}(\mathbb{R}^n))$. From Theorem 2.6, $R^* : \dot{S}_{q-m}(Z_n) \to \dot{S}_{q-m}(\mathbb{R}^n)$ is a strict morphism. Then, by Proposition 3 on p. 263 of [**Hor66**], $R(\dot{S}'_{q-m}(\mathbb{R}^n))$ is weakly closed in $\dot{S}'_{q-m}(Z_n)$ and, therefore, $\mathrm{Ran}\,R = (\mathrm{Ker}\,R^*)^\perp$. $\square$

4. Transition range theorems for the Radon transform

The following theorem was proved in [**Kat97**].

THEOREM 4.1. *Fix any integer $m \geq 0$. Let $\mathcal{X}_m$ be the space of functions $f \in C^\infty(\mathbb{R}^n)$ which admit an asymptotic expansion*

$$(4.1) \qquad f(r\beta) \sim \sum_{k=m}^{\infty} \frac{f_k(\beta)}{r^{n+k}}, \quad r \to \infty, \quad f_k \in C^\infty(S^{n-1}),$$

and the expansion can be differentiated with respect to $x = r\beta$ any number of times. Let $\mathcal{Z}_m$ be the space of even functions $\mu \in C^\infty(Z_n)$, such that

1. *μ satisfies the first m moment conditions, $j = 0, 1, \ldots, m-1$; and*
2. *μ admits an asymptotic expansion*

$$(4.2) \qquad \mu(\alpha, p) \sim \sum_{k=m}^{\infty} \frac{Q_k(\alpha) + b_k(\alpha)}{p^{k+1}}, \quad p \to +\infty,$$

where

$$(4.3) \qquad b_k \in C^\infty(S^{n-1}), \quad b_k(-\alpha) = (-1)^{k+1} b_k(\alpha),$$

Q_k is a homogeneous polynomial of degree k, and the expansion can be differentiated with respect to p any number of times.
Then $R(\mathcal{X}_m) = \mathcal{Z}_m$.

Now we obtain an analogous theorem for the range of the Radon transform on distributions. Fix $0 < q < 1$ and an integer m (not necessarily positive). Let $R_k \in C^\infty([0, \infty))$ be functions which satisfy

$$(4.4) \qquad R_k(r) = r^{-k}, \quad r > 1.$$

We say that a distribution $f \in \dot{S}'_{q-m}(\mathbb{R}^n)$ admits an asymptotic expansion at infinity with respect to the asymptotic sequence $\{r^{-(k+n)}\}_{k \geq m}$ and write

$$(4.5) \qquad f(r\beta) \sim \sum_{k \geq m} \frac{f_k(\beta)}{r^{k+n}}, \quad r \to \infty, \quad f_k \in \mathcal{D}'(S^{n-1}),$$

if

$$(4.6) \qquad f(r\beta) - \sum_{k=m}^{N-1} f_k(\beta) R_{k+n}(r) \in \dot{S}'_{q-N}(\mathbb{R}^n),$$

for all $N \geq m+1$. In a similar fashion, let $A_k, B_k \in C^\infty(\mathbb{R})$ be functions which satisfy

(4.7) $$A_k(p) = B_k(p) = \frac{1}{p^{k+1}}, \ p > 1; \int A_k p^j dp = \int B_k p^j dp = 0, 0 \leq j \leq k-1;$$
$$A_k(-p) = (-1)^k A_k(p), \ B_k(-p) = (-1)^{k+1} B_k(p), \quad p \in \mathbb{R}.$$

We say that a distribution $\mu \in \dot{S}'_{q-m}(Z_n)$ admits an asymptotic expansion at infinity with respect to the asymptotic sequence $\{p^{-(k+1)}\}_{k \geq m}$ and write

(4.8) $$\mu(\alpha, p) \sim \sum_{k \geq m} \frac{\mu_k(\alpha)}{p^{k+1}}, \ p \to +\infty, \ \mu_k \in \mathcal{D}'(S^{n-1}),$$

if

(4.9) $$\mu(\alpha, p) - \sum_{k=m}^{N-1} \eta_k(\alpha, p) \in \dot{S}'_{q-N}(Z_n),$$

for all $N \geq m+1$, where $\eta_k \in \dot{S}'_{q-k}(Z_n)$ are arbitrary distributions, which satisfy $\eta_k(\alpha, p) = \mu_k(\alpha)/p^{k+1}, \alpha \in S^{n-1}, p > 1$. Since η_k's are even, this actually means

(4.10) $$\eta_k(\alpha, p) = (\mu_k(\alpha) + (-1)^k \mu_k(-\alpha))A_k(p) + (\mu_k(\alpha) + (-1)^{k+1} \mu_k(-\alpha))B_k(p)$$

for $\alpha \in S^{n-1}$ and $|p| > 1$.

THEOREM 4.2. *Fix $0 < q < 1$ and an integer $m \geq 0$. Consider the subspace $U_m \subset \dot{S}'_{q-m}(\mathbb{R}^n)$ of distributions admitting an asymptotic expansion at infinity (4.5). Then the range of R on U_m is the subspace $V_m \subset \dot{S}'_{q-m}(Z_n)$ of even distributions which satisfy the first m moment conditions and admit an asymptotic expansion at infinity of the form:*

(4.11) $$\mu(\alpha, p) \sim \sum_{k=m}^{\infty} \frac{Q_k(\alpha) + b_k(\alpha)}{p^{k+1}}, \ p \to +\infty,$$
$$b_k \in \mathcal{D}'(S^{n-1}), \ b_k(-\alpha) = (-1)^{k+1} b_k(\alpha),$$

and Q_k is a homogeneous polynomial of degree k.

PROOF. Fix $N \geq m+1, f \in U_m$, and write

(4.12) $$f(r\beta) = \sum_{k=m}^{N-1} f_k(\beta) R_{k+n}(r) + g_N(r\beta), \ g_N \in \dot{S}'_{q-N}(\mathbb{R}^n).$$

We have $f_k(\beta) = \sum_{l=0}^{\infty} f_{kl} Y_l(\beta)$, where f_{kl} increase at most polynomially as $l \to \infty$. Obviously, in the sense of weak convergence in $\dot{S}'_{q-k}(\mathbb{R}^n)$,

(4.13) $$\sum_{l=0}^{L} f_{kl} Y_l(\beta) R_{k+n}(r) \to f_k(\beta) R_{k+n}(r), \ L \to \infty.$$

Using equation (2.3) in [**Kat97**] and integral 2.21.2.5 in [**PBM88**] (see also (7.312.2) in [**GR94**]) we have

(4.14)
$$R[Y_l(\beta)R_{k+n}(r)] = c_{kl}Y_l(\alpha)/p^{k+1}, \ p > 1,$$
$$c_{kl} = \begin{cases} 0, & k < l \text{ and } k + l \text{ is even,} \\ \neq 0, & \text{for all other } k, l, \end{cases}$$

and $c_{kl} = O(l^{-(k+(n/2))})$, $l \to \infty$, whenever $c_{kl} \neq 0$. Thus,

(4.15) $\quad \mu_{k,L}(\alpha,p) := R\left[\sum_{l=0}^{L} f_{kl}Y_l(\beta)R_{k+n}(r)\right] = \dfrac{Q_{k,L}(\alpha) + b_{k,L}(\alpha)}{p^{k+1}}, \ p > 1,$

where $Q_{k,L}$ are homogeneous polynomials of degree k, $b_{k,L}$ are smooth and $b_{k,L}(-\alpha) = (-1)^{k+1}b_{k,L}(\alpha)$. Since $R: \dot{S}'_{q-m}(\mathbb{R}^n) \to \dot{S}'_{q-m}(Z_n)$ is continuous, $\mu_{k,L}$ converges weakly as $L \to \infty$. Moreover, $Q_{k,L}$ and $b_{k,L}$ are of different parities and $\deg(Q_{k,L}) \leq k$. Hence, $Q_{k,L} \to Q_k$ and $b_{k,L} \to b_k$ in $\mathcal{D}'(S^{n-1})$ as $L \to \infty$, where Q_k and b_k have the properties stated in the theorem. Therefore,

(4.16) $\quad R[f_k(\beta)R_{k+n}(r)] = Q_k(\alpha)A_k(p) + b_k(\alpha)B_k(p), \ |p| > 1.$

By Theorem 3.6, $Rg_N \in \dot{S}'_{q-N}(Z_n)$. This implies $R(U_m) \subset V_m$.

To prove the reverse inclusion, pick $\mu \in V_m$ and write, according to (4.8)–(4.11),

(4.17)
$$\mu(\alpha,p) = \sum_{k=m}^{N-1} \eta_k(\alpha,p) + \nu_N(\alpha,p), \ \eta_k \in \dot{S}'_{q-k}(Z_n), \ \nu_N(\alpha,p) \in \dot{S}'_{q-N}(Z_n),$$
$$\eta_k(\alpha,p) = Q_k(\alpha)A_k(p) + b_k(\alpha)B_k(p).$$

The proof will consist of four steps.

Step 1. By adjusting η_k we can assume that ν_N satisfies the first N moment conditions. Indeed, similarly to [**Kat97**], define

(4.18)
$$\check{\eta}_k(\alpha,p) := \eta_k(\alpha,p) + P_k(p) \cdot \int_{-\infty}^{\infty} \left(\mu(\alpha,t) - \sum_{j=m}^{k-1} \check{\eta}_j(\alpha,t) - \eta_k(\alpha,t)\right) t^k dt, \ k \geq m,$$

and the summation on the right-hand side of (4.18) disappears if $k = m$. Here $P_k \in C_0^{\infty}([-1,1])$ is any function with properties

(4.19) $\quad P_k(-p) = (-1)^k P_k(p), \ \displaystyle\int_{-\infty}^{\infty} P_k(p)p^j dp = \begin{cases} 0, & j = 0, 1, \ldots, k-1, \\ 1, & j = k. \end{cases}$

Clearly, $\eta_k(\alpha,p) = \check{\eta}_k(\alpha,p), |p| > 1$. Moreover, one can see by induction (cf. [**Kat97**], p. 855) that if η_k's are replaced by $\check{\eta}_k$ in (4.17), then ν_N satisfies the first N moment conditions. Therefore, this will be assumed in what follows.

Step 2. Let us find distributions $g_k \in \dot{S}'_{q-k}(\mathbb{R}^n)$ such that $Rg_k = \widehat{g}_k = \eta_k, |p| > 1$. Expanding the given Q_k and b_k in spherical harmonics and solving equations (4.16) with the help of (4.14), we find $G_k \in \mathcal{D}'(S^{n-1}), k \geq m$, and then set $g_k(r\beta) := G_k(\beta)R_{k+n}(r) \in \dot{S}'_{q-k}(\mathbb{R}^n), k \geq m$. Clearly, g_k's have the required property.

Step 3. Consider distributions $h_{kl}(r\beta) := Y_l(\beta)/r^{k+n}$, where $k+l$ is even, $l > k$, defined by

$$(4.20) \quad (h_{kl}, \phi) = \lim_{\epsilon \to 0^+} \int_\epsilon^\infty \frac{1}{r^{k+n}} \left(\int_{S^{n-1}} Y_l(\beta)\phi(r\beta)d\beta \right) r^{n-1}dr, \ \phi \in \dot{S}_{q-k}(\mathbb{R}^n).$$

As is easily seen, $h_{kl} \in \dot{S}'_{q-k}(\mathbb{R}^n)$. A simple calculation based on the Fourier Slice Theorem shows that $\widehat{h}_{kl}(\alpha, p) = d_{kl}\delta^{(k)}(p)Y_l(\alpha)$, where $d_{kl} \neq 0$, $d_{kl} = O(l^{-(k+n/2)})$ as $l \to \infty$.

Step 4. Define constants ψ_{kjl} from the relation

$$(4.21) \quad \int (\eta_k(\alpha, p) - \widehat{g}_k(\alpha, p))p^j dp = \sum_{l \geq 0} \psi_{kjl} Y_l(\alpha).$$

Using distributions h_{kl} constructed in Step 3, define

$$(4.22) \quad h_k(r\beta) := \sum_{j=k}^{N-1} \sum_{\substack{l > j \\ l+j \text{ even}}} \frac{\psi_{kjl}}{d_{jl}(-1)^j j!} \frac{Y_l(\beta)}{r^{j+n}}.$$

Here we have used that ψ_{kjl}/d_{jl} grow at most polynomially as $l \to \infty$. From the results of Step 3 and (4.22),

$$(4.23) \quad \begin{aligned} \int \widehat{h}_k(\alpha, p)p^a dp &= \sum_{j=k}^{N-1} \sum_{\substack{l > j \\ l+j \text{ even}}} \frac{\psi_{kjl}}{(-1)^j j!} Y_l(\alpha)(\delta^j(p), p^a) \\ &= \sum_{\substack{l > a \\ l+a \text{ even}}} \psi_{kal} Y_l(\alpha), \ k \leq a \leq N-1. \end{aligned}$$

By construction, η_k satisfies k moment conditions (this follows from (4.17), (4.7) and (4.18), (4.19)). Since $g_k, h_k \in \dot{S}'_{q-k}(\mathbb{R}^n)$, $\widehat{g}_k, \widehat{h}_k$ satisfy k moment conditions. Comparing (4.21) and (4.23), we conclude that the difference $\eta_k - (\widehat{g}_k + \widehat{h}_k)$ satisfies N moment conditions. Moreover, $\eta_k - \widehat{g}_k$ and $\widehat{h}_k$ are compactly supported. Note also that h_k can be written as $h_k(r\beta) = \sum_{j=k}^{N-1} H_{kj}(\beta)/r^{j+n}$, which is compatible with expansion (4.5), and the coefficients H_{kj} are independent of N. Combining with (4.17) we have

$$(4.24) \quad \mu = \sum_{k=m}^{N-1}(\widehat{g}_k + \widehat{h}_k) + \check{\nu}_N, \ \check{\nu}_N = \nu_N + \sum_{k=m}^{N-1}(\eta_k - (\widehat{g}_k + \widehat{h}_k)).$$

By construction, $\check{\nu}_N \in \dot{S}'_{q-N}(Z_n)$ and satisfies N moment conditions. By Theorem 3.6, there exists $f_N \in \dot{S}'_{q-N}(\mathbb{R}^n)$ such that $\widehat{f}_N = \nu_N$. Let $f = \sum_{k=m}^{N-1}(g_k + h_k) + f_N$. By construction, $\widehat{f} = \mu$. Therefore, f is independent of N. Obviously, $f \in \dot{S}'_{q-m}(\mathbb{R}^n)$. Taking into account the properties of g_k and h_k, we see that f satisfies (4.6). Note that the terms in the expansion of h_k are of the type $H_{kj}(\beta)r^{-(j+n)}, k \leq j \leq N-1$, and by increasing N we do not change the already existing terms in the expansion of f. Since $N \geq m+1$ is arbitrary, $f \in U_m$ and the proof is finished. $\square$

5. Auxiliary results

LEMMA 5.1. *Fix $q < 1$ and consider a function $f(x,y) \in C([-1,1] \times \mathbb{R})$ such that $(1+|y|)^q|f(x,y)| \to 0$ uniformly with respect to $x \in [-1,1]$ as $|y| \to \infty$. One has*

$$(5.1) \qquad g(\lambda) := \int_{-1}^{1} f(x, \lambda x) dx = o(|\lambda|^{-q}), \ |\lambda| \to \infty.$$

PROOF. Fix $\epsilon > 0$ and find $R > 0$ such that

$$(5.2) \qquad (1+|y|)^q|f(x,y)| \le \epsilon, \ x \in [-1,1], \ |y| \ge R.$$

One has

$$(5.3) \quad \begin{aligned}(1+|\lambda|)^q g(\lambda) =& (1+|\lambda|)^q \int_{\substack{x \in [-1,1] \\ |x| < R/|\lambda|}} f(x, \lambda x) dx \\ & + (1+|\lambda|)^q \int_{\substack{x \in [-1,1] \\ |x| > R/|\lambda|}} f(x, \lambda x) dx =: I_1(x) + I_2(x), \ \lambda \ne 0.\end{aligned}$$

Clearly,

$$(5.4) \qquad |I_1(\lambda)| \le \frac{2R(1+|\lambda|)^q}{|\lambda|} \max_{|x| \le 1, |y| \le R} |f(x,y)| \to 0, \ |\lambda| \to \infty.$$

Furthermore, if $|\lambda| > R$,

$$(5.5) \qquad |I_2(\lambda)| \le 2\epsilon(1+|\lambda|)^q \int_{R/|\lambda|}^{1} \frac{dx}{(1+|\lambda|x)^q} \le 2\epsilon \frac{1+|\lambda|^{-1}}{1-q}.$$

Combining (5.3)–(5.5) proves (5.1). □

LEMMA 5.2. *Fix $q < 1$, an integer $m > 0$, and consider a function $f(x,y) \in C^m([-1,1] \times \mathbb{R})$ such that*
1. *$(1+|y|)^{q+m}|\partial_x^l f(x,y)| \to 0$ uniformly with respect to $x \in [-1,1]$ as $|y| \to \infty$ for $0 \le l \le m$, and*
2. *$\int f(x,y) y^j dy \equiv 0$ for $0 \le j \le m-1$.*

Then

$$(5.6) \qquad g(\lambda) := \int_{-1}^{1} f(x, \lambda x) dx = o(|\lambda|^{-(q+m)}), \ |\lambda| \to \infty.$$

PROOF. Define

$$(5.7) \qquad F(x,y) = \int_{-\infty}^{y} \frac{(y-t)^{m-1}}{(m-1)!} f(x,t) dt.$$

Clearly,

$$(5.8) \quad \begin{aligned} & \partial_y^m F(x,y) = f(x,y), \\ & (1+|y|)^{q+j} \partial_x^l \partial_y^j F(x,y) \to 0 \text{ as } |y| \to \infty, \ 0 \le j, l \le m. \end{aligned}$$

Since $(d/dx)F(x,\lambda x) = \partial_x F(x,\lambda x) + \lambda \partial_y F(x,\lambda x)$, we have using (5.8):

$$g(\lambda) = \int_{-1}^{1} \partial_y^m F(x,\lambda x) dx$$

(5.9)
$$= \frac{1}{\lambda} \int_{-1}^{1} \frac{d}{dx}\left[\partial_y^{m-1} F(x,\lambda x)\right] dx - \frac{1}{\lambda} \int_{-1}^{1} \partial_x \partial_y^{m-1} F(x,\lambda x) dx$$

$$= o(|\lambda|^{-(q+m)}) - \frac{1}{\lambda}\int_{-1}^{1} \partial_x \partial_y^{m-1} F(x,\lambda x) dx, \ |\lambda| \to \infty.$$

Repeating this process m times we get

(5.10) $$g(\lambda) = o(|\lambda|^{-(q+m)}) + \frac{1}{(-\lambda)^m} \int_{-1}^{1} \partial_x^m F(x,\lambda x) dx, \ |\lambda| \to \infty.$$

Applying Lemma 5.1 to the integral in the last equation and using (5.8) with $l = m, j = 0$ we finish the proof. □

LEMMA 5.3. *Fix $m \geq 0$ and consider a sequence of functions f_n which satisfy conditions of either Lemma 5.1 if $m = 0$ or Lemma 5.2 if $m > 0$. Suppose*

(5.11) $$\max_{(x,y) \in [-1,1] \times \mathbb{R}} (1+|y|)^{q+m} |\partial_x^l f_n(x,y)| \to 0, \ n \to \infty, \ 0 \leq l \leq m.$$

Define g_n by formula (5.1). Then

(5.12) $$\max_\lambda (1 + |\lambda|)^{q+m} |g_n(\lambda)| \to 0, \ n \to \infty.$$

PROOF. Suppose first $m = 0$. Fix any $R > 0$. Obviously, $\max_{|\lambda| \leq R} |g_n(\lambda)| \to 0, \ n \to \infty$. Moreover, (5.11) implies that the quantities

(5.13) $$\max_{n \geq N} \max_{|x| \leq 1, |y| \leq R} |f_n(x,y)|, \ \max_{n \geq N} \max_{|x| \leq 1, |y| \geq R} (1+|y|)^q |f_n(x,y)|$$

can be made as small as we like by choosing N sufficiently large. The assertion now follows from (5.3)–(5.5). If $m > 0$, define F_n by (5.7). Then

(5.14) $$\max_{(x,y) \in [-1,1] \times \mathbb{R}} (1+|y|)^{q+j} |\partial_x^l \partial_y^j F_n(x,y)| \to 0, \ n \to \infty, \ 0 \leq j, l \leq m.$$

The assertion now follows from (5.9), (5.10), and the first part of the lemma. □

LEMMA 5.4. *Let $t > 0$. Then the Radon transform $R: \dot{S}_{t+(n-1)}(\mathbb{R}^n) \to \dot{S}_t(Z_n)$ is continuous.*

PROOF. Fix $f \in \dot{S}_{t+(n-1)}(\mathbb{R}^n)$. Integrating in the plane $\alpha \cdot x = p$ in polar coordinates we have for $\mu = Rf$:

(5.15) $$\mu(\alpha, p) = \int_0^\infty \int_{\substack{|\omega|=1 \\ \alpha \cdot \omega = 0}} f(p\alpha + s\omega) d\omega \, s^{n-2} ds.$$

Since $f(x) = o(|x|^{-(t+n-1)}), |x| \to \infty$, the integral in (5.15) converges absolutely and $\mu(\alpha, p) = o(|p|^{-t}), p \to \infty$, uniformly with respect to $\alpha \in S^{n-1}$. For convenience, choose a coordinate system such that $\alpha_0 = (1, 0, \ldots, 0)$. Similarly to [**Kat97**], p. 856, we have for α close α_0:

(5.16) $$\mu(\alpha, p) = \frac{1}{\sqrt{1-|\omega|^2}} \int_{\mathbb{R}^{n-1}} f\left(\frac{p - \omega \cdot y}{\sqrt{1-|\omega|^2}}, y\right) dy,$$

where $\alpha = (\sqrt{1-|\omega|^2}, \omega), |\omega| < 1, \omega \in \mathbb{R}^{n-1}$, and (cf. (2.3))

$$(5.17) \qquad \partial_\alpha^\vartheta \mu(\alpha = \alpha_0, p) = \sum_{m \leq |\vartheta|} \int_{\mathbb{R}^{n-1}} \mathcal{P}_m(p,y) \partial_p^m f(p,y) dy,$$

where $\mathcal{P}_m$'s are homogeneous polynomials of $(p,y) \in \mathbb{R}^n$, $\deg(\mathcal{P}_m) = m$. Since $\partial_x^\gamma f(x) = o(|x|^{-(t+|\gamma|+n-1)})$, all the integrals in (5.17) are absolutely convergent. By the same reason we can differentiate with respect to p on both sides of (5.17) any number of times. Hence, $\mu \in C^\infty(Z_n)$ and

$$(5.18) \qquad \partial_p^l \partial_\alpha^\vartheta \mu(\alpha = \alpha_0, p) = \sum_{l \leq m \leq |\vartheta|+l} \int_{\mathbb{R}^{n-1}} \mathcal{Q}_{m-l}(p,y) \partial_p^m f(p,y) dy,$$

where $\mathcal{Q}_m$'s are homogeneous polynomials, $\deg(\mathcal{Q}_m) = m$. This implies

(5.19)
$$|\partial_p^l \partial_\alpha^\vartheta \mu(\alpha = \alpha_0, p)| \leq c \sum_{l \leq m \leq |\vartheta|+l} \sum_{j=0}^{m-l} |p|^{m-l-j} \int_0^\infty \frac{c_m(\sqrt{p^2+u^2})u^j}{(1+\sqrt{p^2+u^2})^{t+n-1+m}} u^{n-2} du$$
$$\leq c_m(|p|) O(|p|^{-(t+l)}) = o(|p|^{-(t+l)}), \ |p| \to \infty,$$

where $c > 0$ and $c_m(r) = \max_{|\gamma|=m, |x| \geq r}(1+|x|)^{q+m}|\partial_x^\gamma f(x)|$. Clearly, estimates (5.19) hold uniformly for all $\alpha_0 \in S^{n-1}$. Hence $\mu \in \dot{S}_t(Z_n)$. Our argument also proves that $R : \dot{S}_{t+(n-1)}(\mathbb{R}^n) \to \dot{S}_t(Z_n)$ is sequentially continuous. The continuity now follows from the closed graph theorem. $\square$

The author wants to use this opportunity to correct an unfortunate mistake in his earlier paper [**Kat97**]. The first and third displayed formulas on p. 856 in [**Kat97**] should be replaced by equations (5.16) and (5.17), respectively. However, all the conclusions made from these equations are still valid.

LEMMA 5.5. *Let $0 < q < 1$. Then the map $\mathcal{H}_j : \dot{S}_q(\mathbb{R}^n) \to \dot{S}_q(\mathbb{R}^n), 1 \leq j \leq n$, is continuous. Moreover, if $n = 1$, $\mathcal{H}$ is an isomorphism onto $\dot{S}_q(\mathbb{R})$.*

PROOF. Assume first that $n = 1$. Take any $f \in \dot{S}_q(\mathbb{R})$ and show $\mathcal{H}f \in \dot{S}_q(\mathbb{R})$. Denote

$$(5.20) \qquad c_m(r) = \max_{|x| \geq r}(1+|x|)^{q+m}|f^{(m)}(x)|.$$

By construction, c_m is continuous, decreasing, and $c_m(r) \to 0$ as $r \to +\infty$. Furthermore,

(5.21)
$$\pi \partial_x^m \mathcal{H} f(x) = \int_{-\infty}^{\infty} \frac{f^{(m)}(x-y)}{y} dy$$
$$= \int_{-x/2}^{x/2} \frac{f^{(m)}(x-y) - f^{(m)}(x)}{y} dy$$
$$+ \left(\sum_{k=1}^{m} (-1)^{k+1} (k-1)! \frac{f^{(m-k)}(x/2)}{(x/2)^k} + (-1)^m m! \int_{x/2}^{\infty} \frac{f(x-y)}{y^{m+1}} dy \right)$$
$$+ \left(\sum_{k=1}^{m} (k-1)! \frac{f^{(m-k)}(3x/2)}{(x/2)^k} + (-1)^m m! \int_{-\infty}^{-x/2} \frac{f(x-y)}{y^{m+1}} dy \right)$$
$$=: I_1(x) + I_2(x) + I_3(x).$$

Suppose, for example, $x \to +\infty$. By (5.20),

(5.22)
$$|I_1(x)| \le x \max_{y \in [x/2, 3x/2]} |f^{(m+1)}(y)| \le \frac{x}{(1+(x/2))^{q+m+1}} c_{m+1}(x/2)$$
$$= o(x^{-(q+m)}), \quad x \to +\infty.$$

Furthermore,

(5.23)
$$|I_2(x)| \le \sum_{k=1}^{m} \frac{(k-1)! c_{m-k}(x/2)}{(1+(x/2))^{q+m-k}(x/2)^k} + m! \int_{x/2}^{\infty} \frac{c_0(|x-y|)}{y^{m+1}(1+|x-y|)^q} dy$$
$$\le o(x^{-(q+m)}) + \frac{m!}{x^{q+m}} \int_{1/2}^{\infty} \frac{c_0(x|1-s|)}{s^{m+1}|1-s|^q} ds, \quad x \to \infty.$$

Consider the last integral in (5.23). We have to show that it is $o(1), x \to +\infty$. Find $\delta, 0 < \delta < 1/2$, such that

(5.24)
$$c_0(0) \int_{1-\delta}^{1+\delta} \frac{ds}{s^{m+1}|1-s|^q} < \frac{\epsilon}{3}.$$

Since $c_m(x) \to 0$ as $x \to \infty$, we can find $R > 0$ such that

(5.25)
$$\int_{1+\delta}^{\infty} \frac{c_0(x|1-s|)}{s^{m+1}|1-s|^q} ds \le c_0(\delta x) \int_{1+\delta}^{\infty} \frac{ds}{s^{m+1}|1-s|^q} ds < \frac{\epsilon}{3}, \quad x > R,$$
$$\int_{1/2}^{1-\delta} \frac{c_0(x|1-s|)}{s^{m+1}|1-s|^q} ds \le c_0(\delta x) \int_{1/2}^{1-\delta} \frac{ds}{s^{m+1}|1-s|^q} ds < \frac{\epsilon}{3}, \quad x > R.$$

Combining (5.24) and (5.25) gives

(5.26)
$$\int_{1/2}^{\infty} \frac{c_0(x|1-s|)}{s^{m+1}|1-s|^q} ds < \epsilon, \quad x > R.$$

The quantity $I_3(x)$ can be estimated in the same fashion.

Combining (5.21)–(5.26) proves that $\mathcal{H}(\dot{S}_q(\mathbb{R})) \subset \dot{S}_q(\mathbb{R})$. Sequential continuity of $\mathcal{H}$ follows from (5.22), (5.23), and a similar estimate for $I_3(x)$. The continuity follows from the closed graph theorem. By applying the Fourier transform we see that $\mathcal{H}$ is injective. Since the injection $C_0^{\infty}(\mathbb{R}) \hookrightarrow \dot{S}_q(\mathbb{R})$ has a dense image and $-\mathcal{H}^2$ is the identity on $C_0^{\infty}(\mathbb{R})$, we conclude that $\mathcal{H}$ is an isomorphism onto $\dot{S}_q(\mathbb{R})$.

Suppose now $n > 1$. For any $f \in \dot{S}_q(\mathbb{R}^n)$ we have

(5.27)
$$\frac{\pi^{(n+1)/2}}{\Gamma((n+1)/2)} \partial_x^\gamma \mathcal{H}_j f(x) = \int_{|y|<|x|/2} \frac{y_j(\partial_x^\gamma f(x-y) - \partial_x^\gamma f(x))}{|y|^{n+1}} dy$$
$$+ \int_{|y|>|x|/2} \frac{y_j \partial_x^\gamma f(x-y)}{|y|^{n+1}} dy =: I_1(x) + I_2(x).$$

Rewriting $I_1(x)$ in spherical coordinates and then estimating it similarly to (5.22), we find:

(5.28)
$$I_1(x) = \int_{S^{n-1}} \left(\int_0^{|x|/2} \frac{\alpha_j(\partial_x^\gamma f(x-r\alpha) - \partial_x^\gamma f(x))}{r} dr \right) d\alpha,$$
$$|I_1(x)| \leq |S^{n-1}| \frac{|x|}{2} \frac{c_{|\gamma|+1}(|x|/2)\sqrt{n}}{(1+|x|/2)^{q+|\gamma|+1}} = o(|x|^{-(q+|\gamma|)}), \quad |x| \to \infty,$$

where $c_m(r) = \max_{|\gamma|=m, |x| \geq r}(1+|x|)^{q+m}|\partial_x^\gamma f(x)|$. To estimate I_2, we integrate by parts first (say, with respect to y_1 if $\gamma_1 > 0$):

(5.29)
$$I_2(x) = \int_{|y|=|x|/2} \frac{y_j \partial_x^{(\gamma_1-1, \gamma_2, \ldots, \gamma_n)} f(x-y)}{|y|^{n+1}} \frac{y_1}{|y|} dy$$
$$+ \int_{|y|>|x|/2} \partial_x^{(\gamma_1-1, \gamma_2, \ldots, \gamma_n)} f(x-y) \partial_{y_1}\left(\frac{y_j}{|y|^{n+1}}\right) dy =: I_{21}(x) + I_{22}(x).$$

Just as in (5.23), the boundary term is $o(|x|^{-(q+|\gamma|)})$. Integrating by parts $|\gamma|$ times, we will get a collection of boundary terms of order $o(|x|^{-(q+|\gamma|)})$ and a volume integral

(5.30)
$$I_3(x) := \int_{|y|>|x|/2} f(x-y) \partial_y^\gamma \left(\frac{y_j}{|y|^{n+1}}\right) dy.$$

Since $\partial_y^\gamma \left(\frac{y_j}{|y|^{n+1}}\right) = A(\alpha) r^{-(n+|\gamma|)}$ for some $A \in C^\infty(S^{n-1})$, we have

(5.31)
$$I_3(x) = \int_{S^{n-1}} A(\alpha) \left(\int_{|x|/2}^\infty \frac{f(x-r\alpha)}{r^{|\gamma|+1}} dr \right) d\alpha.$$

The integral with respect to r can be estimated as follows:

(5.32)
$$\left| \int_{|x|/2}^\infty \frac{f(x-r\alpha)}{r^{|\gamma|+1}} dr \right| \leq \int_{|x|/2}^\infty \frac{c_0(||x|-r|)}{r^{|\gamma|+1}(1+||x|-r|)^q} dr,$$

and the same argument as in the case $n = 1$ shows that the right-hand side of the last equation is $o(|x|^{-(q+|\gamma|)})$. The continuity is now obvious. $\square$

LEMMA 5.6. *Fix $t, 1 < t < n, n \geq 2$. Then the map $\mathcal{I}: \dot{S}_t(\mathbb{R}^n) \to \dot{S}_{t-1}(\mathbb{R}^n)$ is continuous.*

PROOF. One has

(5.33)
$$\frac{2\pi^{n/2}\Gamma(1/2)}{\Gamma((n-1)/2)} \partial_x^\gamma (\mathcal{I}f)(x) = \int_{|y|<|x|/2} \frac{\partial_x^\gamma f(x-y)}{|y|^{n-1}} dy + \int_{|y|>|x|/2} \frac{\partial_x^\gamma f(x-y)}{|y|^{n-1}} dy$$
$$=: I_1(x) + I_2(x).$$

Similarly to (5.28), we establish $I_1(x) = o(|x|^{-(t-1+|\gamma|)})$. Further,

(5.34)
$$I_2(x) = \int_{|z|<|x|/2} \frac{\partial_z^\gamma f(z)}{|x-z|^{n-1}} dz + \int_{\substack{|y|>|x|/2 \\ |x-y|>|x|/2}} \frac{\partial_x^\gamma f(x-y)}{|y|^{n-1}} dy$$
$$=: I_{21}(x) + I_{22}(x).$$

The first integral can be estimated as follows. Suppose first $\gamma = 0$.

(5.35)
$$|I_{21}(x)| \leq \left(\frac{2}{|x|}\right)^{n-1} \int_{|z|<|x|/2} |f(z)|dz \leq \left(\frac{2}{|x|}\right)^{n-1} |S^{n-1}| \int_0^{|x|/2} \frac{c_0(r)}{(1+r)^t} r^{n-1} dr.$$

Since

(5.36)
$$\int_0^{|x|/2} \frac{c_0(r)}{(1+r)^t} r^{n-1} dr \leq c_0(0) \int_0^{\sqrt{|x|/2}} r^{n-1-t} dr + c_0\left(\sqrt{|x|/2}\right) \int_0^{|x|/2} r^{n-1-t} dr$$
$$\leq \frac{c_0(0)}{n-t} \left(\frac{|x|}{2}\right)^{(n-t)/2} + \frac{c_0\left(\sqrt{|x|/2}\right)}{n-t} \left(\frac{|x|}{2}\right)^{n-t},$$

combining with (5.35) gives $I_{21}(x) = o(|x|^{-(t-1)})$. If $\gamma \neq 0$, integrating by parts similarly to (5.29), (5.30), we obtain a collection of boundary terms and a volume integral of the type:

(5.37)
$$I_{bdry}(x) = \int_{|z|=|x|/2} (\partial_z^{\gamma_1} f(z)) \left(\partial_z^{\gamma_2} \frac{1}{|x-z|^{n-1}}\right) \frac{z_j}{|z|} dz,$$
$$\gamma_1 + \gamma_2 + (0,\ldots,\overset{j^{\text{th}}\text{place}}{1},\ldots,0) = \gamma,$$
$$I_{vol}(x) = \int_{|z|<|x|/2} f(z) \left(\partial_z^\gamma \frac{1}{|x-z|^{n-1}}\right) dz.$$

Clearly, $I_{bdry}(x) = o(|x|^{-(t-1+|\gamma|)})$. Since

(5.38) $\qquad \left(\partial_z^\gamma \frac{1}{|x-z|^{n-1}}\right) = O(|x|^{-(n-1+|\gamma|)}), \; |z| < |x|/2, \; |x| \to \infty,$

the same argument as in (5.35), (5.36), proves $I_{vol}(x) = o(|x|^{-(t-1+|\gamma|)})$.

The second integral in (5.34) can be estimated similarly to [**Hel99**], p. 167:

(5.39)
$$|I_{22}(x)| \leq \frac{c_\gamma(|x|/2)}{|x|^{t-1+|\gamma|}} \int_{\substack{|u|>1/2 \\ |x_0-u|>1/2}} \frac{du}{(1+|x_0-u|)^{t+|\gamma|}|u|^{n-1}}, \; x_0 = x/|x| \in S^{n-1}.$$

The integral in (5.39) is convergent because $t > 1$ by assumption. Therefore, $I_2(x) = o(|x|^{-(t-1+|\gamma|)})$. Combining (5.33)–(5.39) proves sequential continuity of $\mathcal{I}$. By the closed graph theorem $\mathcal{I}$ is continuous. $\square$

LEMMA 5.7. *Fix $q, 0 < q < 1$, and $0 \leq l \leq m$. The operator Λ^m can be extended to a continuous map*

(5.40) $\qquad \Lambda^m : \dot{S}_{q-l}(\mathbb{R}^n) \to \dot{S}_{q-l+m}(\mathbb{R}^n).$

PROOF. Using (2.7) we can formally represent Λ^m as follows

$$\Lambda^m = \sum_{\substack{|\alpha|=l \\ |\alpha+\beta|=m}} \mathcal{Q}_\beta(\partial_x)\mathcal{H}^{\alpha+\beta}\mathcal{P}_\alpha(\partial_x), \quad \mathcal{H}^\gamma := \mathcal{H}_1^{\gamma_1}\ldots\mathcal{H}_n^{\gamma_n}, \tag{5.41}$$

where $\mathcal{P}_\alpha$ and $\mathcal{Q}_\beta$ are certain homogeneous polynomials of degree $|\alpha|$ and $|\beta|$, respectively. On $C_0^\infty(\mathbb{R}^n)$ the relation (5.41) coincides with the classical definition of Λ^m because the Riesz transforms and differentiation commute. Since $C_0^\infty(\mathbb{R}^n)$ is dense in $\dot{S}_{q-l}(\mathbb{R}^n)$ and the maps

$$\begin{aligned} \mathcal{P}_\alpha(\partial_x) : \dot{S}_{q-l}(\mathbb{R}^n) \to \dot{S}_q(\mathbb{R}^n), \quad \mathcal{H}_j : \dot{S}_q(\mathbb{R}^n) \to \dot{S}_q(\mathbb{R}^n), \\ \mathcal{Q}_\beta(\partial_x) : \dot{S}_q(\mathbb{R}^n) \to \dot{S}_{q+m-l}(\mathbb{R}^n), \end{aligned} \tag{5.42}$$

are continuous, the desired assertion follows. □

LEMMA 5.8. *Let $f \in \dot{S}_q(\mathbb{R}^n), g \in \dot{S}_{q_1}(\mathbb{R}^n), 0 < q, q_1 < 1; q + q_1 > 1$. Then $(\Lambda^{n-1}f, g) = (f, \Lambda^{n-1}g)$.*

PROOF. Find $f_k, g_k \in C_0^\infty(\mathbb{R}^n)$ such that $f_k \to f$ in $\dot{S}_q(\mathbb{R}^n)$ and $g_k \to g$ in $\dot{S}_{q_1}(\mathbb{R}^n)$. By assumption,

$$|f_k(x)| \leq c_1(1+|x|)^{-q}, \tag{5.43}$$

$$|g_k(x)| \leq c_2(1+|x|)^{-q_1}, \tag{5.44}$$

for some $c_{1,2} > 0$ and all $k \geq 1$. Since $\Lambda^{n-1} : \dot{S}_q(\mathbb{R}^n) \to \dot{S}_{q+n-1}(\mathbb{R}^n)$ is continuous,

$$|(\Lambda^{n-1}f_k)(x)| \leq c_3(1+|x|)^{-(q+n-1)}, \tag{5.45}$$

$$|(\Lambda^{n-1}g_k)(x)| \leq c_4(1+|x|)^{-(q_1+n-1)}, \tag{5.46}$$

for some $c_{3,4} > 0$ and all $k \geq 1$. Since $q + q_1 > 1$, (5.44), (5.45), and the Lebesgue dominated convergence theorem imply

$$(\Lambda^{n-1}f, g) = \lim_{k\to\infty}(\Lambda^{n-1}f_k, g_k). \tag{5.47}$$

Similarly, (5.43), (5.46), and the Lebesgue dominated convergence theorem imply

$$(f, \Lambda^{n-1}g) = \lim_{k\to\infty}(f_k, \Lambda^{n-1}g_k). \tag{5.48}$$

For compactly supported function $(\Lambda^{n-1}f_k, g_k) = (f_k, \Lambda^{n-1}g_k)$, and the result follows from (5.47), (5.48). □

References

[GGV66] I.M. Gelfand, M.I. Graev, and N.Ya. Vilenkin, *Generalized functions. Volume 5: Integral geometry and representation theory*, Academic Press, New York, 1966.

[Gon84] F. B. Gonzalez, *Radon transforms on Grassmann manifolds*, Ph.D. thesis, M.I.T., 1984.

[Gon87] F. B. Gonzalez, *Radon transforms on Grassmann manifolds*, J. Funct. Anal. **71** (1987), 339–362.

[GR94] I. S. Gradshteyn and I. M. Ryzhik, *Table of integrals, series, and products*, 5th ed., Academic Press, Boston, 1994.

[Hel65] S. Helgason, *The Radon transform on Euclidean spaces, compact two-point homogeneous spaces and Grassmann manifolds*, Acta Mathematica **113** (1965), 153–179.

[Hel82] S. Helgason, *Ranges of Radon transforms*, Proceedings of Symposia in Applied Mathematics, Vol. 27 (Providence, RI) (L. Shepp, ed.), Amer. Math. Soc., 1982, pp. 63–70.

[Hel99] S. Helgason, *The Radon transform*, 2nd ed., Birkhauser, Boston, 1999.

[Her83] A. Hertle, *Continuity of the Radon transform and its inverse on Euclidean spaces*, Math. Z. **184** (1983), 165–192.

[Her84]   A. Hertle, *On the range of the Radon transform and its dual*, Math. Ann. **267** (1984), 91–99.
[Hor66]   J. Horvath, *Topological vector spaces and distributions. volume i*, Addison-Wesley, Reading, Mass., 1966.
[Hus65]   Taqdir Husain, *The open mapping and closed graph theorems in topological vector spaces*, Oxford University Press, 1965.
[Kat97]   A. Katsevich, *Range of the Radon transform on functions which do not decay fast at infinity*, SIAM Journal of Mathematical Analysis **28** (1997), no. 4, 852–866.
[Kat01]   A. Katsevich, *New range theorems for the dual Radon transform*, Transactions of the American Mathematical Society **353** (2001), 1089–1102.
[Lan93]   S. Lang, *Real and functional analysis*, 3rd ed., Springer-Verlag, New York, 1993.
[Lou84]   A. K. Louis, *Orthogonal function series expansions and the null space of the Radon transform*, SIAM J. Math. Anal. **15** (1984), 621–633.
[LP70]   P. Lax and R. Phillips, *The Paley-Wiener theorem for the Radon transform*, Comm. Pure Appl. Math. **23** (1970), 409–424.
[Lud60]   D. Ludwig, *The Radon transform on Euclidean spaces*, Comm. Pure Appl. Math. **23** (1960), 49–81.
[PBM88]   A. P. Prudnikov, Yu. A. Brychkov, and O. I. Marichev, *Integrals and series. Volume 2. Special functions*, Gordon and Breach, New York, 1988.
[Ram95]   A.G. Ramm, *The Radon transform is an isomorphism between $L^2(B)$ and $H_e(Z_a)$*, Appl. Math. Lett. **8** (1995), 25–29.
[Ram96]   A.G. Ramm, *Inversion formula and singularities of the solution for the backprojection operator in tomography*, Proc. Amer. Math. Soc. **124** (1996), 567–577.
[RK96]   A. Ramm and A. Katsevich, *The Radon transform and local tomography*, CRC Press, Boca Raton, Florida, 1996.
[SM88]   D. C. Solmon and W. Madych, *A range theorem for the Radon transform*, Proceedings of the Amer. Math. Soc. **104** (1988), 79–85.
[Sol87]   D.C. Solmon, *Asymptotic formulas for the dual Radon transform and applications*, Math. Z. **195** (1987), 321–343.
[SSW77]   K. Smith, D. Solmon, and S. Wagner, *Practical and mathematical aspects of the problem of reconstructing objects from radiographs*, Bull of Amer. Math. Soc. **83** (1977), 1227–1270.
[Tre67]   F. Treves, *Topological vector spaces, distributions, and kernels*, Academic Press, New York, 1967.

DEPARTMENT OF MATHEMATICS, UNIVERSITY OF CENTRAL FLORIDA, ORLANDO, FL 32816
E-mail address: akatsevi@pegasus.cc.ucf.edu

Moment Conditions *Indirectly* Improve Image Quality

S. K. Patch

ABSTRACT. Projecting axial computed tomography (CT) data onto the range of the Radon transform does not improve image quality. However, by monitoring the degree to which CT data satisfies the Helgason-Ludwig range conditions it may be possible to detect failing equipment before serious image artifacts are noticeable.

Introduction

Clinical computed tomography (CT) scanners correct for many sources of error prior to reconstruction. Post-processing is used as well to correct for other measurement errors. Nevertheless, image artifacts sometimes remain. Furthermore, system "down time" is extremely expensive for hospitals. Therefore, it is critical that failing equipment be diagnosed as soon as possible so that repair can be done before image quality (IQ) becomes unacceptable.

Unfortunately, we see in Section 1 that without *a priori* information, enforcing the Helgason-Ludwig moment conditions [1] does nothing to improve IQ, even though measurement errors may be reduced. In Section 2 we see that the same H-L conditions may aid technicians in diagnosing hardware failures. Numerical results are presented in Section 3.

We use the following definitions and notation, restricting ourselves to clinical CT data sets. The imaging object's linear attenuation coefficient (LAC), f, is real-valued, bounded, and of compact support. Parallel beam data (more precisely the $\ln$ of the data) is a line integral of the LAC:

$$(0.1) \qquad Rf(\theta, s) = \int_{x \cdot \theta = s} f(x) dx \quad \text{where} \quad \theta \equiv (cos\phi, sin\phi)$$

See Figure 1 and notice that $Rf \in L^2(S^1 \times \mathbb{R}^1)$ and also that R is even,

$$(0.2) \qquad Rf(\theta, s) = Rf(-\theta, -s)$$

The H-L Range Conditions constrain moments of Rf, defined as follows:

$$(0.3) \qquad \text{Mom}(m, k) \equiv \int_0^{2\pi} e^{ik\phi} \int_\mathbb{R} s^m Rf(\theta, s) \, ds \, d\phi$$

1991 *Mathematics Subject Classification.* Primary 44a12; Secondary 94a08.
Key words and phrases. tomography, range conditions.

© 2001 American Mathematical Society

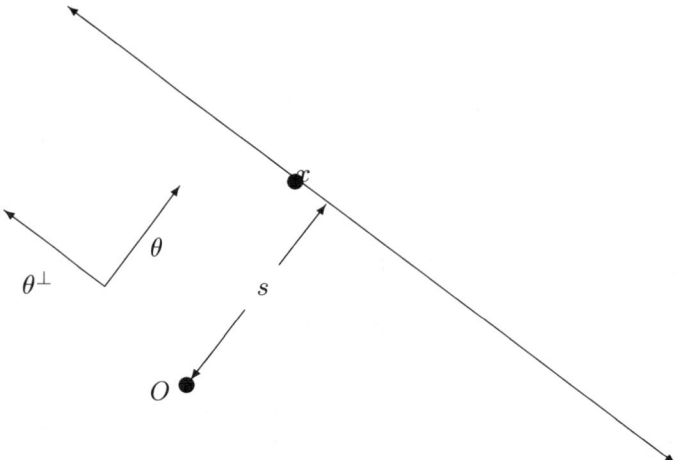

FIGURE 1. $Rf(\theta, s)$ is the line integral of f along the line $\{x \mid x = s\theta + t\theta^\perp, t \in \mathbb{R}^1\}$.

Evenness forces every other moment to be identically zero

$$\begin{aligned}
\text{Mom}(m, k) &\equiv \int_0^{2\pi} e^{ik\phi} \int_{\mathbb{R}} s^m Rf(\theta, s) \, ds \, d\phi \\
&= \int_0^{2\pi} e^{ik(\phi+\pi)} \int_{\mathbb{R}} (-s)^m Rf(-\theta, -s) \, ds \, d\phi \\
&= (-1)^{(m+k)} \text{Mom}(m, k) \quad \text{which implies}
\end{aligned}$$

(0.4) $\quad \text{Mom}(m, k) \equiv 0 \quad \text{for } (m + k) \text{ odd}$

Additionally, the H-L range conditions imply that the Fourier serious expansion of the mth moment of Rf with respect to s has only $(2m + 1)$ nonzero Fourier coefficients. Equivalently, for $|k| > m \geq 0$,

(0.5) $\quad\quad\quad \text{Cond}(m, k) \quad\quad 0 \equiv \text{Mom}(m, k)$

These conditions are sufficient for clinical CT data, which is always real-valued. Finally, note that we need only concern ourselves with moments for $k \geq 0$, since the moments are Hermitian symmetric with respect to k,

(0.6) $\quad\quad\quad \text{Mom}(m, -k) = conj(\text{Mom}(m, k))$

1. Enforcing Range Conditions Does Not Directly Improve IQ

We assume no *a priori* knowledge about our measured data. A natural approach is to make data consistent by enforcing the H-L conditions [2]. We explain why this has not been successful in practice [3]. Readers familiar with this may prefer to skip ahead to Section 2 where we show how moment conditions can be used to monitor CT systems for faulty detector channels.

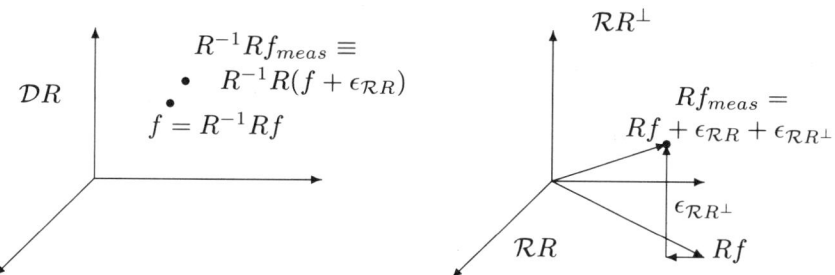

FIGURE 2. The Radon transform has a unique inverse acting upon $\mathcal{R}R$ However, measured CT data contains errors, $\epsilon = \epsilon_{\mathcal{R}R} + \epsilon_{\mathcal{R}R^\perp}$ R^{-1} kills $\epsilon_{\mathcal{R}R^\perp}$ so only measurements errors $\epsilon_{\mathcal{R}R}$ contribute to errors in reconstructing f. Both sides of this figure depict infinite dimensional function spaces; the vertical axis on the right denotes $\mathcal{R}R^\perp$

Break measured CT data into three components: true data, errors in the range of the Radon transform, and errors outside the range of the Radon transform.

$$
\begin{align}
(1.1)\quad Rf_{meas} &= \text{measured CT data } \in L^2(S^1 \times \mathbb{R}^1) \\
(1.2)\quad &= Rf_{true} + \epsilon \quad \text{where} \\
(1.3)\quad \epsilon &= \epsilon_{\mathcal{R}R} + \epsilon_{\mathcal{R}R^\perp} \\
(1.4)\quad \epsilon_{\mathcal{R}R} &\in \mathcal{R}R \\
(1.5)\quad \epsilon_{\mathcal{R}R^\perp} &\in \mathcal{R}R^\perp
\end{align}
$$

where $\mathcal{R}$ denotes the range. See Figure 2. Enforcing the H-L conditions projects Rf_{meas} onto $\mathcal{R}R$, removing the $\epsilon_{\mathcal{R}R^\perp}$ error term. However, measurement errors which were in the range of the Radon transform, $\epsilon_{\mathcal{R}R} \in \mathcal{R}R$, remain so we are left to reconstruct

$$(1.6)\qquad Rf_{recon} = Rf_{true} + \epsilon_{\mathcal{R}R}$$

Fact: Orthogonally projecting measured data onto the range of the Radon transform by enforcing the H-L range conditions leaves the reconstructed image invariant [3].

Details. This follows from the fact that $\mathcal{R}(R)^\perp = \mathcal{N}(R^*)$ Although clinical CT data is typically reconstructed by a filter-backprojection method, the orders may be reversed. In this case, the invariance is made obvious. The backprojection of each line through a point $\mathbf{x}$ can be rewritten as a convolution.

$$(1.7)\qquad R^*Rf(\mathbf{x}) = \int_{S^1} Rf(\mathbf{x}\cdot\theta,\theta)\, d\theta = \int_{\mathbb{R}^2} \frac{f(\mathbf{x}+\mathbf{y})}{|\mathbf{y}|}\, d\mathbf{y}$$

which can be pulled apart by using a Riesz potential and the pseudo-differential operator Λ, satisfying

$$\Lambda \equiv \frac{1}{2\pi}\sqrt{-\Delta} \quad \text{equivalently,} \quad (\Lambda f)\widehat{}(\psi) \equiv |\psi|\hat{f}(\psi)$$

Notice that Λ is just a filter that emphasizes high frequency components. The point is that convolving a function with R_{pot} is the inverse of hitting that function with Λ. An inversion formula for 2D is written below

$$\frac{\Lambda}{2} R^* R f(\mathbf{x}) = f(\mathbf{x}) \tag{1.8}$$

Furthermore, backprojection annihilates the very same errors that we would remove by enforcing the H-L conditions since $\mathcal{N} R^* = \mathcal{R} R^\perp$ Therefore, the reconstructed image is not improved by enforcing range conditions.

$$\frac{\Lambda}{2} R^* R f_{meas}(\mathbf{x}) = \frac{\Lambda}{2} R^* \left(R f_{true} + \epsilon_{\mathcal{R}R} + \epsilon_{\mathcal{R}R^\perp} \right)(\mathbf{x}) \tag{1.9}$$

$$= \frac{\Lambda}{2} R^* \left(R f_{true} + \epsilon_{\mathcal{R}R} \right)(\mathbf{x}) \tag{1.10}$$

$$= \frac{\Lambda}{2} R^* R f_{recon}(\mathbf{x}) \tag{1.11}$$

2. Monitoring Using Moment Conditions

Although the moment conditions are not useful for improving IQ, they have potential as a monitoring tool. The technique described below is *not* currently used by GE Medical Systems, but is covered in [4]. The parallel-beam formulation of the moment conditions is recast into our fan-beam geometry in Section 2.1. The effect of a single bad detector channel is shown to carry a clear "signature" in Section 2.2.

2.1. Fan-Beam Parametrization. The change of variables from parallel-beam to fan-beam geometries is depicted in Figure 3. In fan-beam geometry, we often refer to the view angle as β (rather than θ). The fan angle, γ is analogous to the distance parameter s. The xray focal spot for any particular view is denoted by $\mathbf{S}(\beta)$ with source-to-isocenter distance $D = |\mathbf{S}|$ The change of variables from $(s, \theta(\phi))$ to (β, γ) as shown in Figure 3 is simply

$$\begin{aligned} s &= D\, sin\gamma &\text{where } (s, \phi) \in \mathbb{R}^1 \times [0, 2\pi) \\ \phi &= \pm \gamma & (\beta, \gamma) \in [0, 2\pi) \times [-\pi/2, \pi, 2] \end{aligned} \tag{2.1}$$

with Jacobian

$$J = \begin{vmatrix} D\, cos\gamma & 0 \\ 1 & 1 \end{vmatrix} = D\, cos\gamma \tag{2.2}$$

Therefore, the H-L moments can be rewritten in fan-beam notation for $|k| > m \geq 0$

$$\text{Mom}(k, m) = \int_0^{2\pi} e^{ik\phi} \int_{\mathbb{R}} s^m Rf(\theta(\phi), s)\, ds\, d\phi \tag{2.3}$$

$$= \int_0^{2\pi} e^{ik(\beta+\gamma)} \int_{\mathbb{R}} (D\, sin\gamma)^m\, Rf(D\, sin\gamma, \beta+\gamma)\, (D\, cos\gamma)\, d\gamma\, d\beta$$

$$= D^{m+1} \int_0^{2\pi} e^{ik\beta} \int_{\mathbb{R}} (sin\gamma)^m\, e^{ik\gamma}\, Tf(\beta, \gamma)\, d\gamma\, d\beta$$

where Tf is the "fan beam" data multiplied by the scaling factor $cos\gamma$

$$Tf(\beta, \gamma) \equiv cos\gamma\, Rf(D\, sin\gamma, \beta+\gamma) \tag{2.4}$$

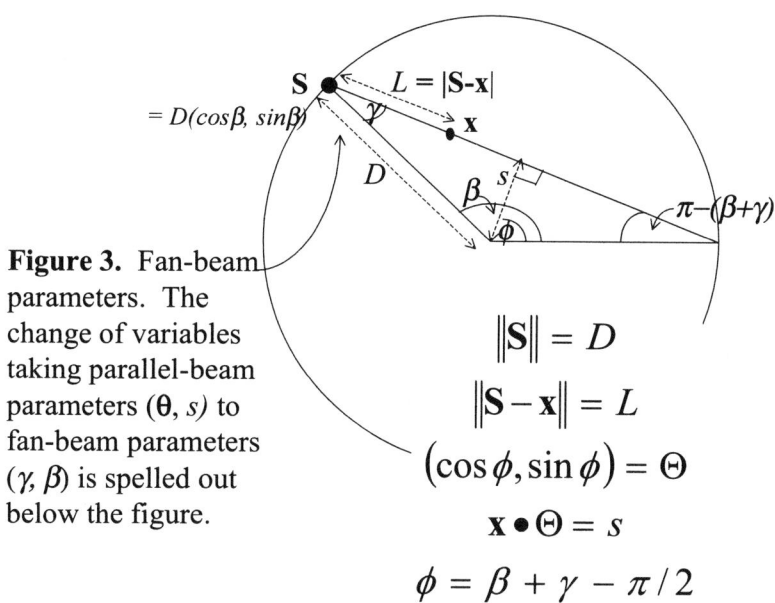

Figure 3. Fan-beam parameters. The change of variables taking parallel-beam parameters (θ, s) to fan-beam parameters (γ, β) is spelled out below the figure.

2.2. Signature of a Faulty Detector.
Suppose that $\hat{T}f$ is measured data, where everthing is correct *except* for one detector at fan position γ_0. We model this in the continuous case as

(2.5) $$\hat{T}f(\beta, \gamma) = Tf(\beta, \gamma) + g(\beta)\, cos(\gamma)\, \delta_{\gamma_0}(\gamma)$$

where Tf satisfies the H-L conditions and g is an unknown, but well behaved, function of β. If 2.5 is a good model of the measured data, then $\hat{T}f$ violates the conditions in a very systematic way. The conditions become

$$\hat{\text{Mom}}(k, m) = \text{Mom}(k, m) +$$
$$D^{m+1} \int_0^{2\pi} e^{ik\beta} \int_{\mathbb{R}} (sin\gamma)^m\, e^{ik\gamma}\, g(\beta)\, cos\gamma\, \delta_{\gamma_0}(\gamma)\, d\gamma\, d\beta$$

(2.6) $$= \text{Mom}(k, m) + G_k\, (D\, sin\gamma_0)^m$$

where $G_k \equiv D\, cos\gamma_o\, e^{ik\gamma_0} \left(\int_0^{2\pi} e^{ik\beta} g(\beta)\, d\beta \right)$ is again some unknown constant, depending only upon k.

Although we do not know the values of $\{G_k\}_{k=1}^\infty$, we do know that ratios of the conditions are related. For example, if we define

(2.7) $$E \equiv D\, sin\gamma_0 \quad \text{and}$$
(2.8) $$\text{Rat}_{k;m_1,m_2} = \frac{\text{Cond}(k, m_1)}{\text{Cond}(k, m_2)}$$

then

(2.9) $\text{Rat}_{k;m_1,m_2} = E^{(m_1-m_2)} \quad \forall |k| > max(m_1,m_2)$
and whenever $(k+m_1), (k+m_2)$ are both odd

Furthermore, $\dfrac{\text{Cond}(k_1,m)}{\text{Cond}(k_2,m)} = \dfrac{G_{k_1}}{G_{k_2}}$ for $m = 0, 1, \ldots, min(|k_1|, |k_2|)$

The H-L moments which *should* be identically zero, but are not are represented below:

(2.10)

	$m=0$	$m=1$	$m=2$	$m=3$	$m=4$	...
$k=0$		G_0E		G_0E^3		...
$k=1$	G_1		G_1E^2		G_1E^4	...
$k=2$	G_2	G_2E		G_2E^3		...
$k=3$	G_3	G_3E	G_3E^2		G_3E^4	...
$k=4$	G_4	G_4E	G_4E^2	G_4E^3		...
$k=5$	G_5	G_5E	G_5E^2	G_5E^3	G_5E^4	...
...	...	...	...	...	...	

The point is that although the $\{G_k\}$'s and E are unknown, the form of the matrix is known. If a CT system's background monitoring were capable of revealing that axial scan data violates the H-L conditions in this way it could automatically notify the operator (and perhaps the GE Medical Systems Service Center!!) that the detector located at

(2.11) $$\gamma_0 = asin(E/D)$$

is suspect. As we shall see in the next section, however, system hardware limitations currently prevent us from monitoring patient data in the background. It still may be possible, however, to diagnose faulty-detectors by periodically scanning a test phantom.

3. Numerical Results

Estimation of the location of a bad detector using equation 2.11 is limited by the accuracy with which we can compute the moments. In Section 3.1 we describe the phantoms and their sinograms used to obtain our results. Sinogram #0 is a discretely sampled Radon transform; sinogram#1 contains partial volume errors, one of many errors unavoidably incurred in physical measurements, no matter how "good" the CT system. See Moments computed from the discrete, but otherwise ideal, sinogram #0 roughly indicate the severity quadrature errors incurred computing the moments. From these we see that partial volume effects are larger - too large to permit *background* monitoring. However, by taking advantage of the fact that the unavoidable measurement errors are roughly constant over time we estimate γ_o from sinograms with as little as one percent multiplicative Gaussian error in a single detector cell.

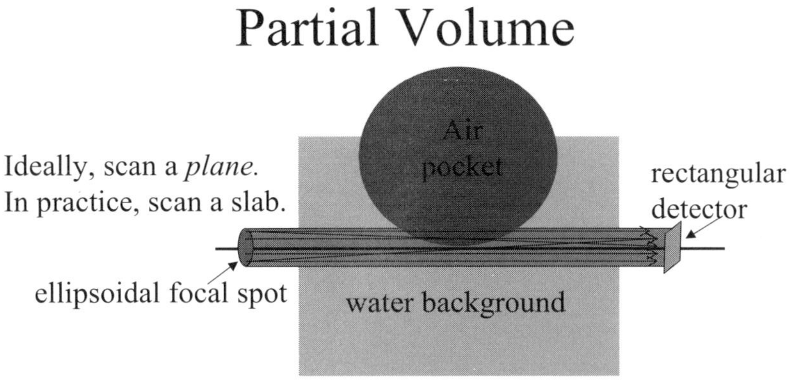

Partial Volume

Ideally, scan a *plane*.
In practice, scan a slab.

Air pocket

rectangular detector

ellipsoidal focal spot

water background

Figure 4. The air pocket in sinogram #1 is depicted above. In the ideal case, line integrals along the infinitesimally thin scan plane would not detect the air pocket. Arrows represent some lines whose integrals are averaged by a CT scanner. The average of line integrals in this "tube" detects the air pocket, albeit imperfectly.

3.1. Numerical Phantom. Numerical results presented here were generated by processing *simulated* sinograms. Sinogram #0 discretely samples "perfect" data; sinogram #1 contains only errors due to partial voluming. Sinogram #0 is used only to indicate difficulties in computing the moments from discrete data. "Partial volume" errors occur because neither the detector cells nor the x-ray focal spot which define the lines along which the linear attenuation coefficient (LAC) are points. In fact, they are $\sim 1mm$ in diameter. Furthermore, detector response is not uniform across the detector cell surface and the x-ray emission rate varies across the surface of the focal spot. Therefore, measured CT data represents an average of many line integrals along lines lying in a tube connecting the x-ray focal spot and detector. See Figure 4.

Scanner geometry for both sinograms is that of *Lightspeed*, a current 3rd generation GE scanner and includes 1/4 detector offset. Source-to-isocenter distance was normalized to $D = 1$. Data are sampled at equidistant points in both the β and γ directions. *Lightspeed* scanners have 888 detector channels in the γ direction covering a fan angle of roughly ±27.5 degrees and takes (as many as) 984 views during a single rotation, $\beta \in [0, 2\pi)$. The trapezoidal rule was used to compute both β and γ integrals, for the sake of simplicity and also to avoid problems created by the fact that the sinogram is not continuously differentiable. (Higher order quadrature techniques for equidistant sampling require derivatives!) The linear attenuation coefficient (LAC) of water is set at 0.19, spine at $0.19 + 0.076 = 0.266$, and ribs at $0.19 + 0.114 = 0.304$.

Analytic Sinogram

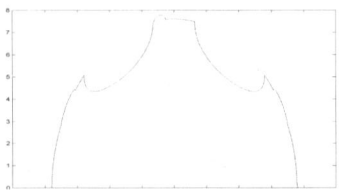

Figure 5 Projection corresponding to gantry angle $\beta = 0$ shown upper right. Entire sinogram on lower right. Below, reconstructed image at (window, level) = (300, 0)

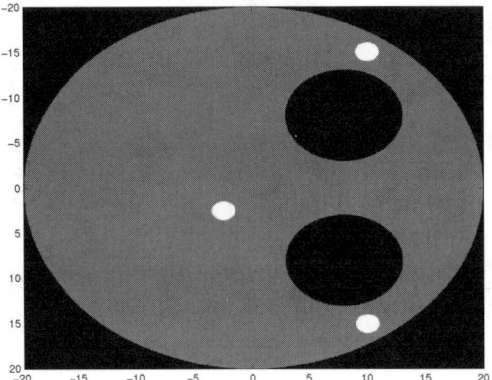
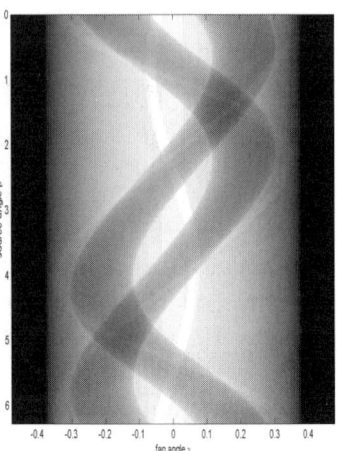

Sinogram #0 is made up solely of constant density circular objects and is therefore simpler than sinogram #1. It was analytically computed assuming a point source and point detector. Three anatomically incorrect ribs lie between two lungs and the outer edge of the water filled phantom's body. See Figure 5 for the entire sinogram, plots of a single projection, and its reconstructed image. One can see that although the sinogram is continuous it contains several "square root singularities" where the first derivative is not defined. H-L moments computed from this sinogram roughly indicate the levels of quadrature errors.

G. Besson wrote the simulation code which generated sinogram #1, which is also used for volumetric simulations. Each element of the sinogram was computed by averaging integrals along lines connecting five points on the focal spot with eight points on the detector cell. Therefore, the curvature of inclusions in the direction normal to the imaging plane is relevant, as will be seen by comparison with a data set simulated for a planar phantom. The object is piecewise constant, a water bottle with ellipsoidal cross-section containing one air pocket, one spine, and seven ribs. The water bottle has a long axis of 40 cm and short axis of 26 cm. The air pocket is an ellipsoid tangent to the imaging plane, and is therefore a large source of partial volume errors. The spine is a cylinder perpendicular to the imaging plane with circular cross-section of diameter 3.4 cm. Each rib is a cylinder with elliptical cross-section with long axis 2 cm and short axis 0.75 cm. The ribs are tilted at varying angles with respect to the imaging plane. See Figures 6 and 7.

Partial Volume Sinogram

Figure 6 Projection corresponding to gantry angle $\beta = 0$ shown upper right. Entire sinogram on lower right. Half-scan image below displayed at (window,level) = (100, 0).

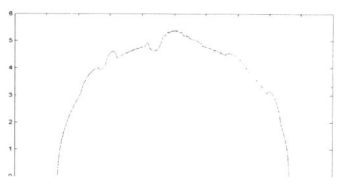

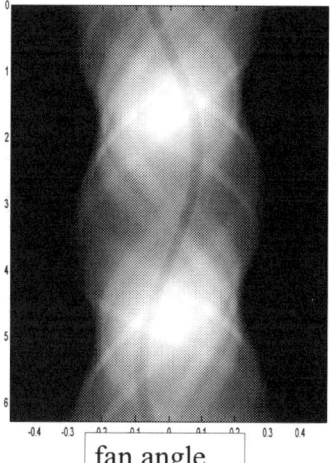

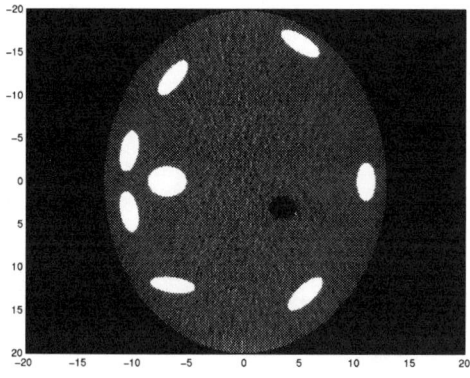

3.2. Quadrature Errors.
Magnitudes of the first few H-L moments from the analytically computed sinogram #1 are listed below:

(3.1) $| Mom\#0(0:4, 0:4) | =$

k	$m = 0$	$m = 1$	$m = 2$	$m = 3$	$m = 4$
0	2.43e+1	1.82e−4	8.68e−1	2.55e−5	6.16e−2
1	2.04e−5	2.49e−1	6.21e−7	9.40e−3	2.66e−8
2	1.50e−6	2.96e−7	8.89e−4	1.76e−8	9.90e−5
3	2.45e−5	2.78e−6	7.09e−7	2.42e−3	4.67e−8
4	5.12e−5	8.22e−7	7.80e−7	2.09e−7	3.69e−4
5	3.90e−5	8.00e−7	3.07e−7	8.32e−8	3.37e−8

Moments corresponding to $(k+m)$ *odd* and $|k| > m$ *should be* identically zero and are in fact several orders of magnitude smaller than moments which the H-L conditions do *not* force to be identically 0. Unfortunately, even a perfectly designed and manufactured system measures only an approximation to the Radon transform. Effects such as partial voluming degrade the moment computations, as can be seen

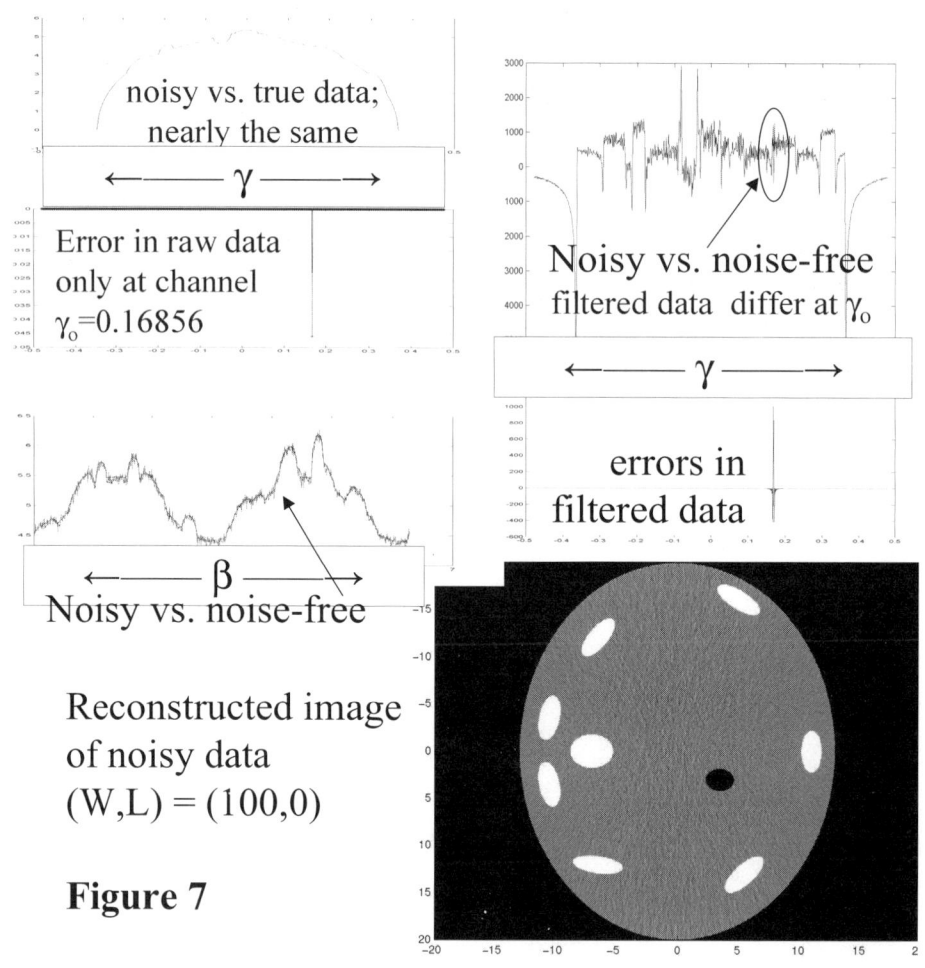

Figure 7

in the moments computed from the partial volumed sinogram #1:

(3.2) $|Mom\#1(0:4, 0:4)| =$

k	$m=0$	$m=1$	$m=2$	$m=3$	$m=4$
0	$1.85e+1$	$2.26e-6$	$4.53e-1$	$6.12e-8$	$2.37e-2$
1	$1.15e-4$	$1.66e-2$	$1.31e-6$	$4.15e-4$	$1.14e-7$
2	$2.65e-4$	$4.76e-5$	$9.01e-2$	$3.97e-6$	$8.82e-3$
3	$7.09e-5$	$1.11e-5$	$1.10e-6$	$1.48e-4$	$1.04e-7$
4	$1.53e-4$	$6.66e-6$	$3.03e-7$	$4.59e-7$	$8.90e-4$
5	$2.74e-5$	$2.78e-6$	$1.50e-6$	$3.50e-7$	$1.26e-7$

Errors due to partial volume are roughly one order of magnitude larger than the quadrature errors suffered in computing |Mom#0|. Therefore we may be forced to sidestep partial voluming by restricting ourselves to scanning test phantoms which are invariant along the patient axis and therefore generate no partial volume errors, or by periodically comparing moments generated from *any* phantom to moments computed from the same phantom immediately after detector calibration, when the detector is known to be in good working order. We explore the latter possibility in the next section.

3.3. γ_o Estimates. Errors due to inconsistent data from a noisy detector must override quadrature errors as well as other errors due to inconsistent data, such as partial voluming. Before proceeding we define first in words

Mom_o^c	array of estimates of H-L moments computed from clean data immediately after detector calibration
ϵ_o^q	array of quadrature errors when computing Mom_o^c
ϵ_o^{pv}	array of unavoidable errors from measuring physical data
Mom^c	array of estimates of H-L moments computed during a periodic system calibration
ϵ^q	array of quadrature errors when computing Mom^c
ϵ^{pv}	array of partial volume errors when computing Mom^c

i.e.,

(3.3) $\quad Mom_o^c = Mom + \epsilon_o^q + \epsilon_o^{pv} \quad$ and
$\quad\quad\quad\; Mom^c = Mom + \epsilon^q + \epsilon^{pv} \quad$ when all detectors function properly

However, when a single detector at γ_o fails, the elements of Mom^c take the form

(3.4) $\quad Mom^c(k,m) = Mom(k,m) + G_k \, (D \, sin\gamma_0)^m + \epsilon^q(k,m) + \epsilon^{pv}(k,m)$

Under reasonable operating conditions, quadrature and partial volume errors remain roughly constant over time, *i.e.*, $\epsilon^q \sim \epsilon_o^q$ and $\epsilon^{pv} \sim \epsilon_o^{pv}$ so subtracting the calibration moments from current moments yields

$$\begin{aligned}
\text{MomDiffs}(k,m) &\equiv Mom^c - Mom_o^c \\
&= (G_k \, (D \, sin\gamma_0)^m + \epsilon^q(k,m) + \epsilon^{pv}(k,m)) \\
&\quad - (\epsilon_o^q(k,m) + \epsilon_o^{pv}(k,m)) \\
\text{(3.5)} \quad &\sim G_k \, (D \, sin\gamma_0)^m
\end{aligned}$$

The pattern becomes all the more obvious when we look at the ratios

$$\begin{aligned}
\text{MomDiffsRatios}(k,m) &\equiv \text{MomDiffs}(k,m+1)/\text{MomDiffs}(k,m) \\
\text{(3.6)} \quad &\sim D \, sin\gamma_0
\end{aligned}$$

We see that this holds with both sinograms when only 1% random noise is added to detector channel #222 at $\gamma_o = -0.24030$ radians as follows

(3.7) $\quad\quad\quad\quad Rf_{noisy}(\beta,\gamma) = (1 + 0.01\delta_{\gamma_o}(\gamma)X)Rf(\beta,\gamma)$

where $X \sim N(0,1)$ and δ_{γ_o} is really a Kroneker δ, since we evaluate at discrete detector locations. MATLAB's random number generator "randn" was used to create such a data set, which generated the following ratios:

(3.8) MomDiffsRatios#0 =

$$\begin{array}{cccc}
-0.238 & -0.238 & -0.238 & -0.238 \\
-0.238+2e{-}10i & -0.238-8e{-}11i & -0.238+9e{-}11i & -0.238-1e{-}10i \\
-0.238-8e{-}11i & -0.238-1e{-}10i & -0.238+1e{-}10i & -0.238+3e{-}11i \\
-0.238-7e{-}11i & -0.238+1e{-}11i & -0.238-2e{-}11i & -0.238+3e{-}11i \\
-0.238+9e{-}12i & -0.238-5e{-}12i & -0.238-1e{-}11i & -0.238+2e{-}11i \\
-0.238-7e{-}11i & -0.238+8e{-}12i & -0.238-6e{-}11i & -0.238+4e{-}11i
\end{array}$$

revealing that the detector at $\gamma_o = asin(-0.238) = -0.24030$ is suspect. Even in the less ideal case where partial volume errors play role, we get very good results:

(3.9) MomDiffsRatios#0 =

$$\begin{array}{cccc}
-0.238 & -0.238 & -0.238 & -0.238 \\
-0.238+9e{-}11i & -0.238+3e{-}11i & -0.238-4e{-}11i & -0.238+2e{-}12i \\
-0.238+9e{-}11i & -0.238+1e{-}11i & -0.238+2e{-}11i & -0.238-5e{-}12i \\
-0.238+1e{-}11i & -0.238-1e{-}10i & -0.238+1e{-}10i & -0.238-7e{-}11i \\
-0.238+3e{-}11i & -0.238-2e{-}11i & -0.238-6e{-}12i & -0.238+2e{-}11i \\
-0.238-7e{-}13i & -0.238+6e{-}12i & -0.238-4e{-}11i & -0.238+5e{-}11i
\end{array}$$

Several different realizations of the noisy sinogram yielded comparable results. This level of noise is minor, producing no discernible errors in the reconstructed image, even when displayed at a window width of 100 Hounsfield units. See Figure 7.

4. Conclusion

Although conventional filtered-backprojection reconstructions of axial CT data are not directly improved by enforcing the H-L moment conditions, the conditions could be used to improve overall system performance. Monitoring for a single bad detector channel is clearly one possible application of the H-L moment conditions, but it is by no means the *only* application. (Please contact the author if you find another good application!)

Thanks to J. Hsieh and T. Pan for pointing me toward the right application and to G. Besson for sharing simulator, as well as his extensive knowledge of the literature.

References

[1] Helgason, S., *The Radon Transform*, Birkhauser, (1980).
[2] Glover, G. H., Eisner, R. L., "Consistent Projection Sets: Fan-Beam Geometry," Computer Aided Tomography and Ultrasonics and Medicine, Raviv, *et. al.* (eds.), North-Holland Publishing Co., pp. 235-251, (1979).
[3] Medoff, B. P., "Inner Product Framework for Image Reconstruction," Proceedings - ICASSP, IEEE International Conference on Acoustics, Speech and Signal Processing, pp. 1073-1076, (1985).
[4] Patch, S. K., "Background Monitoring of CT Data for Existence and Location of a Bad Detector," RD-26,226, filed with the U.S. Patent Office by GE Corporate Research and Development, Oct. 1999.

[5] Basu, S. K., "Range Spaces and Consistency in Linear Inverse Problems and their Application to the Radon Transform," *preprint*.

APPLIED SCIENCE LAB, GE MEDICAL SYSTEMS, PO BOX 414, MILWAUKEE, WI 53201
E-mail address: **sarah.patch@med.ge.com**

Principles of Reconstruction Filter Design in 2D-Computerized Tomography

Andreas Rieder

ABSTRACT. The filtered backprojection algorithm is the most commonly used algorithm for tomographic reconstruction. The quality of the reconstructed density distribution depends heavily on the used reconstruction filter. We discuss general principles to design reconstruction filters with prescribed properties and give concrete examples which are good alternatives to the commonly used Shepp-Logan filter. Moreover, we investigate the proper scaling of the reconstruction filters in relation to the discretization step size in the filtered backprojection algorithm. Numerical experiments illustrate our theoretical results.

1. Introduction

The analytic basis of X-ray computerized tomography in 2D is the reconstruction formula

$$(1.1) \qquad f = (2\pi)^{-1} \mathbf{R}^* \Lambda \mathbf{R} f$$

where f is the searched-for density distribution and $\mathbf{R}$ denotes the *Radon transform* mapping a function to its line integrals, see, e.g., Natterer [**Nat86**]. The operator $\mathbf{R}^*$ is adjoint to $\mathbf{R}$ with respect to suitable L^2-spaces. Formally, Λ is the square root of the 1D Laplacian $-\Delta$, $\Lambda = (-\Delta)^{1/2}$, and it acts on the first variable of $\mathbf{R}f$. All operators involved will be defined explicitly in the next section.

As Λ amplifies high frequencies instabilities appear very likely in reconstructing f from noisy data $\mathbf{R}f$ using (1.1) directly. Therefore, an algorithmic realization of tomographic reconstruction is based on ($\star$ denotes convolution and $\star_1$ denotes convolution with respect to the first variable of $\mathbf{R}f$)

$$(1.2) \qquad f \star e_\gamma = \mathbf{R}^*(v_\gamma \star_1 \mathbf{R}f), \quad e_\gamma = \mathbf{R}^* v_\gamma,$$

where $e_\gamma(x) = e(x/\gamma)/\gamma^2$, $\gamma > 0$, and $e = e_1$ is a smooth function with normalized mean value. The convolution of the Radon data $\mathbf{R}f$ with the *reconstruction kernel* or *reconstruction filter* v_γ implements a low pass filtered version of $\Lambda \mathbf{R}f$. Hence,

---

2000 *Mathematics Subject Classification.* Primary 65R20.

Key words and phrases. computerized tomography, filtered backprojection, mollifier, reconstruction kernel.

The author thanks the American Mathematical Society and the Deutsche Forschungsgemeinschaft (grant RI 975/2-1) for support to attend this conference.

(1.2) is a stabilized or regularized version of (1.1) which lets us recover only $f \star e_\gamma$, a smoothed or mollified approximation to f. Accordingly, e is called a *mollifier*.

A careful discretization of (1.2) leads to the filtered backprojection algorithm (FBA) which is the most frequently used reconstruction algorithm in computerized tomography, see, e.g., Natterer [**Nat86**, Chap. V] or [**Nat99**].

In this paper we pursue two objectives. First, we propose a scheme to design mollifiers and reconstruction kernels with useful features. Second, we solve the scaling problem, that is, we answer the following question. What is a 'good' value for γ in relation to the discretization step size in the FBA?

The scaling problem has been reported probably for the first time by Smith in [**Smi82**, p. 20]. His numerical experiments showed a strong sensitivity of the tomographic reconstructions to the choice of γ. This calls for a criterion to adjust γ. A heuristic rule based on numerical experiments has been suggested by Smith and Keinert [**SK85**, Sec. VI]. Looking at the scaling problem from sampling theory lead Natterer to a choice of γ for reconstructing essentially band-limited functions, see [**Nat86**, Chap. V].

We begin this paper by introducing some definitions and notations. Based on the reconstruction formula of the Radon transform we are able to compute explicitly three different types of new reconstruction filters (Section 3). Our main result is presented in Section 4. A Fourier analysis of the reconstruction error yields an equation to determine the scaling parameter γ relative to a fixed discretization step size in the FBA. We analyze this equation in Section 5 and give numerical values of valid scaling parameters for the new reconstruction kernels from Section 3. The scale factors of the classical Shepp-Logan reconstruction filter can also be obtained via our technique as we demonstrate in Section 6. In the final section we compare tomographic reconstructions with respect to different filters including the Shepp-Logan filter.

2. Preliminaries

Here we give detailed definitions of the operators and operations from the previous section.

The *Radon transform* $\mathbf{R}$ maps a function $f \in L^2(\Omega)$ to its line integrals. By Ω we denote the unit ball in $\mathbb{R}^2$ centered about the origin. We have that

$$\mathbf{R}f(s,\vartheta) := \int_{L(s,\vartheta) \cap \Omega} f(x) \, \mathrm{d}\sigma(x)$$

where $L(s,\vartheta) = \{\tau\, \omega^\perp(\vartheta) + s\, \omega(\vartheta) \mid \tau \in \mathbb{R}\}$, $s \in]-1,1[$, $\omega(\vartheta) = (\cos \vartheta, \sin \vartheta)^t$ and $\omega^\perp(\vartheta) = (-\sin \vartheta, \cos \vartheta)^t$ for $\vartheta \in]0,\pi[$. This parameterization of lines intersecting Ω gives rise to the *parallel scanning geometry*.

The Radon transform is injective and maps $L^2(\Omega)$ boundedly to $L^2(Z)$ where $Z :=]-1,1[\times]0,\pi[$, see, e.g., Natterer [**Nat86**]. Let $\mathbf{R}^* : L^2(Z) \to L^2(\Omega)$ be the adjoint of $\mathbf{R}$ which is called *backprojection*. Then,

$$\mathbf{R}^* g(x) := \int_0^\pi g(x^t\, \omega(\vartheta), \vartheta) \, \mathrm{d}\vartheta.$$

REMARK 2.1. Please observe that our backprojection $\mathbf{R}^*$ is slightly different from the backprojection operator $\mathbf{R}^\#$ used by Natterer [**Nat86**]. Natterer considers $\mathbf{R}$ as a mapping from $L^2(\Omega)$ to $L^2(]-1;1[\times]0,2\pi[)$. As a consequence, statements from [**Nat86**] concerning $\mathbf{R}^*$ have to be adjusted to the present situation. This can be done easily enough since $2\mathbf{R}^*g = \mathbf{R}^\# g$ whenever $g(s,\vartheta) = g(-s, \vartheta + \pi)$.

In the sequel we denote the Fourier transform of a function $f \in L^1(\mathbb{R}^d)$ by $\widehat{f}(y) := (2\pi)^{-d/2} \int_{\mathbb{R}^d} f(x) \exp(-\imath y^t x)\,\mathrm{d}x$. The Λ-operator is defined by the Fourier transform via

$$\widehat{\Lambda f}(\xi) := \|\xi\|\,\widehat{f}(\xi). \tag{2.1}$$

We introduce the Sobolev spaces $H_0^\alpha(\Omega)$, $\alpha \in \mathbb{R}$, to be the closure of $\mathcal{C}_0^\infty(\Omega)$, the space of infinitely differentiable functions with compact support in Ω, with respect to the norm $\|f\|_\alpha^2 = \int_{\mathbb{R}^2} \left(1 + \|\xi\|^2\right)^\alpha |\widehat{f}(\xi)|^2\,\mathrm{d}\xi$.

3. Obtaining reconstruction kernels from mollifiers

Our approach is based on the reconstruction formula (1.1) which holds true for $f \in H_0^{1/2}(\Omega)$. With $e_\gamma \in H_0^{1/2}(\Omega)$ we define

$$v_\gamma := \Lambda \mathbf{R} e_\gamma / (2\pi) \tag{3.1}$$

and obtain $e_\gamma = \mathbf{R}^* v_\gamma$ such that (1.2) applies. Assuming sufficient smoothness as well as radial symmetry of e then v_γ only depends on s and (3.1) may be expressed as

$$v_\gamma(s) = \frac{1}{\pi} \int_0^\infty \sigma\, \widehat{e_\gamma}(\sigma, 0)\, \cos(s\,\sigma)\,\mathrm{d}\sigma, \tag{3.2}$$

compare Natterer [**Nat86**, (1.5) on p. 103]. Set $v = v_1$. Then, $v_\gamma(s) = v(s/\gamma)/\gamma^2$.

The following asymptotic result (3.3) follows readily from (3.1) and an explicit representation of Λ due to Faridani et al. [**FFRS97**, Formula (2.1)]. We provide a rather elementary proof using integration by parts.

LEMMA 3.1. *Let $e \in H_0^\alpha(\Omega)$, $\alpha > 1$, be radially symmetric with $\int_\Omega e(x)\,\mathrm{d}x = 1$. Then,*

$$\lim_{|s| \to \infty} s^2\, v(s) = -\frac{1}{2\,\pi^2}. \tag{3.3}$$

Especially, the kernel v cannot be compactly supported.

PROOF. Our assumptions on e make the integral (3.2) converge and we are allowed to apply integration by parts two times:

$$\begin{aligned}
s^2\,v(s) &= -\frac{1}{\pi} \int_0^\infty \sigma\,\widehat{e}(\sigma, 0)\,\frac{\mathrm{d}^2}{\mathrm{d}\sigma^2} \cos(s\,\sigma)\,\mathrm{d}\sigma \\
&= \frac{1}{\pi}\Big((\widehat{e}(\sigma,0) + \sigma\,\widehat{e}'(\sigma,0))\,\cos(s\,\sigma)\Big|_{\sigma=0}^{\sigma=\infty} \\
&\qquad - \int_0^\infty \frac{\mathrm{d}^2}{\mathrm{d}\sigma^2}(\sigma\,\widehat{e}(\sigma,0))\,\cos(s\,\sigma)\,\mathrm{d}\sigma\Big) \\
&= \frac{1}{\pi}\Big(-\widehat{e}(0,0) - \int_0^\infty \frac{\mathrm{d}^2}{\mathrm{d}\sigma^2}(\sigma\,\widehat{e}(\sigma,0))\,\cos(s\,\sigma)\,\mathrm{d}\sigma\Big).
\end{aligned}$$

Letting s go to infinity the statement follows by the Riemann-Lebesgue Lemma as $\frac{\mathrm{d}^2}{\mathrm{d}\sigma^2}(\sigma\,\widehat{e}(\sigma,0))$ is an integrable function. $\square$

We now give examples for mollifier/reconstruction kernel pairs.

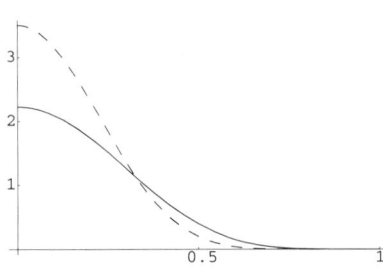 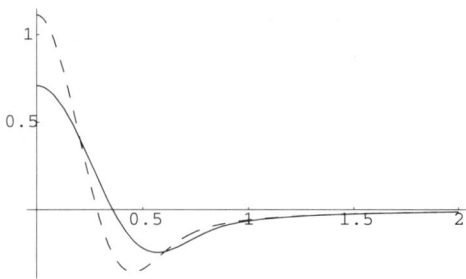

FIGURE 1. Radial parts of the mollifiers e^n from (3.4) on the left and corresponding reconstruction kernels v^n from (3.7) on the right for $n = 6$ (solid curves) and $n = 10$ (dashed curves).

EXAMPLE 3.2. We define a family $\{e^n\}_{n\in\mathbb{N}}$ of radial mollifiers compactly supported in Ω:

$$(3.4) \quad e^n(x) := \frac{n+1}{\pi} \, p_n(\|x\|) \quad \text{where} \quad p_n(t) = \begin{cases} (1-t^2)^n & : \quad |t| \leq 1 \\ 0 & : \quad \text{otherwise} \end{cases}.$$

The smoothness of e^n increases with n (to be precise: $e^n \in H_0^s(\Omega)$ for any $s < n + 1/2$) and $\int_{\mathbb{R}^2} e^n(x)\,dx = 1$.

In view of (3.2) we first need to calculate the Fourier transform of e^n. By the radial symmetry and compact support of e^n we find that

$$(3.5) \quad \widehat{e^n}(\rho\omega) = \frac{n+1}{\pi} \int_0^1 r\, p_n(r)\, \mathrm{J}_0(r\rho)\, dr, \quad \omega \in S^1,$$

where

$$(3.6) \quad \mathrm{J}_\nu(z) = \left(\frac{z}{2}\right)^\nu \sum_{k=0}^\infty \frac{(-1)^k}{k!\,\Gamma(\nu+k+1)} \left(\frac{z}{2}\right)^{2k}$$

is the *Bessel function of the first kind of order* ν with Γ denoting *Euler's Gamma function*. Plugging the series expansion of J_0 into (3.5) and interchanging integration and summation we obtain

$$\widehat{e^n}(\rho\omega) = \frac{(n+1)!}{2\pi} \sum_{k=0}^\infty (-1)^k \frac{\rho^{2k}}{4^k\,k!\,\Gamma(k+n+2)}$$

$$\stackrel{(3.6)}{=} \frac{2^n\,(n+1)!}{\pi} \rho^{-(n+1)}\, \mathrm{J}_{n+1}(\rho).$$

Finally we found

$$v^n(s) = \frac{2^n\,(n+1)!}{\pi^2} \int_0^\infty \sigma^{-n}\, \mathrm{J}_{n+1}(\sigma)\, \cos(s\,\sigma)\, d\sigma$$

$$(3.7) \quad = \frac{1}{2\pi^2} \begin{cases} 2(n+1)\, {}_2\mathrm{F}_1(1, -n;\, 1/2;\, s^2) & : \quad |s| \leq 1 \\ -\, {}_2\mathrm{F}_1(1, 3/2;\, n+2;\, 1/s^2)/s^2 & : \quad |s| > 1 \end{cases}.$$

The latter equality comes from formula (6.699) by Gradshteyn and Ryzhik [**GR80**] where ${}_2\mathrm{F}_1$ is the *hypergeometric series*. Figure 1 displays the graphs of e^n (3.4) and v^n (3.7) for $n = 6, 10$.

REMARK 3.3. Nothing prevents us to allow real $n > 1/2$ in (3.4). All computations which lead to (3.7) remain valid. The special choice $n = m + 1/2$, $m \in \mathbb{N}$, leads to the so called *logarithmic kernels* propagated by Smith and co-workers, see, e.g., [**Smi82**, eq. (6.8)]. However, they have not been aware of the connection to the hypergeometric series.

EXAMPLE 3.4. We introduce a second family of mollifiers closely related to the first one:

$$(3.8) \qquad e^n(x) := \frac{(n+2)(n+3)}{\pi} \left(\frac{2}{n+3} - \|x\|^2 \right) p_n(\|x\|)$$

with p_n from (3.4). Again, $\int_{\mathbb{R}^2} e^n(x) \, dx = 1$ and $e^n \in H_0^s(\Omega)$ for $s < n + 1/2$. Moreover,

$$\int_{\mathbb{R}^2} x^\alpha \, e^n(x) \, dx = 0 \quad \text{for all pairs } \alpha \in \mathbb{N}^2 \text{ with } 1 \leq |\alpha| \leq 3.$$

Additional vanishing moments of e^n improve the convergence speed of $f \star e_\gamma^n$ to f as $\gamma \to 0$ provided f has sufficient smoothness:

$$\|f - f \star e_\gamma^n\|_{H^r(\mathbb{R}^2)} \leq C \, \|f\|_{H^{r+3}(\mathbb{R}^2)} \, \gamma^3.$$

This feature might be useful if only few or noisy Radon data are available requiring a relatively large γ.

For the Fourier transform of e^n we obtain ($\omega \in S^1$)

$$\widehat{e^n}(\rho\,\omega) = \frac{(n+2)(n+3)}{\pi} \int_0^1 r \left(\frac{2}{n+3} - r^2 \right) p_n(r) \, J_0(r\rho) \, dr$$

$$= \frac{2^n \, (n+2)!}{\pi} \, \rho^{-(n+2)} \left(2\, J_{n+2}(\rho) + \rho \, J_{n+3}(\rho) \right).$$

The second equality may be verified using the series representation (3.6) of the Bessel functions. By formula (6.699) from [**GR80**] we conclude that

$$(3.9) \qquad v^n(s) =$$

$$\begin{cases} \dfrac{n+2}{\pi^2} \left({}_2F_1\left(1, -(n+1); 1/2; s^2\right) + {}_2F_1\left(2, -(n+1); 1/2; s^2\right) \right) & : |s| \leq 1 \\[1ex] \dfrac{1}{2\pi^2 s^2} \left(\dfrac{3 \, {}_2F_1\left(2, 5/2; n+4; s^{-2}\right)}{2(n+3) s^2} - {}_2F_1\left(1, 3/2; n+3; s^{-2}\right) \right) & : |s| > 1 \end{cases}$$

Again, these mollifiers and kernels are defined for real $n > 1/2$. Figure 2 shows the graphs of e^n (3.8) and v^n (3.9) for $n = 6, 10$.

EXAMPLE 3.5. We will now briefly recall the reconstruction kernel which belongs to the Gaussian mollifier

$$(3.10) \qquad G(x) = \frac{18}{\pi} \, \exp\left(-18 \, \|x\|^2 \right).$$

Clearly, G is not supported in Ω. However, since $\int_\Omega G(x) \, dx \approx 1 - 7 \cdot 10^{-7}$ and $\widehat{G}(\omega) \approx 9 \cdot 10^{-8}$ for all $\omega \in S^1$ we may consider G a mollifier in $H_0^{1/2}(\Omega)$. In [**RS00**, Eq. (5.14)] we computed the reconstruction kernel v to be

$$(3.11) \qquad v(s) = \frac{18}{\pi^2} \left(1 + \imath \, \sqrt{18\pi} \, s \, \exp\left(-18 s^2\right) \, \mathrm{erf}\left(\imath \sqrt{18}\, s\right) \right)$$

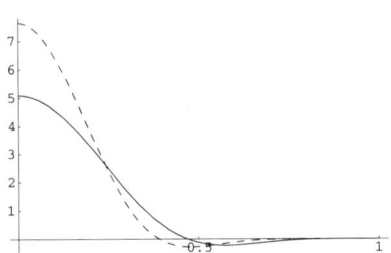 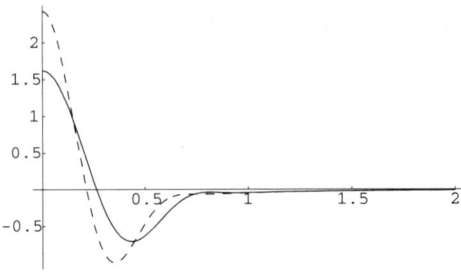

FIGURE 2. Radial parts of the mollifiers e^n from (3.8) on the left and corresponding reconstruction kernels v^n from (3.9) on the right for $n = 6$ (solid curves) and $n = 10$ (dashed curves).

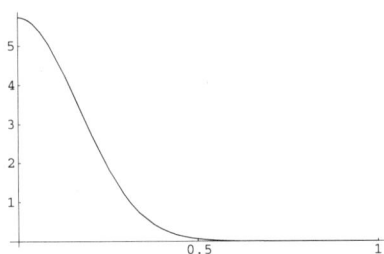 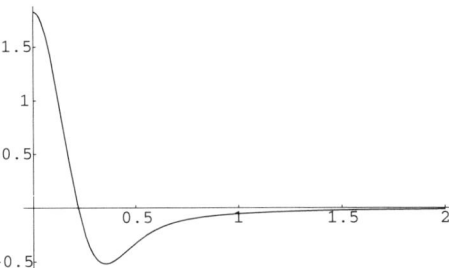

FIGURE 3. Radial part of the Gaussian mollifier (3.10) on the left and corresponding reconstruction kernel (3.11) on the right.

where $\mathrm{erf}(t) = (2/\sqrt{\pi}) \int_0^t \exp(-z^2)\,\mathrm{d}z$ is the error function. The graphs of G and v can be viewed in Figure 3. First computations with this kernel have been reported by Dietz [**Die99**].

REMARK 3.6. The classical way to obtain reconstruction kernels is to choose $\widehat{e}$ in (3.2) as a low pass filter, that is, $\widehat{e}(\xi) = 0$ for $\|\xi\| > 1$. For instance, the ideal low pass filter, $\widehat{e}(\xi) = 1$ for $\|\xi\| \leq 1$, gives the Ram-Lak reconstruction kernel [**RL71**]. The kernel due to Shepp and Logan [**SL74**] comes from $\widehat{e}(\xi) = \mathrm{sinc}(\pi\|\xi\|/2)$ with $\mathrm{sinc}(t) := \sin(t)/t$. More examples of this kind are presented by Chang and Herman [**CH80**]. All reconstruction kernels obtained in this way correspond to mollifiers e which are *not* compactly supported. In particular Lemma 3.1 does not apply. We will comment on the Shepp-Logan filter in more detail in Section 6.

4. Fourier analysis of the reconstruction error

The following analysis of the reconstruction error is inspired by a result of Faridani [**Far90**] for local tomography, see also [**RDS00**].

To avoid technicalities we assume the searched-for density distribution f to be in $\mathcal{C}_0^\infty(\Omega)$. We further consider only that part of the FBA which discretizes the convolution $\star_1$ in (1.2). We assume to know the semi-discrete Radon data $g(s_k, \vartheta) := \mathbf{R}f(s_k, \vartheta)$ for $\vartheta \in [0, \pi[$ and $s_k = h\,k$ where $k = -q, \ldots, q$ and $h = 1/q$, $q \in \mathbb{N}$, is the discretization step size. Our model of the FBA reconstructs f_R by

(4.1) $$f_R(x) = \mathbf{R}^* \mathrm{I}_h (w \overset{h}{\star} g)(x), \quad x \in \Omega,$$

where $\overset{h}{\star}$ denotes the *discrete convolution*,

$$\left(w \overset{h}{\star} g(\cdot, \vartheta)\right)_l = h \sum_{k=-q}^{q} w_{l-k}\, g(s_k, \vartheta), \quad l \in \mathbb{Z}, \tag{4.2}$$

and $\{w_k\}$ is a sequence associated with the chosen kernel v_γ such that $\left(w \overset{h}{\star} g(\cdot, \vartheta)\right)_l \approx (v_\gamma \star g(\cdot, \vartheta))(s_l)$ for $l \in \mathbb{Z}$. For instance,

$$w_k = v_\gamma(s_k), \quad k \in \mathbb{Z}. \tag{4.3}$$

Other weight sequences $\{w_k\}$ are proposed by Smith in [**Smi82**, Sec. 6].

In (4.1), the operator $\mathrm{I}_h : \mathbb{R}^{2q+1} \to L^2(\mathbb{R})$ represents piecewise linear *interpolation* with respect to the step size h:

$$\mathrm{I}_h a(s) = \sum_{k=-q}^{q} a_k\, B_h(s - hk), \quad B_h(s) = B(s/h), \tag{4.4}$$

with B being the linear B-spline supported in $[-1, 1]$.

We will now investigate the difference

$$\delta f := f_R - \mathbf{R}^*\!\left(\chi v_\gamma \star_1 g\right) = \mathbf{R}^*\!\left(\mathrm{I}_h(w \overset{h}{\star} g) - \chi v_\gamma \star_1 g\right) \tag{4.5}$$

where χ denotes the indicator function of the interval $[-2, 2]$. Please notice that δf coincides with the reconstruction error $f_R - f \star e_\gamma$ in the region of interest Ω. This is because $\mathbf{R}^*(\chi v_\gamma \star_1 g)(x) = \mathbf{R}^*(v_\gamma \star_1 g)(x)$ for $x \in \Omega$.

THEOREM 4.1. *Let f be in $C_0^\infty(\Omega)$ and let $\{w_k\}_{k \in \mathbb{Z}}$ be the even real weight sequence used in (4.2). Then, we have the error splitting*

$$\delta f(x) = E_1(x) + E_2(x)$$

with

$$|E_1(x)| \le \int_{\mathbb{R}^2} \frac{|\widehat{f}(\xi)|}{\|\xi\|}\, |\Delta(\|\xi\|)|\, \mathrm{d}\xi \tag{4.6}$$

$$|E_2(x)| \le \pi \sup_{\omega \in S^1} \int_{\mathbb{R}} \left[1 - \mathrm{sinc}^2(h\sigma/2)\right] |\mathrm{T}_h w(\sigma)|\, |\widehat{f}(\sigma\omega)|\, \mathrm{d}\sigma \tag{4.7}$$

where $\mathrm{sinc}(t) = \sin(t)/t$,

$$\Delta(\sigma) = \mathrm{sinc}^2(h\sigma/2)\, \mathrm{T}_h w(\sigma) - \sqrt{2\pi}\, \widehat{\chi v_\gamma}(\sigma),$$

and

$$\mathrm{T}_h w(\sigma) = h\left(w_0 + 2 \sum_{k=1}^{2q} w_k\, \cos(hk\sigma)\right).$$

PROOF. We will need the Poisson summation formula

$$h \sum_{l \in \mathbb{Z}} e^{-\imath l h \sigma}\, v(lh) = \sqrt{2\pi} \sum_{l \in \mathbb{Z}} \widehat{v}(\sigma - 2\pi l/h), \quad \sigma \in \mathbb{R},\ h > 0,$$

which holds true for $v \in \mathcal{S}(\mathbb{R})$, the Schwartz space of rapidly decaying functions, see, e.g., Natterer [**Nat86**, Chap. VII.1]. Moreover, we will rely on the identity

$$\widehat{\mathbf{R}^* \mathrm{w}}(\xi) = \sqrt{\frac{\pi}{2}}\, \|\xi\|^{-1} \left(\widehat{\mathrm{w}}(\|\xi\|, \arg \xi) + \widehat{\mathrm{w}}(-\|\xi\|, \pi + \arg \xi)\right) \tag{4.8}$$

see, e.g., Natterer [**Nat86**, Chap. II, Th. 1.4]. In (4.8) we have a 2D Fourier transform on the left and a 1D Fourier transform (with respect to s) on the right hand side. By $\arg\xi$ we denote the argument of $\xi \in \mathbb{R}^2$, that is, $\xi = \|\xi\|\,\omega(\arg\xi)$.

We start by computing the Fourier transform of δf. In view of (4.5) and (4.8) we only need to compute the 1D Fourier transforms of $\mathrm{I}_h(w \overset{h}{\star} g)$ and $\chi v_\gamma \star_1 g$ with respect to s. First, we take the Fourier transform of $\mathrm{I}_h(w \overset{h}{\star} g)$:

$$\widehat{\mathrm{I}_h(w \overset{h}{\star} g)}(\sigma) = h\,\widehat{B_h}(\sigma) \sum_{k=-q}^{q} (w \overset{h}{\star} g)_k\, \mathrm{e}^{-\imath k h \sigma}$$

$$= h^2\,\widehat{B}(h\sigma) \sum_{k=-q}^{q} \sum_{l=-q}^{q} w_{k-l}\, g(h\,l, \vartheta)\, \mathrm{e}^{-\imath k h \sigma}$$

$$= h^2\,\widehat{B}(h\sigma) \sum_{l \in \mathbb{Z}} \sum_{r=-2q}^{2q} w_r\, g(h\,l, \vartheta)\, \mathrm{e}^{-\imath (r+l) h \sigma}$$

$$= 2\pi h\,\widehat{B}(h\sigma) \sum_{r=-2q}^{2q} w_r\, \mathrm{e}^{-\imath r h \sigma} \sum_{l \in \mathbb{Z}} \widehat{f}((\sigma - 2\pi l/h)\cdot\omega(\vartheta))$$

where, in the last step, we used the Poisson summation formula and then the projection slice theorem,

$$(4.9) \qquad \widehat{\mathbf{R}f}(\sigma, \vartheta) = \sqrt{2\pi}\,\widehat{f}(\sigma\cdot\omega(\vartheta)),$$

see, e.g., Natterer [**Nat86**, Chap. II.1, Th. 1.1]. By $\widehat{B}(h\sigma) = \mathrm{sinc}^2(h\sigma/2)/\sqrt{2\pi}$ and by the symmetry of $\{w_k\}$ we find that

$$(4.10) \quad \widehat{\mathrm{I}_h(w \overset{h}{\star} g)}(\sigma) = \sqrt{2\pi}\,\mathrm{sinc}^2(h\sigma/2)\,\mathrm{T}_h w(\sigma) \sum_{l\in\mathbb{Z}} \widehat{f}((\sigma - 2\pi l/h)\cdot\omega(\vartheta)).$$

Second, we compute the Fourier transform of $\chi v_\gamma \star_1 g$. We have

$$(4.11) \quad \widehat{(\chi v_\gamma \star_1 g(\cdot,\vartheta))}(\sigma) = \sqrt{2\pi}\,\widehat{\chi v_\gamma}(\sigma)\,\widehat{g}(\sigma,\vartheta)$$
$$\overset{(4.9)}{=} 2\pi\,\widehat{\chi v_\gamma}(\sigma)\,\widehat{f}(\sigma\cdot\omega(\vartheta)).$$

Combining (4.5), (4.8), (4.10), and (4.11) we found that

$$(4.12\mathrm{a}) \qquad \widehat{\delta f} = \widehat{E}_1 + \widehat{E}_2$$

with

$$(4.12\mathrm{b}) \qquad \widehat{E}_1(\xi) = 2\pi\,\frac{\widehat{f}(\xi)}{\|\xi\|}\,\Delta(\|\xi\|),$$

$$(4.12\mathrm{c}) \qquad \widehat{E}_2(\xi) = 2\pi\,\mathrm{sinc}^2(h\|\xi\|/2)\,\mathrm{T}_h w(\|\xi\|)$$
$$\times \sum_{l \in \mathbb{Z}\setminus\{0\}} \frac{\widehat{f}((\|\xi\| - 2\pi l/h)\,\xi/\|\xi\|)}{\|\xi\|}.$$

Both $\widehat{E}_1$ and $\widehat{E}_2$ are in $L^1(\mathbb{R}^2)$. Thus, $|E_i(x)| \le (2\pi)^{-1}\int |\widehat{E}_i(\xi)|\,\mathrm{d}\xi$, $i=1,2$, which readily yields the bound for E_1. The bound for E_2 requires further attention. Using polar coordinates and $\sum_{l\in\mathbb{Z}} \operatorname{sinc}^2(z+\pi l) = 1$ independent of $z \in \mathbb{R}$ we have

$$|E_2(x)| \leq \frac{1}{2} \sum_{l \in \mathbb{Z}\setminus\{0\}} \int_{S^1} \int_{\mathbb{R}} |\widehat{f}((\sigma - 2\pi l/h)\,\omega)| \operatorname{sinc}^2(h\,\sigma/2) \, |T_h w(\sigma)| \, \mathrm{d}\sigma \, \mathrm{d}\omega$$

$$= \frac{1}{2} \int_{S^1} \int_{\mathbb{R}} |\widehat{f}(\sigma\,\omega)| \sum_{l\in\mathbb{Z}\setminus\{0\}} \operatorname{sinc}^2(h\,\sigma/2 + \pi l) \, |T_h w(\sigma)| \, \mathrm{d}\sigma \, \mathrm{d}\omega$$

$$= \frac{1}{2} \int_{S^1} \int_{\mathbb{R}} |\widehat{f}(\sigma\,\omega)| \left[1 - \operatorname{sinc}^2(h\,\sigma/2)\right] |T_h w(\sigma)| \, \mathrm{d}\sigma \, \mathrm{d}\omega.$$

A straight forward estimate of the latter expressions gives the bound for E_2. $\square$

Now we discuss the magnitude of the two errors (4.6) and (4.7). To this end we recall typical properties of density functions f in tomography. Since f has a compact support its Fourier transform cannot be band-limited. However, we may assume that f is essentially b-band-limited, i.e., frequencies larger than b do not contribute much to f:

$$(4.13) \qquad \int_{\|\xi\| \geq b} |\widehat{f}(\xi)| \, \mathrm{d}\xi \approx 0.$$

Further, f is usually non-negative so that $|\widehat{f}|$ has a sharply peaked maximum at the origin.

Hence, the error E_1 will be small if and only if Δ is small, especially near the origin. We therefore need to study how $\Delta(\|\xi\|)$ behaves as $\|\xi\|$ becomes small.

LEMMA 4.2. *For v_γ and $\{w_k\}$ even we have that*

$$(4.14) \qquad \left| \operatorname{sinc}^2(h\,\sigma/2) \, T_h w(\sigma) - \sqrt{2\pi} \, \widehat{\chi v_\gamma}(\sigma) \right| = \mathcal{O}(\sigma^2) \quad \text{as } \sigma \to 0$$

if and only if

$$(4.15) \qquad h \left(w_0 + 2 \sum_{k=1}^{2q} w_k \right) = \int_{-2}^{2} v_\gamma(s) \, \mathrm{d}s.$$

PROOF. The only-if-direction is trivial. Now assume that (4.15) holds. By the triangle inequality we find

$$\left| \operatorname{sinc}^2(h\,\sigma/2) \, T_h w(\sigma) - \sqrt{2\pi} \, \widehat{\chi v_\gamma}(\sigma) \right|$$

$$\leq \ |\operatorname{sinc}^2(h\,\sigma/2) - 1| \, |T_h w(\sigma)| + |T_h w(\sigma) - T_h w(0)|$$

$$+ \underbrace{|T_h w(0) - \sqrt{2\pi} \, \widehat{\chi v_\gamma}(0)|}_{\stackrel{(4.15)}{=} 0} + \sqrt{2\pi} \, |\widehat{\chi v_\gamma}(0) - \widehat{\chi v_\gamma}(\sigma)|.$$

All non-zero terms of the above right hand side are of order $\mathcal{O}(\sigma^2)$ as $\sigma \to 0$ which implies the assertion (4.14). For instance,

$$\sqrt{2\pi} \, |\widehat{\chi v_\gamma}(0) - \widehat{\chi v_\gamma}(\sigma)| \leq \int_{-2}^{2} |v_\gamma(s)| \, |1 - \cos(\sigma\,s)| \, \mathrm{d}s$$

where $1 - \cos(\sigma\,s) = \sigma^2 s^2 + \mathcal{O}(\sigma^4)$ uniformly in $s \in [-2, 2]$ as $\sigma \to 0$. $\square$

Condition (4.15) is necessary for E_1 to be small since then $\Delta(\sigma)/|\sigma|$ is bounded and we may estimate

$$|E_1(x)| \le \int_{\|\xi\|<b} |\widehat{f}(\xi)|\,\mathrm{d}\xi \sup_{|\sigma|<b}\left|\frac{\Delta(\sigma)}{\sigma}\right| + \int_{\|\xi\|\ge b} |\widehat{f}(\xi)|\,\mathrm{d}\xi \sup_{|\sigma|\ge b}\left|\frac{\Delta(\sigma)}{\sigma}\right|.$$

The second term on the right is negligible by (4.13) and by $\lim_{|\sigma|\to\infty}\Delta(\sigma)=0$. So, E_1 is dominated by the discretization error $\sup_{|\sigma|<b}|\Delta(\sigma)/\sigma|$ which is small when $\mathrm{sinc}^2(h\,\sigma/2)\,\mathrm{T}_h w(\sigma)$ approximates $\sqrt{2\pi}\,\widehat{\chi v_\gamma}(\sigma)$ well in $[-b,b]$.

Let us now focus on E_2. If h is fixed but smaller than the critical sampling rate π/b (Nyquist rate) and (4.15) applies then both factors $1-\mathrm{sinc}^2(h\,\sigma/2)$ and $|\mathrm{T}_h w(\sigma)|$ are small exactly where $|\widehat{f}|$ is large, and vice versa. We conclude that the error E_2 will be small.

We summarize: under (4.15) and (4.13) the reconstruction error δf is caused only by discretization provided $h \le \pi/b$. No artefacts will show up due to low frequency inconsistencies of the discrete and the continuous reconstruction kernels.

REMARK 4.3. We consider the weight sequences given by (4.3). Here, condition (4.15) seems to contradict the *kernel sum rule*: only filter functions v_γ with a *vanishing* kernel sum, that is, $\sum_{k\in\mathbb{Z}} v_\gamma(s_k)=0$, should be used, see, Smith [**Smi82**, Eq. (6.10)] and Natterer [**Nat99**, p. 114]. However, there is no contradiction since, for h and γ *small*, the kernel sum rule is a fairly good approximation to (4.15). We emphasize that (4.15) is numerically easier accessible than the kernel sum rule, see the next section.

There are two obvious ways to satisfy (4.15). First, we may define the w_k's as local averages of v_γ:

$$w_k := \frac{1}{h}\int_{s_k-h/2}^{s_k+h/2} v_\gamma(s)\,\mathrm{d}s,\ k=1-2q,\ldots,2q-1,\ \ w_{-2q}=w_{2q}:=\frac{1}{h}\int_{2-h/2}^{2} v_\gamma(s)\,\mathrm{d}s.$$

This choice was already motivated by Smith [**Smi82**, (6.12)] from a different point of view. Here γ has to be adjusted before computing the w_k's. For instance, determine γ such that v_γ attains its minimum at h or $2h$. We will not pursue this approach any further in the present paper.

The second way to guarantee (4.15) is presented in detail in the following section.

5. Computing γ

In this section we restrict ourselves to the w_k's from (4.3). For a given reconstruction kernel v and a fixed discretization step size h we consider (4.15) as a nonlinear equation for computing γ. We are therefore seeking zeroes of the *mean value function* $\mathrm{M}(\gamma,h)$,

(5.1)
$$\mathrm{M}(\gamma,h) := \frac{\int_{-2}^{2} v_\gamma(s)\,\mathrm{d}s - \mathrm{T}_h w(0)}{\int_{-2}^{2} v_\gamma(s)\,\mathrm{d}s} = 1 - \frac{\mathrm{T}_h w(0)}{\int_{-2}^{2} v_\gamma(s)\,\mathrm{d}s}$$

$$\stackrel{(4.3)}{=} 1 - \frac{h\left(v(0)/2 + \sum_{k=1}^{2q} v(h\,k/\gamma)\right)}{\gamma \int_0^{2/\gamma} v(s)\,\mathrm{d}s}.$$

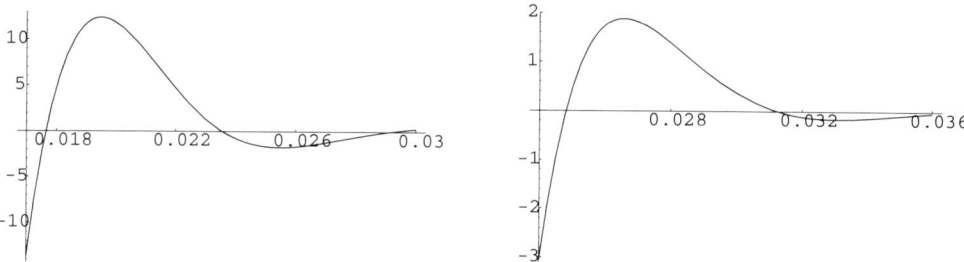

FIGURE 4. Mean value function $M(\cdot, 0.01)$ with respect to v^n from (3.7) with $n = 6$ (left) and $n = 10$ (right). The three zeroes are the respective smallest ones (at least from a computational point of view). See Table 1 for the numerical values of the zeroes.

One easily sees that

(5.2) $$\lim_{\gamma \to 0} M(\gamma, h) = -\infty \quad \text{and} \quad \lim_{\gamma \to \infty} M(\gamma, h) = -h/4.$$

From the typical shape of a reconstruction kernel, see Figures 1, 2 and 3, we may expect the mean value function to have zeroes (actually, an even number of zeroes). Indeed, in the concave and convex parts of the reconstruction kernels the trapezoidal rule $T_h w(0)$ produces quadrature errors with different sign. This leaves the possibility to select γ such that the positive and negative errors cancel each other. Our heuristic argument can be made more rigorous by the Peano representation of the quadrature error, see, e.g., Davis and Rabinowitz [**DR75**].

For the three examples from Section 3 we are able to express the integral $\int_0^{2/\gamma} v(s)\,ds$ by special functions (generalized hypergeometric series in Examples 3.2 and 3.4, error function in Example 3.5). These functions are implemented in the mathematical software environment *Mathematica*. Using *Mathematica* we can efficiently and accurately compute roots of the mean value function $M(\cdot, h)$.

TABLE 1. Three smallest zeroes of the mean value function $M(\cdot, h)$ with respect to the reconstruction kernel family (3.7) for $h = 1/100$ and $h = 1/200$.

v^6	v^8	v^{10}
\multicolumn{3}{c}{$h = 0.01$}		
0.0176529903910958	0.0212575642667490	0.0248099877825188
0.0235717697184353	0.0274457574145479	0.0312331705302724
0.0291182560577191	0.0330879866453031	0.0369605816087359
\multicolumn{3}{c}{$h = 0.005$}		
0.0088265533879969	0.0106289271020307	0.0124053500164855
0.0117854167076440	0.0137213281958366	0.0156119089136376
0.0145613503797639	0.0165528124292464	0.0185115846242859

Numerical computations indicated that the mean value functions relative to the families (3.7) and (3.9) have many zeroes. For three reconstruction kernels from

TABLE 2. Three smallest zeroes of the mean value function $M(\cdot, h)$ with respect to the reconstruction kernel family (3.9) for $h = 1/100$ and $h = 1/200$.

$h = 0.01$		
v^6	v^8	v^{10}
0.0210528823403499	0.0246196221516816	0.0281644779349004
0.0272675397543600	0.0310708882538176	0.0348088528885206
0.0329521791079712	0.0369054141050471	0.0407715560315404
$h = 0.005$		
0.0105264658013230	0.0123098661681931	0.0140823598625049
0.0136335980640357	0.0155349632785071	0.0174031202152876
0.0164767690344411	0.0184551543945099	0.0203936526140549

TABLE 3. The smallest zero of the mean value function $M(\cdot, h)$ with respect to the Gaussian reconstruction kernel (3.11) for $h = 1/100$ and $h = 1/200$.

$h = 0.01$	$h = 0.005$
0.0510222102639069701	0.0267205478436277552

each family we computed the three smallest zeroes, see Tables 1 and 2. We only have numerical evidence based on the left relation in (5.2) that these zeroes are in fact the smallest ones. Figure 4 shows the graphs of the mean value function $M(\cdot, 0.01)$ with respect to the kernels v^6 and v^{10} from (3.7).

The mean value function for the Gaussian reconstruction kernel (3.11) seems to possess exactly two zeroes. Again we only have numerical evidence based on both relations in (5.2). Only the smaller zero is meaningful for tomographic reconstruction (the Gaussian mollifier (3.10) scaled with the larger zero has Ω as effective support which is too large). See Table 3 for numerical values.

6. Scaling the Shepp-Logan filter

We will apply the results from Section 4 to the Shepp-Logan filter [**SL74**]

$$v^{\mathrm{sl}}(s) = \frac{1}{\pi^3} \frac{\pi/2 - s \sin s}{\pi^2/4 - s^2}.$$

It is obtained from (3.2) by $\widehat{e}(\xi) = \mathrm{sinc}(\pi \|\xi\|/2)$, $\|\xi\| \leq 1$, and $\widehat{e}(\xi) = 0$, otherwise, see Remark 3.6. The Shepp-Logan filter is probably the most frequently used reconstruction filter in computerized tomography. Figure 5 shows the graph of $v^{\mathrm{sl}}_{h/\pi}$ with $h = 0.01$.

Scaling the Shepp-Logan filter with the roots of its corresponding mean value function $M(\cdot, h)$ yields disastrous reconstructions. Here our scaling theory fails since v^{sl} violates the asymptotic relation (3.3). However, it is well known that the scale factor $\gamma_h = h/\pi$ works perfectly for v^{sl}, see, e.g., Natterer [**Nat86**, Chap. V.1.1]. How does all this fit into our framework?

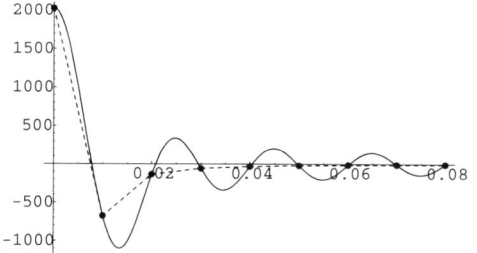
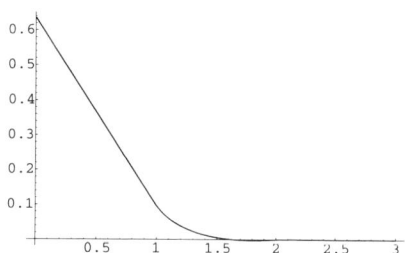

FIGURE 5. Left: Shepp-Logan filter v^{sl} scaled by $0.01/\pi$ (solid line), sampled points (6.1) (black dots), and kernel $\widetilde{v}_{0.01}^{\mathrm{sl}}$ (6.2) (dashed line), right: radial part of the mollifier $\widetilde{e}^{\mathrm{sl}}$ (6.4) belonging to $\widetilde{v}^{\mathrm{sl}}$.

To answer this question we look at the discrete values ($s_k = h\,k$)

$$(6.1) \quad v_{\gamma_h}^{\mathrm{sl}}(s_k) = \frac{1}{\gamma_h^2}\,v^{\mathrm{sl}}\!\left(\frac{h}{\gamma_h}k\right) = \frac{\pi^2}{h^2}\,v^{\mathrm{sl}}(\pi k) = \frac{2}{\pi^2\,h^2}\,\frac{1}{1-4\,k^2}, \quad k \in \mathbb{Z},$$

which are indicated by black dots in Figure 5 (left). The sampled values (6.1) satisfy a discrete version of the asymptotic relation (3.3):

$$\lim_{|k|\to\infty} s_k^2\,v_{\gamma_h}^{\mathrm{sl}}(s_k) = -\frac{1}{2\pi^2}.$$

So it seems as if the samples belong to a reconstruction kernel obtained from (3.2) with a smooth and compactly supported mollifier e_{γ_h}. Therefore, following Smith [**Smi82**] we consider the kernel

$$(6.2) \qquad \widetilde{v}^{\mathrm{sl}} := \mathrm{I}_1 v_{1/\pi}^{\mathrm{sl}}$$

which attains the values $v_{1/\pi}^{\mathrm{sl}}(k)$ at the integers k and which is linear in between (see (4.4) for I_1). Please observe that $\widetilde{v}_h^{\mathrm{sl}}(s_k) = v_{\gamma_h}^{\mathrm{sl}}(s_k)$, see Figure 5 (left). In this framework a modification of the weights $\{w_k\}$ will prove appropriate being slightly different to (4.3):

$$w_k = \begin{cases} \widetilde{v}_\gamma^{\mathrm{sl}}(s_k) & : \quad k = 1-2q,\ldots,2q-1 \\ \widetilde{v}_\gamma^{\mathrm{sl}}(s_k)/2 & : \quad k \in \{-2q,2q\} \\ 0 & : \quad \text{otherwise} \end{cases}.$$

This re-definition of the weights, however, does not affect the value of the discrete convolution (4.2) as $g(-1,\vartheta) = g(1,\vartheta) = 0$ for all ϑ. The corresponding mean value function is

$$\widetilde{\mathrm{M}}(\gamma,h) = 1 - \frac{h\left(\left(\widetilde{v}^{\mathrm{sl}}(0) + \widetilde{v}^{\mathrm{sl}}(2/\gamma)\right)/2 + \sum_{k=1}^{2q-1}\widetilde{v}^{\mathrm{sl}}(h\,k/\gamma)\right)}{\gamma \int_0^{2/\gamma} \widetilde{v}^{\mathrm{sl}}(s)\,\mathrm{d}s}.$$

The roots of $\widetilde{\mathrm{M}}(\cdot,h)$ are easy to spot. The piecewise linear function $\widetilde{v}_\gamma^{\mathrm{sl}}$ is integrated exactly by the trapezoidal rule with step size h whenever the scale factor γ is a multiple of h:

$$(6.3) \qquad \widetilde{\mathrm{M}}(\gamma,h) = 0 \quad \Longleftrightarrow \quad \gamma \in h\,\mathbb{N}.$$

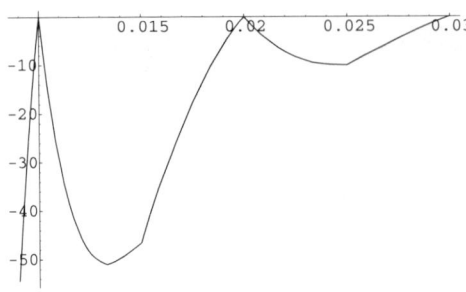

FIGURE 6. Mean value function $\widetilde{M}(\cdot, 0.01)$ for values of the argument near the three smallest zeroes, see (6.3).

See Figure 6 for the graph of $\widetilde{M}(\cdot, 0.01)$ for values of the argument near to the three smallest zeroes.

We just showed that we can handle the Shepp-Logan filter within our unified scaling theory for reconstruction kernels. The optimal scaling factors, which have been deduced before by more or less heuristic arguments from sampling theory, now have an additional analytic foundation.

For illustrational reasons we computed numerically the mollifier $\widetilde{e}^{\mathrm{sl}}$ belonging to $\widetilde{v}^{\mathrm{sl}}$ via the formula (see [**RDS00**])

$$(6.4) \qquad \widetilde{e}^{\mathrm{sl}}(x) = 2 \int_0^{\pi/2} \widetilde{v}^{\mathrm{sl}}(\|x\| \cos \vartheta) \, \mathrm{d}\vartheta.$$

The radial part of $\widetilde{e}^{\mathrm{sl}}$ is plotted in Figure 5 (right).

REMARK 6.1. The reader may have noticed in Tables 1 to 3 that doubling h results roughly in doubling the zeroes of M (5.1). This observation can be explained in the framework of this section. If h is small the piecewise linear interpolate $I_h v_\gamma$ approximates v_γ well in $[-2, 2]$. Moreover, for γ small we have $\mathrm{M}(\gamma, \cdot) \approx \widetilde{\mathrm{M}}(\gamma, \cdot)$. Both facts together lead to $\mathrm{M}(2\gamma^\star, 2h) \approx \mathrm{M}(\gamma^\star, h) = 0$ whenever the zero $\gamma^\star$ and its corresponding discretization step size h are small enough.

7. Numerical simulations

We provide numerical experiments to demonstrate the influence of the scaling and of the underlying reconstruction kernel on the tomographic reconstruction. We do not intend to give a rigorous study for ranking the kernels with respect to the quality of the reconstruction.

The synthetic Radon data were computed from the Shepp-Logan head phantom [**SL74**] which simulates the density relations and the geometry in a human skull, see Figure 7. The phantom is composed of 11 superimposed ellipses.

Our implementation of the filtered backprojection algorithm for the parallel scanning geometry is taken from Natterer [**Nat86**, Chap. V.1.1]. In all our computations below we worked with $p = 300$ directions (angles) and 201 rays per direction, that is, $q = 100$ and $h = 0.01$, respectively. All reconstructions below are shown on a 251×251 grid.

Figure 8 displays six reconstructions of f^{sl} with respect to two different reconstruction kernels. For the same reconstruction kernel they differ in the used scaling parameter γ. The left column of Figure 8 shows reconstructions where we used

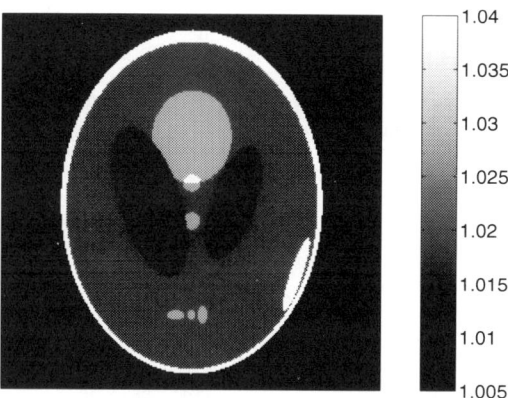

FIGURE 7. Head phantom f^{sl} due to Shepp and Logan [**SL74**].

v^6 of family (3.7). From top to bottom γ runs through the three values listed in Table 1 for v^6 and $h = 0.01$. The right column shows reconstructions where v^{10} of family (3.9) is the reconstruction kernel. Here the γ-values are those from Table 2.

As expected, increasing γ results in smoother reconstructions and larger relative errors as the support of the mollifier increases. The sharpest contrasts can be observed in both top images. In the middle and bottom images on the left the small white ellipse with the large eccentricity merges completely in the scull. In the original phantom (Figure 7) this ellipse is clearly separated from the scull.

Figure 9 allows for a more quantitative evaluation of the reconstruction errors. We plotted both the original and reconstructed values along a column. To be precise we plotted the entries 30 to 229 of the middle column 126. The cross sections are shown at the same locations as the corresponding full reconstructions in Figure 8. Surprisingly, the reconstructions with the smallest overall-errors (top images of Figures 8 and 9) show a shift in the reconstructed values along the cross section. The shift vanishes with the next larger admissible γ-values (middle images) only to be slightly present again for the largest γ-values (bottom images). The behavior just described occurs also in neighboring cross sections. Why does the shift not depend monotonically on the scaling parameter? Which feature of the reconstruction kernels causes the shift? Finding answers yields a better understanding of reconstruction kernels and may finally help to improve the quality of the reconstructed images.

We present reconstructions and corresponding cross sections using the Gaussian kernel (3.11) and the Shepp-Logan kernel (6.1) in Figures 10 and 11, respectively. The Gaussian kernel leads to smooth reconstructions which are robust to noisy data. This robustness has been observed by Dietz [**Die99**].

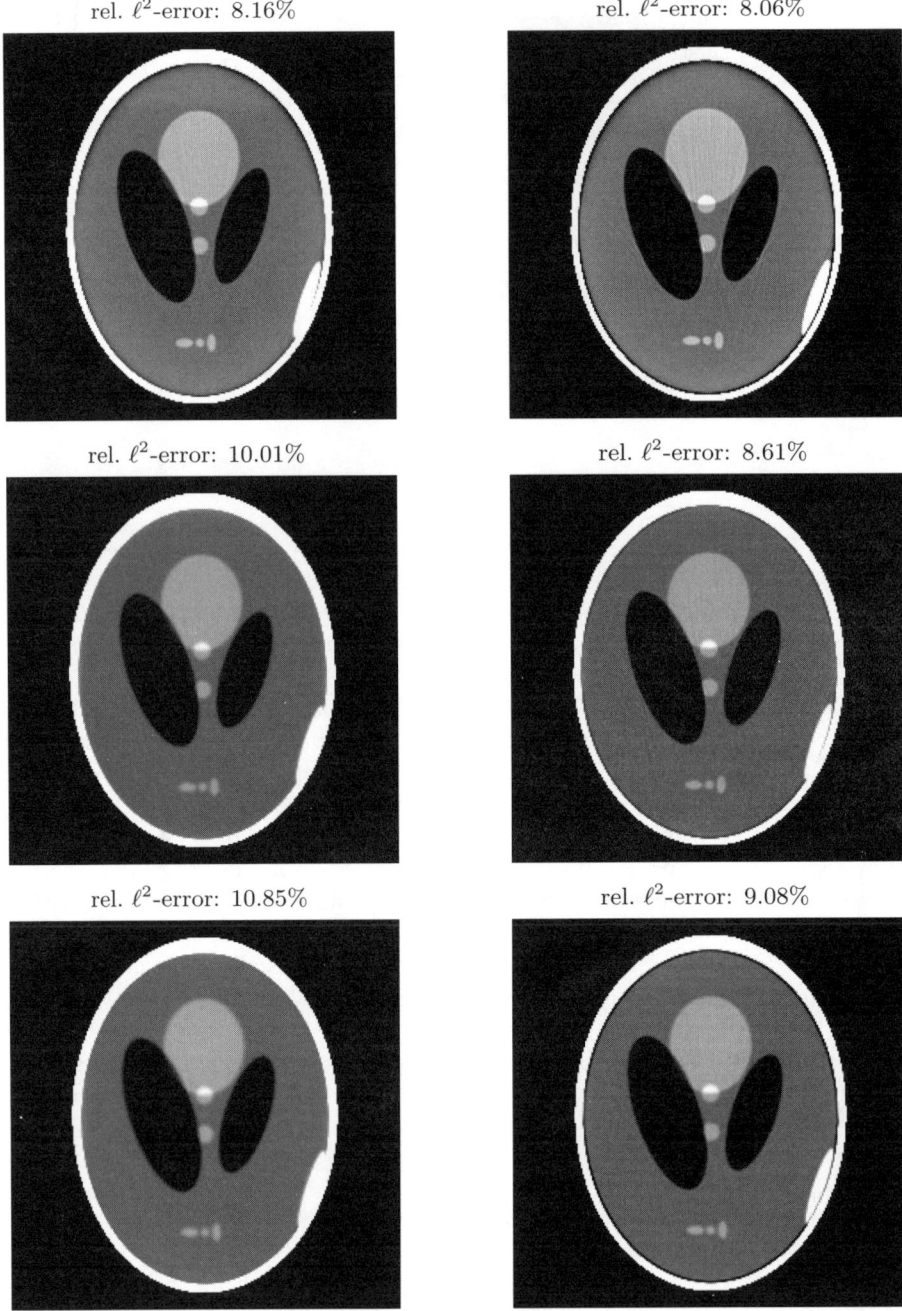

FIGURE 8. Reconstructions of f^{sl}. Left column: reconstruction kernel v^6 from family (3.7), right column: reconstruction kernel v^{10} from family (3.9). The scaling parameter γ of the kernels increases from top to bottom. The values of γ are the respective ones listed in Table 1 and Table 2 for $h = 0.01$. The relative ℓ^2-error in the support of f^{sl} is printed on top of each reconstruction.

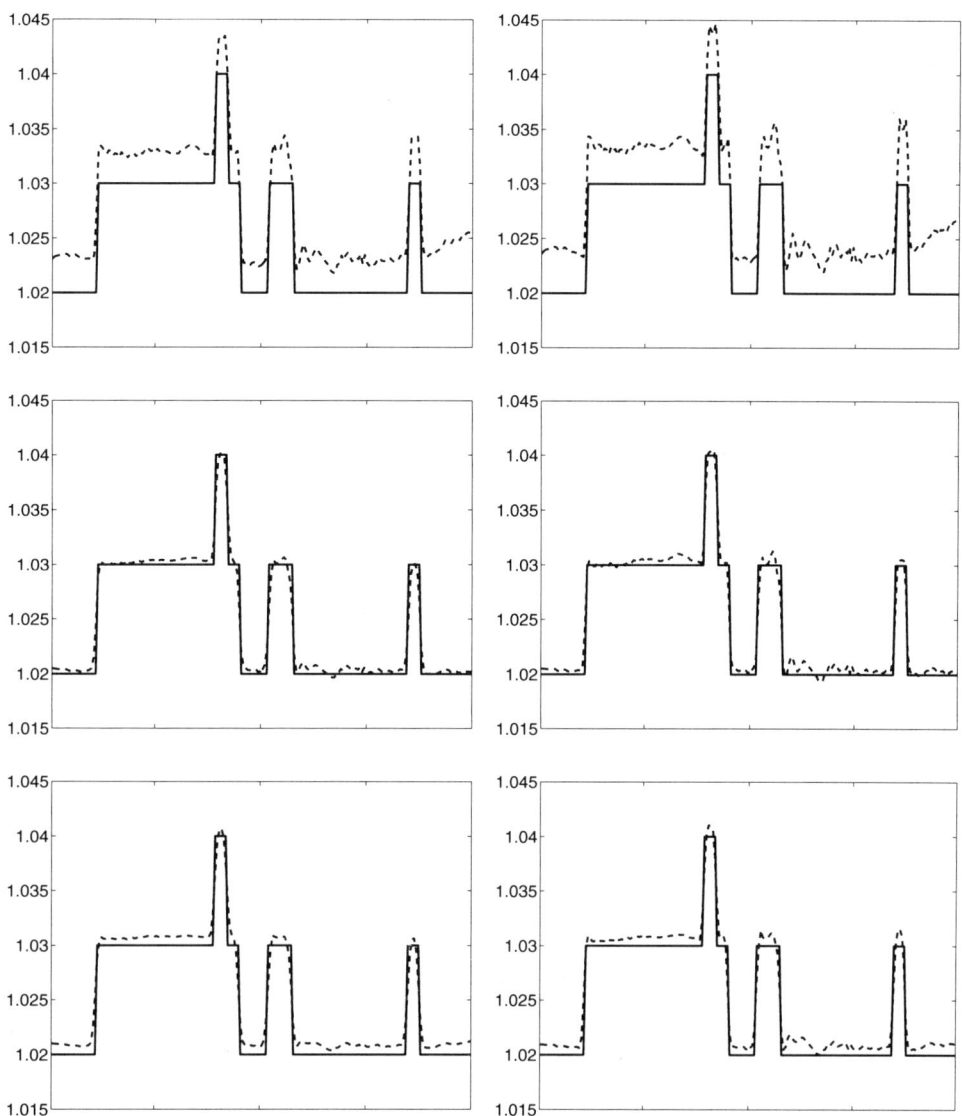

FIGURE 9. Cross sections (dashed lines) of the reconstructions displayed in Figure 8 at the same location. The corresponding cross section of the original $f^{\text{sl}}(29:230,126)$ is drawn using a solid line.

rel. ℓ^2-error: 11.55%

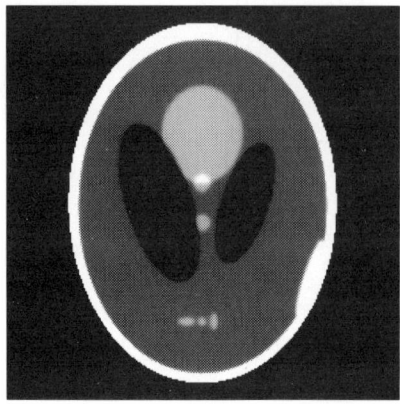 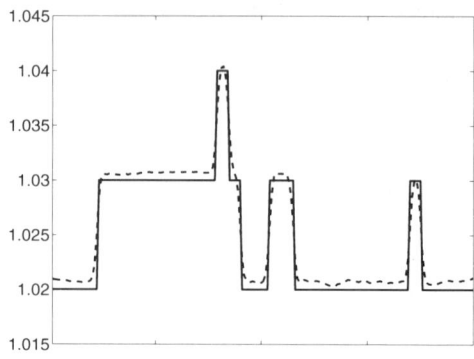

FIGURE 10. Left: reconstruction of f^{sl} using the Gaussian kernel (3.11). The scaling parameter is the one from Table 3. Right: Corresponding cross section as in Figure 9.

rel. ℓ^2-error: 8.62%

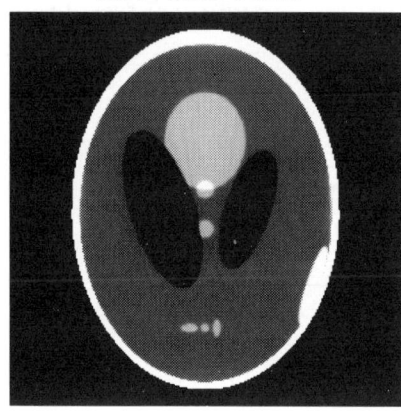 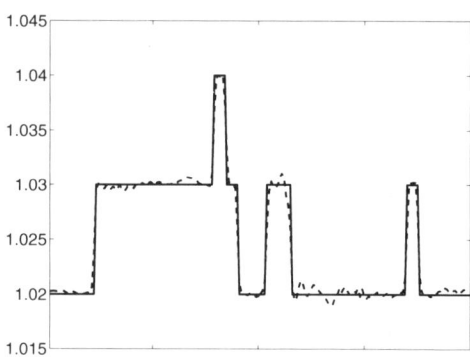

FIGURE 11. Left: reconstruction of f^{sl} using the Shepp-Logan kernel (6.1). Right: Corresponding cross section as in Figure 9.

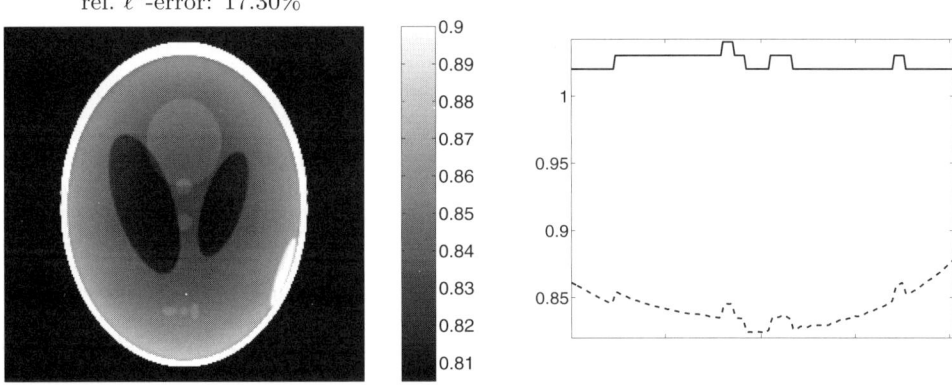

FIGURE 12. Left: reconstruction of f^{sl} using kernel v^6 from (3.7) with wrong scaling parameter $\widetilde{\gamma} = 0.0177$. The reconstruction with the correct scaling $\gamma = 0.017652990\ldots$ is displayed in Figure 8 (top left). Right: Corresponding cross section (dashed line) as in Figure 9 ($f^{\text{sl}}(29:230, 126)$ solid line).

Finally, we demonstrate what happens if the mean value condition (4.15) is not satisfied. To this end we slightly perturbed $\gamma = 0.017652990\ldots$, the smallest zero of $M(\cdot, 0.01)$ with respect to v^6 from (3.7). As perturbed scaling parameter we worked with $\widetilde{\gamma} = 0.0177$ having a relative magnitude of perturbation less than 0.3%. The reconstruction with $v^6_{\widetilde{\gamma}}$ is shown in Figure 12 (left). In perturbing the exact γ the relative ℓ^2-error explodes from 8.16% to 17.30%. This instability in γ can be explained from the graph of $M(\cdot, 0.01)$ near its smallest zero, see Figure 4 (left). As the graph is very steep near $\gamma = 0.017652990\ldots$ the numerical value $M(\widetilde{\gamma}, 0.01) \approx 0.74$ is far from being zero. Hence, (4.15) is strongly violated.

Nevertheless, the wrong reconstruction reveals the correct geometry. All geometric parameters of the 11 ellipses of the Shepp-Logan phantom are resolved. In view of (4.12) we are able to provide an explanation. As (4.15) is violated the reconstruction error δf is dominated by a multiple of $\Lambda^{-1} f$ given by $\widehat{\Lambda^{-1} f}(\xi) = \widehat{f}(\xi)/\|\xi\|$, compare (2.1). Thus, $f_R \approx c\, \Lambda^{-1} f + f \star e_{\widetilde{\gamma}}$ with a number $c \neq 0$, that is, we reconstruct $f \star e_{\widetilde{\gamma}}$ contaminated by a multiple of $\Lambda^{-1} f$. The latter artefact also appears in local tomography when the kernel sum does not vanish, see Faridani [**Far90**]. Since the operator Λ^{-1} can be used to reduce the cupping effect in local tomography it has been studied intensively by Faridani et al. [**FRS92, FFRS97**], see also Smith and Keinert [**SK85**]. For instance, $\Lambda^{-1} f$ is smoother than f (continuous at least) and is cupped in regions where f is constant. The superposition of $f \star e_{\widetilde{\gamma}}$ by $c\, \Lambda^{-1} f$ can be observed clearly in the cross section shown in Figure 12 (right).

8. Concluding remarks

The proposed scheme to scale reconstruction filters for the filtered backprojection algorithm works fine in 2D as long as the parallel scanning geometry is considered. The analytic techniques used to derive the mean value condition (4.15) are limited to this geometry. However, many commercial scanners realize the fan beam geometry to speed up data acquisition (especially in medical imaging). Therefore, an extension of our results to the fan beam geometry is important.

References

[CH80]   T. Chang and G. T. Herman, *A scientific study of filter selection for a fan-beam convolution reconstruction algorithm*, SIAM J. Appl. Math. **39** (1980), 83–105.

[Die99] R. Dietz, *Die Approximative Inverse als Rekonstruktionsmethode in der Röntgen-Computertomographie (approximate inverse as reconstruction method in computerized tomography)*, Ph.D. thesis, Universität des Saarlandes, Fachbereich Mathematik, 66041 Saarbrücken, Germany, 1999.

[DR75]   P. J. Davis and P. Rabinowitz, *Methods of numerical integration*, Computer Science and Applied Mathematics, Academic Press, New York, 1975.

[Far90] A. Faridani, *Praktische Fragen der lokalen Tomographie*, Z. Angew. Math. Mech. **70** (1990), T530–T532.

[FFRS97] A. Faridani, D. Finch, E. Ritman, and K. T. Smith, *Local tomography II*, SIAM J. Appl. Math. **57** (1997), no. 4, 1095–1127.

[FRS92] A. Faridani, E. Ritman, and K. T. Smith, *Local tomography*, SIAM J. Appl. Math. **52** (1992), no. 2, 459–484, 1193–1198.

[GR80]   I. S. Gradshteyn and I. M. Ryzhik, *Table of integrals, series, and products*, Academic Press, San Diego, 1980.

[Nat86] F. Natterer, *The mathematics of computerized tomography*, Wiley, Chichester, 1986.

[Nat99] F. Natterer, *Numerical methods in tomography*, Acta Numerica, Cambridge University Press, New York, 1999, pp. 107–141.

[RDS00] A. Rieder, R. Dietz, and Th. Schuster, *Approximate inverse meets local tomography*, Math. Meth. Appl. Sci. **23** (2000), no. 15, 1373–1387.

[RL71]   G. N. Ramachandran and A. V. Lakshminarayanan, *Three dimensional reconstruction from radiographs and electron micrographs: application of convolutions instead of Fourier transforms*, Proc. Nat. Acad. Sci. USA **68** (1971), 2236–2240.

[RS00]   A. Rieder and Th. Schuster, *The approximate inverse in action with an application to computerized tomography*, SIAM J. Numer. Anal. **37** (2000), no. 6, 1909–1929.

[SK85]   K. T. Smith and F. Keinert, *Mathematical foundations of computed tomography*, Applied Optics **24** (1985), 3950–3957.

[SL74]   L. A. Shepp and B. F. Logan, *The Fourier reconstruction of a head section*, IEEE Trans. Nuc. Sci. **21** (1974), 21–43.

[Smi82] K. T. Smith, *Reconstruction formulas in computed tomography*, Proc. Symp. Appl. Math. **27** (1982), 7–23.

Institut für Wissenschaftliches Rechnen und Mathematische Modellbildung, Universität Karlsruhe, 76128 Karlsruhe, Germany

E-mail address: `andreas.rieder@mathematik.uni-karlsruhe.de`

The k-dimensional Radon Transform on the n-sphere and Related Wavelet Transforms

Boris Rubin and Dmitry Ryabogin

ABSTRACT. Continuous wavelet transforms, associated with the k-dimensional spherical Radon transform Rf on the n-dimensional unit sphere $S^n, n \geq 2$, are introduced. It is assumed that $f \in L^p(S^n)$, $1 \leq p < \infty$, or $f \in C(S^n)$. For the operator R and for its left inverse R^{-1} explicit representations are given in terms of the relevant continuous wavelet transforms.

1. Introduction and main results

Let $S^n \subset \mathbb{R}^{n+1}$ be the unit sphere and let Ξ be the set of all k-dimensional totally geodesic submanifolds of S^n, $1 \leq k \leq n-1$. Given a continuous function f on S^n, consider the k-dimensional spherical Radon transform

$$(1.1) \qquad Rf(\xi) = \int_\xi f(x) dm(x), \qquad \xi \in \Xi,$$

where dm is the natural measure on ξ induced by the Lebesgue measure on S^n and normalized so that $\int_\xi dm(x) = 1$. This transform was studied by different authors (see, e.g., [1], [6], [7], [8], [11], [13], [24]) and plays an important role in geometrical problems for convex bodies ([2], [9], [10], [28]). In the present article we develop a wavelet approach to the inversion problem for (1.1). The following approaches to this problem, which differ from ours, are known in the literature.

S. Helgason [11] suggested two inversion procedures based on the duality principle. The first formula reads ([11], p. 93)

$$(1.2) \qquad f = P(\Delta) R^* Rf,$$

and works only for k even. Here $P(\Delta)$ is a certain polynomial of the Laplace-Beltrami operator on S^n, R^* is the dual transform which designates the average

---

2000 *Mathematics Subject Classification.* 44A12.
Key words and phrases. The spherical Radon transform, continuous wavelet transforms.
The first author was supported in part by the Edmund Landau Center for Research in Mathematical Analysis and Related Areas, sponsored by the Minerva Foundation (Germany).

© 2001 American Mathematical Society

over the set of all ξ containing x. The second inversion formula ([11], p. 99), which works for all $1 \le k \le n-1$, is as follows:

$$(1.3) \qquad f(x) = \frac{c}{2}\Bigl[\Bigl(\frac{d}{d(u^2)}\Bigr)^k \int_0^u (R^*_{\cos^{-1}(v)} Rf)(x) v^k (u^2 - v^2)^{k/2-1} dv \Bigr]\Big|_{u=1}.$$

Here $c^{-1} = (k-1)! \, |S^k|/2^{k+1}$, $|S^k|$ is the area of the unit k-sphere, $R^*_{\cos^{-1}(v)} Rf$ is the average of Rf over the set of all ξ at the distance $p = \cos^{-1}(v)$ from x. This formula is based on the observation that $R_p^* Rf$ can be written as a fractional integral of order $k/2$ of a certain average of f (see [11], pp. 98, 99).

R. Strichartz ([24], p. 725) suggested the following inversion formulas:

$$(1.4) \qquad f = c_k^{-1} \mathcal{R}(-\widetilde{\Delta})^{k/2} E_{-1} R^* Rf = c_k^{-1} \mathcal{R}(-\widetilde{\Delta})^{k/2} E_{k-n} R^* R.$$

Here $c_k = 2^k \Gamma((k+1)/2) \Gamma(n/2) / \sqrt{\pi} \; \Gamma((n-k)/2)^1$, $\mathcal{R}$ denotes the restriction operator from functions on $\mathbb{R}^{n+1}$ to S^n; $E_\lambda f$ is the extension of f to a homogeneous function of degree λ; $\widetilde{\Delta}$ denotes the Laplacian on $\mathbb{R}^{n+1}$. The idea of this approach goes back to V.I. Semyanistyi [22, 23] who studied the case $k = n-1$.

T. Kakehi ([13], p. 319) showed how to use (1.2) in the case of k odd. He constructed an operator L such that LRf is the $(k+1)$-dimensional spherical Radon transform and then applied (1.2) to LRf.

One should mention the papers by I. M. Gelfand, S. G. Gindikin, M. I. Graev [6], S. G. Gindikin [7], E. L. Grinberg [8] (and references therein), where inversion formulas are given in the context of projective spaces. A series of inversion formulas like (1.2), which work for all $1 \le k \le n-1$ and involve fractional integrals associated with Rf, was obtained by B. Rubin [18, 19].

There is a number of remarkable papers by D'Agnolo, F. B. Gonzalez, E. L. Grinberg, T. Kakehi, E. T. Quinto and others devoted to the range characterization of Rf for infinitely smooth f (see [11] for references). In the present paper, we do not concern ourselves with this problem and only recall some facts related to L^p-spaces. In the case $k = n-1$ the range $R(L^2(S^n))$ coincides with the Sobolev space $L^2_{(n-1)/2}(S^n)$ (on even functions). For $R(L^p(S^n))$, $p \ne 2$, it is not so, and one has the proper imbeddings [17]

$$L^p_{\delta, even}(S^n) \subset R(L^p_{even}(S^n)) \subset L^p_{\gamma, even}(S^n),$$

$\delta = (n-1)/2 + |1/p - 1/2|(n-1)$, $\gamma = (n-1)/2 - |1/p - 1/2|(n-1)$, $1 < p < \infty$, with the best possible parameters. Similar results for $k < n-1$ can be found in [24]. In the case $k = n-1$ exact characterization of the ranges $R(L^p(S^n))$, $1 \le p < \infty$ and $R(C(S^n))$ was given in [17] in terms of the relevant wavelet transforms. For $k < n-1$, $p \ne 2$, the characterization of $R(L^p(S^n))$ is not known. In order to give a flavour of the structure of the range $R(C^\infty(S^n))$ (or $R(L^2(S^n))$) for $k < n-1$, consider an orthonormal system of spherical harmonics of even degree on S^n. It can be "lifted" to the relevant Stiefel (or Grassmann) manifold and closed in the corresponding C^∞-topology (or the $L^2_{k/2}$ Sobolev norm). This closure coincides with $R(C^\infty(S^n))$ (or $R(L^2(S^n))$); see [18, 24] for details.

---

[1] In [24] this constant is given with misprints. We present it in the correct form.

Let us return to (1.2)-(1.4). These formulae work pointwise for sufficiently good smooth functions (in (1.4) some orthogonality conditions should be imposed; see [23]). For arbitrary even $f \in L^p(S^n)$ and $f \in C(S^n)$, all "doubtful" operations in (1.2)-(1.4) can be treated somehow in the framework of the distribution theory. This way is not so interesting, and it is natural to ask whether (1.2)-(1.4) are applicable to L^p-functions (or continuous functions) directly. This question was investigated in [19]. The basic difficulties are as follows. It is known [24] that

$$R^*Rf(x) = c_k \int_{S^n} (1 - |x \cdot y|^2)^{(k-n)/2} f(y) dy \stackrel{\text{def}}{=} Q^k f(x), \qquad c_k = \text{const.}$$

By the Funk-Hecke formula, the Fourier-Laplace multiplier of Q^k has the form

$$\frac{\Gamma((j+n-k)/2)\,\Gamma((j+1)/2)}{\Gamma((j+k+1)/2)\,\Gamma((j+n)/2)} \qquad (\sim j^{-k} \quad \text{as} \quad j \to \infty)$$

for j even and 0 for j odd. Owing to this, Q^k is an operator of the potential type acting from $L^p(S^n)$ into the Sobolev space $L^p_k(S^n)$ [15]. Investigation of the potential type operators on S^n in L^p- and Lipschitz spaces was carried out by Vakulov [25, 26] in a full generality. In particular, by the Hardy-Littlewood-Sobolev theorem, Q^k is a linear bounded operator from $L^p(S^n)$ into $L^q(S^n)$, $q = np(n-kp)^{-1}$, for $1 < p < n/k$, and this result is sharp. Thus, if $k < n/p$, a pointwise differentiation in (1.2) may be impossible in principle. Similar difficulties arise in connection with Semyanistyi-Strichartz' formula (1.4) in which $(-\tilde{\Delta})^{k/2}$ is understood in the sense of Φ'-distributions where Φ stands for the space of Schwartz functions orthogonal to all polynomials (see [22, 23] for details). The passage from S^n to $\mathbb{R}^{n+1}$ yields additional problems related to potentials on $\mathbb{R}^{n+1}$. In principle, realization of $(-\tilde{\Delta})^{k/2}$ in the context of L^p-spaces is possible in terms of hypersingular integrals or, more generally, by using wavelet type representations [15, 20], but this way looks too complicated and not so natural by taking into account that the problem was originally stated on S^n rather than on $\mathbb{R}^{n+1}$.

Helgason's formula (1.3) contains an additional averaging parameter $\cos^{-1}(v)$ which gives more freedom. One can show [19] that (1.3) can be applied directly to $f \in L^p(S^n)$, $1 \leq p < \infty$, provided that all derivatives are interpreted in the a.e. sense or in the L^p-norm (with respect to the x-variable). If $f \in C(S^n)$, (1.3) is true in the usual pointwise sense.

In order to overcome the aforementioned difficulties, B. Rubin [17] suggested "a wavelet approach" to the problem and inverted (1.1) for $k = n - 1$ by using suitable wavelet transforms.

Main results. In the present paper we extend the results from [17] to all $1 \leq k \leq n - 1$. Our inversion formula has a simple form and agrees with the general philosophy developed in [16]. Following [6] and [24], we regard (1.1) as a function $Rf(v)$ on the Stiefel manifold $V = V_{n+1,n-k}$ of all orthonormal $(n-k)$-frames in $\mathbb{R}^{n+1}$. Here $v = (v^1, \ldots, v^{n-k}) \in V$ is an $(n+1) \times (n-k)$ matrix with pairwise orthogonal unit column vectors $v^1, \ldots, v^{n-k} \in \mathbb{R}^{n+1}$.

The Radon transform (1.1) can be written as

$$(1.5) \qquad Rf(v) = \frac{1}{|S^k|} \int_{S^k} f(r_v \eta) d\eta, \quad v \in V.$$

Here and on $d\eta$ is the natural Lebesgue measure on S^k, $r_v \in SO(n+1)$ is a rotation such that $r_v v_0 = v$ where $v_0 = (e_{k+2}, \ldots, e_{n+1})$ is the coordinate $(n-k)$-frame, $e_i = (0, \ldots, \overset{i}{1}, 0, \ldots, 0) \in \mathbb{R}^{n+1}$.

For $f \in L^1(S^n)$ and $\varphi \in L^1(V)$, the intertwining continuous wavelet transforms associated with (1.1) are defined by

$$(1.6) \qquad Wf(v,t) = t^{k-n} \int_{S^n} f(x) w(|x \cdot v|/t) dx, \qquad v \in V, \quad t > 0,$$

$$(1.7) \qquad \overset{*}{W}\varphi(x,t) = t^{k-n} \int_{V} \varphi(v) w(|x \cdot v|/t) dv, \qquad x \in S^n, \quad t > 0.$$

Here w is a sufficiently nice "wavelet function" on $\mathbb{R}_+ = [0, \infty)$; $x \cdot v$ is an $(n-k)$-vector defined by

$$x \cdot v = [x_1, \ldots, x_{n+1}] \begin{bmatrix} v_1^1 \ldots v_1^{n-k} \\ \ldots \ldots \ldots \\ v_{n+1}^1 \ldots v_{n+1}^{n-k} \end{bmatrix} = [\sum_{j=1}^{n+1} x_j v_j^1, \ldots, \sum_{j=1}^{n+1} x_j v_j^{n-k}];$$

dx and dv are the corresponding $SO(n+1)$-invariant measures on S^n and V, normalized so that $\sigma_n = |S^n| = 2\pi^{(n+1)/2}/\Gamma((n+1)/2)$ and $|V| = \sigma_n \sigma_{n-1} \cdots \sigma_{k+1}$ (see e.g., [12], [21, p. 208]). For $k = n-1$, transforms (1.6) and (1.7) are identical and coincide with those in [17].

THEOREM A. *Let*

$$(1.8) \qquad \int_0^\infty \tau^{j+n-k-1} w(\tau) d\tau = 0 \quad \forall j = 0, 2, 4, \ldots, 2[k/2],$$

($[k/2]$ *is the integral part of* $k/2$),

$$(1.9) \qquad \int_1^\infty \tau^{\beta+n-k-1} |w(\tau)| d\tau < \infty \qquad \text{for some } \beta > k.$$

Suppose that $\varphi(v) = Rf(v)$, $v \in V$, *where* f *is an even function belonging to* $L^p(S^n)$, $1 \leq p < \infty$, *or* $f \in C(S^n)$. *Then*

$$(1.10) \qquad \int_0^\infty \frac{(\overset{*}{W}\varphi)(x,t)}{t^{k+1}} dt = \lim_{\varepsilon \to 0} \int_\varepsilon^\infty \frac{(\overset{*}{W}\varphi)(x,t)}{t^{k+1}} dt = \alpha \, f(x),$$

$$(1.11) \qquad \alpha = \frac{|V| \Gamma(\frac{n+1}{2})}{\pi^{1/2} \Gamma(\frac{n-k}{2})} \begin{cases} \Gamma(-k/2) \int_0^\infty \tau^{n-1} w(\tau) d\tau & \text{if } k \text{ is odd,} \\ \frac{2(-1)^{1+k/2}}{(k/2)!} \int_0^\infty \tau^{n-1} w(\tau) \log \tau d\tau & \text{if } k \text{ is even.} \end{cases}$$

The limit in (1.10) is understood in the L^p-*norm and in the a.e. sense. For* $f \in C(S^n)$, *the limit in (1.10) is interpreted in the sup-norm.*

The next statement contains an analogue of the Calderón reproducing formula for the k-dimensional spherical Radon transform.

THEOREM B. *Let*

(1.12) $$\int_0^\infty w(\tau)\tau^{n-k-1}d\tau = 0, \qquad \int_0^\infty |w(\tau)\log\tau|\tau^{n-k-1}d\tau < \infty.$$

If $f \in L^p(S^n)$, $1 \leq p < \infty$, *or* $f \in C(S^n)$, *then*

(1.13)
$$\int_0^\infty \frac{(Wf)(v,t)}{t}dt = \lim_{\varepsilon \to 0}\int_\varepsilon^\infty \frac{(Wf)(v,t)}{t}dt = \beta\, Rf(v),$$

$$\beta = \frac{4\pi^{(n-k)/2}}{\Gamma((n-k)/2)}\int_0^\infty w(\tau)\tau^{n-k-1}\log\frac{1}{\tau}d\tau,$$

where the limit is understood in the L^p-norm or in the sup-norm.

REMARK. One can define the Radon transform (1.1) as a function on the dual Stiefel manifold $\tilde{V} = V_{n+1,k+1}$:

$$\tilde{R}f(u) = \frac{1}{|S^k|}\int_{S^k} f(\tilde{r}_u x)d\sigma(x), \quad u \in \tilde{V}.$$

Here $\tilde{r}_u \in SO(n+1)$ so that $\tilde{r}_u u_0 = u$; $u_0 = (e_1, \ldots, e_{k+1}) \in \tilde{V}$ is the coordinate $(k+1)$-frame. It is clear that $\tilde{R}f(u) = Rf(v)$ provided that columns of the $(n+1) \times (n+1)$-matrix (u,v) generate an orthonormal basis of $\mathbb{R}^{n+1}$. One can reformulate Theorems A and B in terms of $\tilde{R}f$. In this case the intertwining continuous wavelet transforms, associated with $\tilde{R}f$, are defined by:

(1.14) $$Uf(u,t) = t^{k-n}\int_{S^n} f(x)w(\sqrt{1-|x\cdot u|^2}/t)dx, \qquad u \in \tilde{V}, \ t > 0,$$

(1.15) $$\overset{*}{U}f(x,t) = t^{k-n}\int_{\tilde{V}} \varphi(u)w(\sqrt{1-|x\cdot u|^2}/t)du, \qquad x \in S^n, \ t > 0.$$

Motivation. Why do we seek new inversion formulae although so many are available? To answer this question we recall that similar situation occurred in *Fractional Calculus*, the branch of analysis which studies fractional integrals and derivatives of functions of one and several variables [15, 20]. The point is that numerous transforms of the Radon type can be immersed in suitable analytic families of fractional integrals generalizing the classical ones like those of Riemann-Liouville on the real line or Riesz potentials on $\mathbb{R}^n$. A remarkable feature of these fractional integrals is that the set of singularities of the corresponding kernel is a manifold of dimension ≥ 1. This observation was used implicitly by I.M. Gelfand (with collaborators), V.I. Semyanistyi, E.E. Petrov, and developed in [16] in the context of L^p-spaces for various transforms of the Radon type. It enables one to apply methods and ideas of fractional calculus to problems of integral geometry.

In fractional calculus one discriminates between the Riemann-Liouville and Marchaud fractional derivative. In fact, they represent different forms of analytic continuation of the same object, namely, the Riemann-Liouville fractional integral

$(I_+^\alpha \psi)(t) = (1/\Gamma(\alpha)) \int_{-\infty}^t \psi(\tau)(t-\tau)^{\alpha-1} d\tau$ (see [20, Section 5] and [15, Section 10] for discussion and further details). The derivative of Marchaud is "more sensitive" to function spaces (L^p, C, Lipschitz or whatever) one deals with, while implementation of the Riemann-Liouville fractional derivative for the same purposes may have some restrictions and needs additional justification. In many dimensions the method of Marchaud leads to hypersingular integrals which proved to be a powerful tool in function theory for characterization of spaces of fractional smoothness and inversion of operators of the potential type (see, e.g., the papers by E.M. Stein, P.I. Lizorkin, S.G. Samko, B. Rubin and others mentioned in [15, 20]). Wavelet type representations of fractional integrals and derivatives generalize Marchaud's constructions and are more flexilble. They are especially important in numerical calculations where it is desirable to have the wavelet w smooth and well localized. The philosophy of such a generalization was developed by B. Rubin [15] who extended this idea to operators of integral geometry [16].

The favour of the wavelet approach to studying singularities of higher dimension was also noticed by other authors. At about the same time continuous wavelet transforms, associated to the Radon transform on $\mathbb{R}^n$, were introduced independently by D.L. Donoho and E.J. Candès [3-5], who gave them a new name *ridgelet transforms* (see also N. Murata [14] for application to analysis of neural networks).

In a sense, formulae (1.2)-(1.4) can be viewed as those of the Riemann-Liouville type, involving additional averaging operators which are inevitable in the integral geometrical set up. The formula (1.10) can be treated as that of the Marchaud type (in wavelet interpretation). An important feature of (1.10) is that one can choose any wavelet function w he likes. Note also that (1.10) has the same form as inversion formulas for many other transforms of Radon type written in the wavelet language [16].

Acknowledgement. We are grateful to the referee and to Eric Todd Quinto for valuable comments and suggestions.

2. Preliminaries

LEMMA 2.1. *Let $f \in L^1(S^n)$, $f \geq 0$. Then*

$$(2.1) \qquad \|Rf\|_{L^1(V)} = \sigma_n^{-1} |V| \, \|f\|_{L^1(S^n)}.$$

PROOF. By (1.5),

$$\|Rf\|_{L^1(V)} = \frac{1}{\sigma_k} \int_{S^k} d\eta \int_V f(r_v \eta) dv = \frac{1}{\sigma_k} \int_{S^k} d\eta \int_V dv \int_{SO(n+1)} f(r_{\gamma v} \eta) d\gamma$$

$$= \frac{|V|}{\sigma_k} \int_{S^k} d\eta \int_{SO(n+1)} f(\alpha \eta) d\alpha = \frac{|V|}{\sigma_n} \|f\|_{L^1(S^n)}. \qquad \square$$

We shall use the bispherical coordinates on S^n (see, e.g., [27], p. 12) defined by

$$(2.2) \qquad x = \xi \cos\theta + \eta \sin\theta,$$

$$\xi \in S^{n-k-1} \subset \mathbb{R}^{n-k}, \qquad \eta \in S^k \subset \mathbb{R}^{k+1}, \qquad 0 \leq \theta \leq \frac{\pi}{2},$$

$$\mathbb{R}^{k+1} = span\ (e_1,\ldots,e_{k+1}); \qquad \mathbb{R}^{n-k} = span\ (e_{k+2},\ldots,e_{n+1}).$$

According to (2.2),
$$dx = \sin^k\theta \cos^{n-k-1}\theta\ d\theta d\xi d\eta.$$

For $v \in V$ and $\theta \in [0,\pi/2]$, we introduce a mean-value operator

(2.3) $$M_{\cos\theta}f(v) = \frac{1}{\sigma_{n-k-1}\sigma_k} \int_{S^{n-k-1}} d\xi \int_{S^k} f(r_v(\xi\cos\theta + \eta\sin\theta))d\eta,$$

so that
$$\int_{S^n} f(x)dx = \sigma_{n-k-1}\sigma_k \int_0^{\pi/2} \sin^k\theta \cos^{n-k-1}\theta\ M_{\cos\theta}f(v)d\theta.$$

In particular, for $\theta = \pi/2$ we have $M_0 f(v) = Rf(v)$.

LEMMA 2.2. *Let $f \in L^p(S^n)$, $1 \leq p < \infty$. Then*

(2.4) $$\sup_{\theta\in[0,\pi/2]} \|M_{\cos\theta}f\|_{L^p(V)} \leq (|V|/\sigma_n)^{1/p} \|f\|_{L^p(S^n)},$$

(2.5) $$\lim_{\theta\to\pi/2} \|M_{\cos\theta}f - Rf\|_{L^p(V)} = 0.$$

If $f \in C(S^n)$, then $M_{\cos\theta}f \to Rf$ as $\theta \to \pi/2$ uniformly on S^n.

PROOF. By (2.3) and Minkowski's inequality,

$$\|M_{\cos\theta}f\|_{L^p(V)} \leq \frac{|V|^{1/p}}{\sigma_{n-k-1}\sigma_k} \int_{S^{n-k-1}} d\xi \int_{S^k} d\eta \left(\int_{SO(n+1)} |f(r_{gv_0}(\xi\cos\theta + \eta\sin\theta))|^p dg \right)^{1/p}$$

$$= (|V|/\sigma_n)^{1/p} \|f\|_{L^p(S^n)}.$$

Furthermore, $\|M_{\cos\theta}f - Rf\|_{L^p(V)}$ does not exceed

$$\frac{|V|^{1/p}}{\sigma_{n-k-1}\sigma_k} \int_{S^{n-k-1}} d\xi \int_{S^k} d\eta \left(\int_{SO(n+1)} |f(r_{gv_0}(\xi\cos\theta + \eta\sin\theta)) - f(r_{gv_0}\eta)|^p dg \right)^{1/p}.$$

The last expression tends to 0 as $\theta \to \pi/2$ provided $f \in C(S^n)$. Since $C(S^n)$ is dense in $L^p(S^n)$, $1 \leq p < \infty$, we have (2.5). The proof of $M_{\cos\theta}f \to Rf$ as $\theta \to \pi/2$ uniformly on S^n is similar with $\|\cdot\|_p$ replaced by the corresponding sup-norm. $\square$

Consider the intertwining operators which were introduced in [18]:
$$Af(v) = \int_{S^n} a(|x\cdot v|)f(x)dx, \qquad \overset{*}{A}\varphi(x) = \int_V a(|x\cdot v|)\varphi(v)dv.$$

LEMMA 2.3. *Let $f \in L^p(S^n)$, $\varphi \in L^p(V)$, $1 \leq p \leq \infty$. Then*

(2.6) $$\|Af\|_{L^p(V)} \leq \sigma_{n-k-1}\sigma_k\,(|V|/\sigma_n)^{1/p} \|a\|\ \|f\|_{L^p(S^n)},$$

(2.7) $$\|\overset{*}{A}\varphi\|_{L^p(S^n)} \leq \sigma_{n-k-1}\sigma_k\,(|V|/\sigma_n)^{1-1/p} \|a\|\ \|\varphi\|_{L^p(V)},$$

where $\|a\| = \int_0^{\pi/2} \sin^k\theta \cos^{n-k-1}\theta\ |a(\cos\theta)|d\theta$.

PROOF. Replacing $x \to r_v x$ and passing to the bispherical coordinates (2.2), we have:

$$(2.8) \qquad Af(v) = \sigma_{n-k-1}\sigma_k \int_0^{\pi/2} a(\cos\theta) \sin^k\theta \cos^{n-k-1}\theta M_{\cos\theta} f(v) d\theta.$$

Now (2.6) follows from (2.8), (2.4) and Minkowski's inequality; (2.7) follows from (2.6) by duality: $\int_{S^n} f(x)\overset{*}{A}\varphi(x)dx = \int_V \varphi(v)Af(v)dv$. □

Given $x \in S^n$ and $t \in (-1,1)$, denote

$$(2.9) \qquad \mathbb{M}_t f(x) = \frac{(1-t^2)^{(1-n)/2}}{\sigma_{n-1}} \int_{\{y \in S^n:\, x \cdot y = t\}} f(y) d\sigma(y),$$

where $x \cdot y = x_1 y_1 + \cdots + x_{n+1} y_{n+1}$ is a usual inner product in $\mathbb{R}^{n+1}$, $d\sigma(y)$ designates the corresponding Lebesgue measure induced by that on S^n. The integral (2.9) is the mean value of f on the planar section of S^n by the hyperplane $x \cdot y = t$.

LEMMA 2.4. (cf. [11], p. 97). *Let $r_x \in SO(n+1)$ be such that $r_x e_{n+1} = x \in S^n$. Suppose that $f \in L^1(S^n)$. Then*

$$(2.10) \qquad \int_{SO(n)} Rf(r_x \rho r_x^{-1} \gamma v_0) d\rho = \frac{1}{\sigma_k} \int_{S^k} \mathbb{M}_{x \cdot \gamma \eta} f(x) d\eta$$

for a. e. $x \in S^n$ and for any $\gamma \in SO(n+1)$.

PROOF. We start with the obvious formula

$$\int_{SO(n)} \varphi(\rho z) d\rho = \mathbb{M}_{z_{n+1}} \varphi(e_{n+1}), \qquad z \in S^n,$$

and set $z = r_x^{-1} \gamma \eta$, $\varphi(y) = f(r_x y)$. This yields

$$(2.11) \qquad \int_{SO(n)} f(r_x \rho r_x^{-1} \gamma \eta) d\rho = \mathbb{M}_{x \cdot \gamma \eta} f(x).$$

Since $Rf(\alpha v) = 1/\sigma_k \int_{S^k} f(\alpha r_v \eta) d\eta \; \forall \alpha \in SO(n+1)$, by (1.5) and (2.11) we obtain

$$\int_{SO(n)} Rf(r_x \rho r_x^{-1} \gamma v_0) d\rho = \frac{1}{\sigma_k} \int_{S^k} d\eta \int_{SO(n)} f(r_x \rho r_x^{-1} \gamma \eta) d\rho = \frac{1}{\sigma_k} \int_{S^k} \mathbb{M}_{x \cdot \gamma \eta} f(x) d\eta.$$

□

3. Proof of Theorem A.

In the following we deal with the Riemann-Liouville fractional integrals

$$(3.1) \qquad (I_{0+}^\alpha \psi)(t) = \frac{1}{\Gamma(\alpha)} \int_0^t \frac{\psi(\tau) d\tau}{(t-\tau)^{1-\alpha}}, \qquad (I_{1-}^\alpha \psi)(t) = \frac{1}{\Gamma(\alpha)} \int_t^1 \frac{\psi(\tau) d\tau}{(\tau-t)^{1-\alpha}},$$

$Re\,\alpha > 0$. The next statement generalizes Lemma 2.3 of [17].

LEMMA 3.1. *Let a be such that $\int_0^1 z^{n-k-1}(1-z^2)^{(k-1)/2}|a(z)|dz < \infty$, and let $f \in L^1(S^n)$ be even. Then for a.e. $x \in S^n$,*

$$(3.2) \quad \overset{*}{A}Rf(x) \equiv \int_V a(|x \cdot v|)Rf(v)dv = c_{k,n} \int_0^1 z^{n-k-1} a(z)(I_{1-}^{k/2} g)(z^2) dz,$$

where $g(u) = (1-u)^{-1/2} \mathbb{M}_{\sqrt{1-u}} f(x)$,

$$(3.3) \quad c_{k,n} = \frac{2|V|\,\Gamma((n+1)/2)}{\sqrt{\pi}\,\Gamma((n-k)/2)}.$$

PROOF. By (2.1) and (2.7) the expression $\overset{*}{A}Rf(x)$ is well-defined for a.e. $x \in S^n$. Fix any such x, and replace $v \to r_x v$, where $r_x \in SO(n+1), r_x e_{n+1} = x$. We get

$$\overset{*}{A}Rf(x) = \int_V a(|e_{n+1} \cdot v|)Rf(r_x v)dv = |V| \int_{SO(n+1)} a(|e_{n+1} \cdot \gamma v_0|)Rf(r_x \gamma v_0)d\gamma.$$

Next we replace $\gamma \to \rho r_x^{-1}\gamma, \rho \in SO(n)$, and integrate in ρ. Since $e_{n+1} \cdot \rho r_x^{-1}\gamma v_0 = e_{n+1} \cdot r_x^{-1}\gamma v_0$, we use (2.10) to obtain

$$\overset{*}{A}Rf(x) = |V| \int_{SO(n+1)} a(|e_{n+1} \cdot r_x^{-1}\gamma v_0|)d\gamma \int_{SO(n)} Rf(r_x \rho r_x^{-1}\gamma v_0)d\rho$$

$$= \frac{|V|}{\sigma_k} \int_{SO(n+1)} a(|e_{n+1} \cdot r_x^{-1}\gamma v_0|)d\gamma \int_{S^k} \mathbb{M}_{x \cdot \gamma \eta} f(x)d\eta \qquad (r_x^{-1}\gamma = \alpha)$$

$$= \frac{|V|}{\sigma_k} \int_{SO(n+1)} a(|e_{n+1} \cdot \alpha v_0|)d\alpha \int_{S^k} \mathbb{M}_{e_{n+1} \cdot \alpha \eta} f(x)d\eta$$

$$= \frac{|V|}{\sigma_k \sigma_n} \int_{S^n} a(|y \cdot v_0|)dy \int_{S^k} \mathbb{M}_{y \cdot \eta} f(x)d\eta$$

$$= \frac{2|V|}{\sigma_k \sigma_n} \int_{S^n} a(|y \cdot v_0|)dy \int_{y \cdot \eta > 0} \mathbb{M}_{y \cdot \eta} f(x)d\eta.$$

Passing to bispherical coordinates $y = \xi'\cos\theta + \eta'\sin\theta$, we have $|y \cdot v_0| = \cos\theta$, $y \cdot \eta = (\eta' \cdot \eta)\sin\theta$. This yields

$$\int_{y\eta > 0} \mathbb{M}_{y \cdot \eta} f(x)d\eta = \int_{\eta \cdot \eta' > 0} \mathbb{M}_{(\eta \cdot \eta')\sin\theta} f(x)d\eta = \sigma_{k-1} \int_0^1 (1-\tau^2)^{k/2-1} \mathbb{M}_{\tau \sin\theta} f(x)d\tau,$$

and therefore $\overset{*}{A}Rf(x)$ is equal to

$$\frac{2|V|\sigma_{k-1}\sigma_{n-k-1}}{\sigma_n} \int_0^{\pi/2} \sin^k\theta \cos^{n-k-1}\theta\, a(\cos\theta)d\theta \int_0^1 (1-\tau^2)^{k/2-1}\mathbb{M}_{\tau\sin\theta}f(x)d\tau$$

$$= \frac{2|V|\sigma_{k-1}\sigma_{n-k-1}}{\sigma_n} \int_0^1 z^{n-k-1}(1-z^2)^{(k-1)/2}a(z)dz \int_0^1 (1-\tau^2)^{k/2-1}\mathbb{M}_{\tau\sqrt{1-z^2}}f(x)d\tau.$$

Finally we replace $\tau\sqrt{1-z^2}$ by $\sqrt{1-u}$ and obtain (3.2). $\square$

Consider the dual wavelet transform (1.7) applied to $\varphi = Rf$. By setting

$$w_1(\tau) = \tau^{(n-k)/2-1}w(\sqrt{\tau}), \qquad h = I_{0+}^{k/2}w_1,$$

owing to Lemma 3.1, we get

COROLLARY 3.2.

(3.4) $$(\overset{*}{W}Rf)(x,t) = \frac{c_{k,n}\, t^{k-2}}{2} \int_0^1 g(u)h\left(\frac{u}{t^2}\right) du,$$

g and $c_{k,n}$ being the same as in Lemma 3.1.

PROOF. By (1.7) and (3.2),

$$(\overset{*}{W}Rf)(x,t) = \frac{c_{k,n}}{t^{n-k}} \int_0^1 z^{n-k-1}w\left(\frac{z}{t}\right)(I_{1-}^{k/2}g)(z^2)dz$$

$$= \frac{c_{k,n}}{\Gamma(k/2)} \int_0^{1/t} s^{n-k-1}w(s)ds \int_{t^2s^2}^1 g(u)(u-t^2s^2)^{k/2-1}du$$

$$= \frac{c_{k,n}}{2\Gamma(k/2)} \int_0^{1/t^2} \tau^{(n-k)/2-1}w(\sqrt{\tau})d\tau \int_\tau^{1/t^2} g(t^2r)(r-\tau)^{k/2-1}dr$$

$$= \frac{c_{k,n}\, t^k}{2} \int_0^{1/t^2} g(t^2r)(I_{0+}^{k/2}w_1)(r)dr = \frac{c_{k,n}\, t^{k-2}}{2} \int_0^1 g(u)h\left(\frac{u}{t^2}\right) du. \quad \square$$

REMARK 3.3. By Lemma 4.12 of [15], $h \in L^1(\mathbb{R}_+)$ provided

(3.5) $$\int_0^\infty \tau^j w_1(\tau)d\tau = 0 \; \forall j = 0, 1, \ldots, m = \begin{cases} k/2 - 1, & \text{if } k/2 \in \mathbb{N}, \\ [k/2], & \text{if } k/2 \notin \mathbb{N}, \end{cases}$$

(3.6) $$\int_1^\infty \tau^{k/2}|w_1(\tau)|d\tau < \infty.$$

PROOF OF THEOREM A. We proceed as in the proof of Theorem 1.1 from [17]. Denote

$$(3.7) \qquad \lambda(s) = \frac{1}{s}(I_{0+}^{k/2+1}w_1)(s), \quad \Lambda_\varepsilon(\tau) = \frac{c_{k,n}}{4\,\sigma_{n-1}} \frac{(1-\tau^2)^{1-n/2}}{\varepsilon^2} \lambda\left(\frac{1-\tau^2}{\varepsilon^2}\right),$$

$$(3.8) \qquad (\mathcal{L}_\varepsilon f)(x) = \int_{S^n} \Lambda_\varepsilon(x \cdot y) f(y) dy,$$

and prove the equality

$$(3.9) \qquad J_\varepsilon R f(x) \overset{\text{def}}{=} \int_\varepsilon^\infty (\overset{*}{W} Rf)(x,t) \frac{dt}{t^{k+1}} = (\mathcal{L}_\varepsilon f)(x).$$

By (3.4),

$$J_\varepsilon Rf = \frac{c_{k,n}}{2} \int_\varepsilon^\infty \frac{dt}{t^3} \int_0^1 g(u) h\left(\frac{u}{t^2}\right) du = \frac{c_{k,n}}{2} \int_0^1 g(u) du \int_\varepsilon^\infty h\left(\frac{u}{t^2}\right) \frac{dt}{t^3}$$

$$= \frac{c_{k,n}}{4} \int_0^{1/\varepsilon^2} \lambda(s) g(\varepsilon^2 s) ds.$$

Using (2.12) from [17] and the definition of g (see Lemma 3.1), we get

$$J_\varepsilon Rf = \frac{c_{k,n}}{4} \int_0^{1/\varepsilon^2} \lambda(s)(1-\varepsilon^2 s)^{-1/2} \mathbb{M}_{\sqrt{1-\varepsilon^2 s}} f ds$$

$$(3.10) \qquad = \frac{c_{k,n}}{4\,\varepsilon^2} \int_{-1}^1 (\mathbb{M}_\tau f)(x) \lambda\left(\frac{1-\tau^2}{\varepsilon^2}\right) d\tau$$

$$= \frac{c_{k,n}}{4\,\sigma_{n-1}} \int_{S^n} f(y) \left[\frac{(1-(x \cdot y)^2)^{1-n/2}}{\varepsilon^2} \lambda\left(\frac{1-(x \cdot y)^2}{\varepsilon^2}\right) \right] dy,$$

which gives (3.9).

To complete the proof it remains to show that $\mathcal{L}_\varepsilon f \to \alpha f$ as $\varepsilon \to 0$ in the required sense. For this purpose one can use a standard machinery of approximation to the identity. As in ([17], p. 212), for each spherical harmonic Y_j, j even, we have

$$(3.11) \qquad \lim_{\varepsilon \to 0} \mathcal{L}_\varepsilon Y_j \to \alpha Y_j, \quad \alpha = \frac{c_{k,n}}{4} \int_0^\infty \lambda(s) ds.$$

By Lemma 2.4 from [17],

$$\int_0^\infty \lambda(s) ds = \begin{cases} \Gamma(-k/2) \int_0^\infty s^{k/2} w_1(s) ds & \text{if } k/2 \notin \mathbb{N}, \\ \frac{(-1)^{1+k/2}}{(k/2)!} \int_0^\infty s^{k/2} w_1(s) \log s\, ds & \text{if } k/2 \in \mathbb{N}, \end{cases}$$

provided

$$\int_0^\infty s^j w_1(s)ds = 0, \quad \forall j = 0, 1, \ldots, [k/2],$$

$$\int_0^\infty s^\beta |w_1(s)|ds < \infty \text{ for some } \beta > k/2.$$

Thus, α in (3.11) has the form (1.11), and the result follows. $\square$

4. Proof of Theorem B

Fix any $t > 0$. By (2.6), $Wf(v,t)$ is well-defined for a.e. $v \in V$. Passing to bispherical coordinates in (1.6), we obtain (cf. (2.8)):

$$Wf(v,t) = \frac{\sigma_{n-k-1}\sigma_k}{t^{n-k}} \int_0^1 (1-\tau^2)^{(k-1)/2} \tau^{n-k-1} w\left(\frac{\tau}{t}\right) M_\tau f(v) d\tau,$$

and therefore

$$\int_\varepsilon^\infty \frac{Wf(v,t)}{t} dt = \sigma_{n-k-1}\sigma_k \int_0^1 (1-\tau^2)^{(k-1)/2} \tau^{n-k-1} M_\tau f(v) d\tau \int_\varepsilon^\infty w\left(\frac{\tau}{t}\right) \frac{dt}{t^{1+n-k}}$$

$$(4.1) \qquad = \sigma_{n-k-1}\sigma_k \int_0^{1/\varepsilon} (1-\varepsilon^2\tau^2)^{(k-1)/2} M_{\varepsilon\tau} f(v) k(\tau) d\tau,$$

$k(\tau) = \tau^{-1} \int_0^\tau w(s) s^{n-k-1} ds$. By (1.12),

$$k \in L^1(\mathbb{R}_+) \quad \text{and} \quad \int_0^\infty k(\tau) d\tau = \int_0^\infty w(s) s^{n-k-1} \log \frac{1}{s} ds$$

(a simple proof of this assertion can be found in ([15], p. 190). Owing to Lemma 2.2, one can pass to the limit in (4.1) in the L^p- and sup-norm as $\varepsilon \to 0$, and the required result follows. $\square$

References

[1] C. A. Berenstein, E. C. Tarabusi, and A. Kurusa, *Radon transform on spaces of constant curvature*, Proc. Amer. Math. Soc. **125** (1997), 455–461.

[2] J. Bourgain and G. Zhang, *On a generalization of the Buseman-Petty problem*, Collection: Convex geometric analysis (Berkeley, CA, 1996), Cambridge Univ. Press, Cambridge, 1999, pp. 65–76.

[3] E.J. Candès, *Ridgelets: Theory and applications*. Ph. D. Thesis, Department of Statistics, Stanford University, 1998.

[4] ———, *Harmonic analysis of neural networks*, Applied and Comp. Harm. Analysis **6** (1999), 197–218.

[5] D.L. Donoho, *Tight frames of k-plane ridgelets and the problem of representing d-dimensional singularities in $\mathbb{R}^n$*, Proc. Nat. Acad. Sci. USA **96** (1999), 1828–1833.

[6] I. M. Gelfand, S. G. Gindikin and M. I. Graev, *Integral geometry in affine and projective spaces*, Izrail M. Gelfand collected papers, vol. III, Springer-Verlag, 1989, pp. 99–227.

[7] S. G. Gindikin, *Real integral geometry and complex analysis*, C. A. Berenstein, et al., Integral Geometry, Radon Transform and Complex Analysis, Lect. Notes in Math., vol. 1684, Springer, Berlin, 1998, pp. 70–98.
[8] E. L. Grinberg, *On images of Radon transforms*, Duke Math. J. **52** (1985), 939–972.
[9] E. L. Grinberg and E. T. Quinto, *Analytic continuation of convex bodies and Funk's characterization of the sphere* (submitted).
[10] E. L. Grinberg and G. Zhang, *Convolutions, transforms, and convex bodies*, Proc. London Math. Soc. (3) **78** (1999), 77–115.
[11] S. Helgason, *The Radon transform*, Birkhäuser, Boston, Second edition, 1999.
[12] A. T. James, *Normal multivariate analysis and the orthogonal group*, Annals of Math. Statistics **25** (1954), 38–75.
[13] T. Kakehi, *Range characterization of Radon transforms on S^n and $P^n R$*, J. Math. Kyoto Univ. **33** (1993), 315–328.
[14] N. Murata, *An integral representation of functions using three-layered networks and their approximation bounds*, Neural Networks **9** (1996), 947–956.
[15] B. Rubin, *Fractional integrals and potentials*, Addison Wesley Longman, Essex, U.K., 1996.
[16] _____, *Fractional Calculus and wavelet transforms in integral geometry*, Fractional calculus and Applied Analysis **1** (1998), No. 2, 193–219.
[17] _____, *Spherical Radon transform and related wavelet transforms*, Appl. and Comp. Harmonic Anal. **5** (1998), 202–215.
[18] _____, *Inversion formulas for Radon transforms on S^n and intertwining fractional integrals*, Hebrew University of Jerusalem, Preprint No. 10 (1999-2000).
[19] _____, *Inversion formulas for totally geodesic Radon transforms on the sphere in L^p-spaces*, Hebrew University of Jerusalem, Preprint (2000).
[20] S.G. Samko, A.A. Kilbas and O.I. Marichev, *Fractional integrals and derivatives. Theory and applications*, Gordon and Breach Sc. Publ., New York, 1993.
[21] L. A. Santalo, *Integral geometry and geometric probability*, Addison-Wesley Publ. Comp., London, 1976.
[22] V.I. Semyanistyi, *Homogeneous functions and some problems of integral geometry in spaces of constant curvature*, Soviet Math. Dokl. **2** (1961), 59–62.
[23] _____, *Some integral transformations and integral geometry in an elliptic space*, Trudy Sem. Vektor. Tenzor. Analizu **12** (1963), 397–441 (Russian).
[24] R. S. Strichartz, *L^p-estimates for Radon transforms in Euclidean and non-euclidean spaces*, Duke Math. J. **48** (1981), 699–727.
[25] B.G. Vakulov, *Theorems of the Hardy-Littlewood-Sobolev type for operators of the potential type in $L_p(S_{n-1};\rho)$*, Rostov. Gos. Univ., Rostov-on-Don, Dep. in VINITI, No. 5435–B86, 1986 (Russian).
[26] B.G. Vakulov, *Operators of the potential type on the sphere in generalized Hölder spaces,*, Rostov. Gos. Univ., Rostov-on-Don, Dep. in VINITI, No. 1563–B86, 1986 (Russian).
[27] N. Ja. Vilenkin, and A. V. Klimyk, *Representations of Lie groups and Special functions, Vol 2*, Kluwer Academic publishers, Dordrecht, 1993.
[28] G. Zhang, *Sections of convex bodies*, Amer. J. Math. **118** (1996), 319-340.

B. RUBIN, INSTITUTE OF MATHEMATICS, HEBREW UNIVERSITY OF JERUSALEM, GIVAT RAM, JERUSALEM 91904, ISRAEL
  E-mail address: boris@math.huji.ac.il,

D. RYABOGIN, DEPARTMENT OF MATHEMATICS, UNIVERSITY OF MISSOURI, COLUMBIA, M0 65211, USA
  E-mail address: ryabs@math.missouri.edu

Reconstruction of High Contrast 2-D Conductivities by the Algorithm of A. Nachman

Samuli Siltanen, Jennifer L. Mueller, and David Isaacson

ABSTRACT. The uniqueness proof of A. Nachman [20] for the 2D inverse conductivity problem outlines a reconstruction algorithm for determining the unknown conductivity in a region Ω from knowledge of the Dirichlet-to-Neumann map. The algorithm is a direct method based on the techniques of inverse scattering. Here we present a practical implementation of the algorithm in Nachman's proof and demonstrate its effectiveness on several radially symmetric conductivities in $C^3(\Omega)$ and $C^4(\Omega)$.

1. Introduction

Let $\Omega \subset \mathbb{R}^2$ be a bounded, simply connected C^∞ domain and $\gamma \in C^2(\Omega)$. We assume that $\gamma(x) \geq c > 0$ and $\gamma \equiv 1$ in a neighborhood of $\partial \Omega$.

Define the Dirichlet-to-Neumann map corresponding to γ by

$$(1.1) \qquad \Lambda_\gamma : H^{1/2}(\partial\Omega) \to H^{-1/2}(\partial\Omega), \qquad \langle \Lambda_\gamma f, g \rangle = \int_\Omega \gamma \nabla u \cdot \nabla v,$$

where v is any $H^1(\Omega)$ function with trace g on the boundary and u is the unique $H^1(\Omega)$ solution of the Dirichlet problem

$$(1.2) \qquad \begin{cases} \nabla \cdot \gamma \nabla u = 0 \text{ in } \Omega, \\ u = f \text{ on } \partial\Omega. \end{cases}$$

The inverse conductivity problem of Calderón [7] is to decide whether γ is uniquely determined by Λ_γ and if so, reconstruct γ from the knowledge of Λ_γ. Physically, u represents the electric potential and γ the conductivity. Knowledge of the Dirichlet-to-Neumann map is tantamount to knowing the resulting current pattern on $\partial\Omega$ corresponding to any prescribed voltage pattern on $\partial\Omega$. The inverse problem has applications in geophysics, nondestructive testing and a medical imaging technique known as Electrical Impedance Tomography (EIT). In EIT the domain Ω is often a cross-section of the body, such as a patient's chest. A basis of current patterns is applied on electrodes attached around the patient's chest and the resulting voltages are measured on the electrodes. Since the tissues and organs in the body have

---

2000 *Mathematics Subject Classification.* 35R30.
The second author was supported by an NSF Mathematical Sciences Postdoctoral Fellowship.

© 2001 American Mathematical Society

different conductivities, plotting the conductivity distribution $\gamma(x)$ yields a 2-D image of a cross-section of the chest. See [9] for a recent survey article on EIT.

The inverse conductivity problem is ill-posed in the sense that large changes in the conductivity can correspond to small changes in the boundary data. Furthermore, the problem is more difficult in dimension $n = 2$ than $n \geq 3$. One can see this formally by considering the Schwartz kernel of the data. The kernel contains $2(n-1)$ degrees of freedom, so in 2-D the problem is formally determined, while in $n \geq 3$ it is overdetermined. Thus in 2-D one must in some sense make use of all of the data.

We briefly discuss some of the milestone results for the inverse conductivity problem, with no claim to completeness. Motivated by geophysical applications, in 1980 Calderón [7] showed how to determine nearly constant conductivities from the Dirichlet-to-Neumann map. In 1985 Kohn and Vogelius [17] proved that if $\partial\Omega$ is C^∞ and γ is piecewise analytic, then Λ_γ determines γ uniquely in dimensions $n \geq 2$. Global uniqueness for $n \geq 3$ with γ in $C^\infty(\bar\Omega)$ and C^∞ $\partial\Omega$ was established by Sylvester and Uhlmann [24]. Nachman [19] gave a reconstruction method for dimensions $n \geq 3$ for $\gamma \in C^{1,1}$ with $\partial\Omega \in C^{1,1}$. The global uniqueness question in 2-D remained open until 1995, when it was resolved for $\partial\Omega$ Lipschitz and $\gamma \in W^{2,p}(\Omega), p > 1$, by Nachman [20]. This result was sharpened in 1997 to $W^{1,p}(\Omega), p > 2$, conductivities by Brown and Uhlmann [5]. For generalizations of the above results to more general spaces and other related work see the references given in [25, 20, 15].

An important feature of Nachman's proof is that it is constructive; it outlines a direct method for solving for γ without iteration. Previous algorithms have relied on iterative techniques such as least squares minimization (see, for example, [10], [16], [6]), linearization ([7, 2, 8]), or layer-stripping ([22], [23]). Nachman's proof is based on the techniques of inverse scattering and the $\bar\partial$ method. The $\bar\partial$ method was first used by Beals and Coifman [3] for the quantum inverse scattering problem in 1-D and was extended to 2-D problems in [4]. It makes use of the Green's function for the Laplacian introduced by Faddeev [11] and the corresponding exponentially growing solutions to the Schrödinger equation. In the $\bar\partial$ approach a nonphysical function not directly measurable in experiments known as the scattering transform $\mathbf{t}(k)$ plays a key role. In [20] an integral equation is derived for determining the scattering transform from the Dirichlet-to-Neumann data. The scattering transform is a function in the $\bar\partial$ equation, which must be solved to obtain the exponentially growing solutions in the complex plane. In the final step, the conductivity γ is recovered by taking the small k limit of the solution to the $\bar\partial$ equation.

In this work we review the reconstruction algorithm set forth in Nachman's proof and the implementation developed in [21]. In [21] the algorithm was tested on two radially symmetric C^∞ conductivities, one low-contrast example with height 1.2 and one high-contrast example with height 4. Here we consider three radially symmetric C^3 examples, with heights approximately 4, 6, and 8, and we study a slightly more complicated C^4 example. These new examples provide insight into the relationship between the conductivity and the scattering transform and demonstrate the effectiveness of the algorithm on less regular and higher contrast conductivities. The C^4 example dips below the boundary value $\gamma = 1$ in region near $\partial\Omega$. Such an example can be viewed as the first step in simulating physically relevant conductivities, which would have features such as a resistive region (corresponding to skin) near the boundary.

2. The reconstruction algorithm

In this section we give a brief outline of the reconstruction algorithm set forth in Nachman's uniqueness proof [20].

First, the conductivity equation (1.2) is transformed to the Schrödinger equation by the following change of variables. Let $q \in C_0^0(\Omega)$ be given by

$$q = \gamma^{-1/2}\Delta\gamma^{1/2}.$$

If u is a solution of $\nabla \cdot \gamma\nabla u = 0$ in Ω, defining $\tilde{u} = \gamma^{1/2}u$ yields

(2.1) $$(-\Delta + q)\tilde{u} = 0 \text{ in } \Omega.$$

While in [20] the first step is to find $\gamma|_{\partial\Omega}$ and $\partial\gamma/\partial\nu|_{\partial\Omega}$ and to continue γ artificially to be one outside a neighborhood of Ω, we omit that step in this work and consider only conductivities that are one in a neighborhood of $\partial\Omega$. This assumption allows us to smoothly extend $\gamma = 1$ and $q = 0$ to the whole plane. We may therefore study equation (2.1) in all of $\mathbb{R}^2$.

The exponentially behaving solutions of (2.1) introduced by Faddeev [11] are the key to the reconstruction. By Theorem 1.1 of [20] for any $k \in \mathbb{C} \setminus 0$ there is a unique solution $\psi(x,k)$ of

(2.2) $$(-\Delta + q)\psi(x,k) = 0 \qquad x \in \mathbb{R}^2$$

satisfying $e^{-ikx}\psi(\cdot,k) - 1 \in W^{1,\tilde{p}}(\mathbb{R}^2)$ for any $2 < \tilde{p} < \infty$. The space $W^{1,\tilde{p}}(\mathbb{R}^2)$ is a special case of the definition

$$W^{m,\rho}(E) = \{f \in L^\rho(E) \,|\, \partial^\alpha f \in L^\rho(E),\, |\alpha| \leq m\}$$

for an arbitrary domain $E \subset \mathbb{R}^n$ and $1 \leq \rho \leq \infty, m \geq 0$.

Denote

(2.3) $$\mu(x,k) := e^{-ikx}\psi(x,k), \qquad x \in \mathbb{R}^2, k \in \mathbb{C} \setminus 0.$$

Here $ikx = i(k_1 + ik_2)(x_1 + ix_2)$. Then μ satisfies

(2.4) $$(-\Delta - 2ik\bar{\partial} + q)\mu = 0$$

where $\bar{\partial} = (\partial/\partial x_1 + i\partial/\partial x_2)/2$. The condition $\mu - 1 \in W^{1,\tilde{p}}$ and the Sobolev imbedding theorem yield that μ is continuous and tends to one asymptotically when $|x| \to \infty$.

The reconstruction of γ is based on the use of an intermediate object called the *non-physical scattering transform* $\mathbf{t}$, which is not directly measurable in experiments:

(2.5) $$\mathbf{t}(k) := \int_{\mathbb{R}^2} e_k(x)\mu(x,k)q(x)dx, \qquad k \in \mathbb{C} \setminus 0,$$

where $e_k(x) := \exp(i(kx + \bar{k}\bar{x}))$. Note that since μ is asymptotically close to one, $\mathbf{t}(k)$ is approximately the Fourier transform of $q(x)$ evaluated at the point $(-2k_1, 2k_2) \in \mathbb{R}^2$.

The direct reconstruction method consists of two main steps:
1. Given Λ_γ, determine the scattering transform $\mathbf{t}(k)$.
2. Determine γ from the knowledge of $\mathbf{t}(k)$.

To obtain $\mathbf{t}(k)$ from the Dirichlet-to-Neumann data, one must first solve an integral equation for the trace on $\partial\Omega$ of the exponentially growing solution ψ. Denote the Dirichlet-to-Neumann map of the homogeneous conductivity 1 by Λ_1 and note that since $\gamma \equiv 1$ near $\partial\Omega$ the maps Λ_γ and the Dirichlet-to-Neumann map Λ_q of the Schrödinger problem are the same. By Theorem 5 of [20] for any $k \in \mathbb{C} \setminus 0$ the following integral equation is uniquely solvable on $H^{1/2}(\partial\Omega)$.

$$\psi(\cdot,k)|_{\partial\Omega} = e^{ikx} - S_k(\Lambda_\gamma - \Lambda_1)\psi(\cdot,k) \tag{2.6}$$

where

$$(S_k\phi)(x) := \int_{\partial\Omega} G_k(x-y)\phi(y)d\sigma(y), \tag{2.7}$$

and $G_k(x)$ is the Faddeev Green's function defined by

$$G_k(x) := e^{ikx}g_k(x), \qquad -\Delta G_k = \delta, \tag{2.8}$$

$$g_k(x) := \frac{1}{(2\pi)^2}\int_{\mathbb{R}^2}\frac{e^{ix\cdot\xi}}{\xi(\bar{\xi}+2k)}d\xi, \qquad (-\Delta - 4ik\bar{\partial})g_k = \delta. \tag{2.9}$$

Now $\mathbf{t}(k)$ can be recovered from the formula

$$\mathbf{t}(k) = \int_{\partial\Omega} e^{i\bar{k}\bar{x}}(\Lambda_\gamma - \Lambda_1)\psi(\cdot,k)d\sigma. \tag{2.10}$$

To determine γ from $\mathbf{t}(k)$, one must first find $\mu(x,k)$ for all $k \in \mathbb{C}\setminus 0$. Namely, Theorem 2.1 of [20] implies that the $\bar{\partial}$ equation

$$\frac{\partial}{\partial\bar{k}}\mu(x,k) = \frac{1}{4\pi\bar{k}}\mathbf{t}(k)e_{-k}(x)\overline{\mu(x,k)}, \quad k \neq 0, \tag{2.11}$$

holds and Theorem 4.1 of [20] shows that (2.11) is uniquely solvable. The solution satisfies the Fredholm integral equation

$$\mu(x,k) = 1 + \frac{1}{(2\pi)^2}\int_{\mathbb{R}^2}\frac{\mathbf{t}(k')}{(k-k')\bar{k'}}e_{-x}(k')\overline{\mu(x,k')}dk_1'dk_2' \tag{2.12}$$

for all $k \in \mathbb{C}\setminus 0, x \in \mathbb{R}^2$. Note that the integral is taken over the k-plane, so to solve (2.12) $\mu(x,k)$ is needed for all values of $k \in \mathbb{C}\setminus 0$. Also note that in the solution of the $\bar{\partial}$ equation x is kept fixed, so the computations can be carried out only in the region of interest.

One can recover the conductivity $\gamma(x)$ from the function $\mu(x,k)$. Formally, since

$$(-\Delta + q)e^{ikx}\mu(x,k) = 0,$$

taking $k=0$ implies

$$(-\Delta + \frac{\Delta\gamma^{1/2}}{\gamma^{1/2}})\mu(x,0) = 0$$

and we see that

$$\gamma^{1/2}(x) = \lim_{k\to 0}\mu(x,k).$$

3. From Λ_γ to $\mathbf{t}(k)$

In this section we describe how we obtain the scattering transform from the Dirichlet-to-Neumann map numerically. In [18] it is shown that the step from Λ_γ to $\mathbf{t}(k)$ has only a logarithmic stability estimate while the step from $\mathbf{t}(k)$ to γ has linear stability.

In [21] it is shown that if $\gamma(x)$ is rotationally symmetric, so is $\mathbf{t}(k)$. In this work we consider radial conductivity distributions and make use of this fact in our construction of the Dirichlet-to-Neumann data and in our approximation of $\mathbf{t}(k)$. In the numerical solution of the $\bar{\partial}$ equation, no radially symmetry of γ or $\mathbf{t}(k)$ is assumed. If $\gamma(x) = \gamma(|x|)$ we know from [23] that the functions $\phi_n(\theta) := (2\pi)^{-1/2} e^{in\theta}$, $n \in \mathbb{Z}$, are eigenfunctions for Λ_γ. Thus Λ_γ can be represented in the trigonometric basis by the collection $\{\lambda_n\}_{n=-\infty}^{\infty}$ of its eigenvalues.

To produce numerical data, we need to approximate these eigenvalues numerically. As explained in [12], we note that if two radial L^∞ conductivities γ and $\tilde{\gamma}$ satisfy $\gamma(x) \leq \tilde{\gamma}(x)$ pointwise in Ω, the eigenvalues satisfy $\lambda_n \leq \tilde{\lambda}_n$. Moreover, for piecewise constant radial conductivities we can compute the eigenvalues explicitly [21]. Thus approximating the C^2 conductivity from above and below with piecewise constant conductivities gives us numerical upper and lower bounds for the eigenvalues. See [21] for more details.

To obtain $\mathbf{t}(k)$ from $\{\lambda_n\}$ numerically, we use the approximation $\psi(x,k)|_{\partial\Omega} \approx e^{ikx}$, as opposed to solving the integral equation (2.6). Expanding e^{ikx} in a Fourier series on the circle $x = e^{i\theta}$ yields [14]

$$e^{ikx} = \sum_{n=-\infty}^{\infty} a_n(k) e^{in\theta} \quad \text{with} \quad a_n(k) = \begin{cases} \frac{(ik)^n}{n!}, & n \geq 0 \\ 0, & n < 0. \end{cases}$$

Substituting this series into formula (2.10) gives

$$(3.1) \qquad \mathbf{t}(k) \approx \mathbf{t}^{\exp}(k) = \sum_{n=1}^{N} (\lambda_n - |n|) \frac{(-1)^n |k|^{2n}}{(n!)^2}.$$

Although one would expect this approximation to be more accurate for small q, we obtained reasonable results in our examples even when q was not small.

4. From $\mathbf{t}(k)$ to γ

By [18] the $\bar{\partial}$ inversion $\mathbf{t} \to \gamma$ is well-posed and even contributes some smoothing. The fact that the $\bar{\partial}$ equation must be solved independently for each x in the region of interest to obtain $\gamma(x)$ suggests the use of parallelization in a numerical method. Here a 2-D adaptation of the method of product integrals presented in [1] in 1-D is used to solve the $\bar{\partial}$ equation in parallel for the x values in the region of interest. The idea of the method is to factor the integrand into its smooth part and its singular part and approximate the smooth part with a simple function, such as an interpolatory polynomial. The new integrand is then computed analytically where possible. We describe the method applied to the weakly singular second-order Fredholm integral equation (2.12) below. We mention that in the numerical solution of the $\bar{\partial}$ equation, no radially symmetry of γ or $\mathbf{t}(k)$ is assumed.

For $s \in \mathbb{C} \setminus 0$ write equation (2.12) as

$$(4.1) \qquad \mu(x,s) = 1 + \frac{1}{4\pi^2} \int_{\mathbb{R}^2} H(s,k) L(x,k) \overline{\mu(x,k)} dk_1 dk_2$$

where $k = k_1 + ik_2$ and

(4.2) $$H(s,k) := \frac{1}{(s-k)} \quad \text{and} \quad L(x,k) := \frac{\mathbf{t}(k)}{\overline{k}} e_{-x}(k).$$

In [21] it is shown that for $\gamma \in C^{2+m}(\Omega), m \geq 1$, $|\mathbf{t}(k)| \leq C|k|^{-m}$ for large $|k|$. Thus, the integrand in (4.1) approaches zero as $|k| \to \infty$, and for numerical purposes we choose $A, C > 0$ sufficiently large and instead solve

$$\mu(x,s) = 1 + \frac{1}{4\pi^2} \int_{-A}^{A} \int_{-C}^{C} H(s,k) L(x,k) \overline{\mu(x,k)} dk_1 dk_2.$$

Next, define a mesh on $[-A, A] \times [-C, C]$ in such a way that $k_1 = 0, k_2 = 0$ is not a mesh point. Let

$$\begin{aligned} u_j &= -A + jh_u^j, \quad j = 0, \ldots, N+1 \\ v_i &= -C + ih_v^i, \quad i = 0, \ldots, N+1. \end{aligned}$$

where $u_{j+1} - u_j = h_u^j > 0, j = 0, \ldots, N$ and $v_{i+1} - v_i = h_v^i > 0, i = 0, \ldots, N$.

Since $\mu \sim 1$ for $|k|$ large, on the set $S := \{[u_j, u_{j+1}] \times [v_i, v_{i+1}] : j \in \{0, N-1\}$ or $i \in \{0, N-1\}\}$ of outer mesh elements we set $\mu \equiv 1$. Then we wish to solve

(4.3) $$\mu(x,s) = g(x,s) + \frac{1}{4\pi^2} \sum_{j=1}^{N-1} \sum_{i=1}^{N-1} \int_{u_j}^{u_{j+1}} \int_{v_i}^{v_{i+1}} H(s,k) L(x,k) \overline{\mu(x,k)} dk_1 dk_2$$

where

(4.4) $$g(x,s) := 1 + \frac{1}{4\pi^2} \int_S H(s,k) L(x,k) dk_1 dk_2.$$

We then approximate the function $f(x,k) := L(x,k)\overline{\mu(x,k)}$ by an interpolatory polynomial. Here, we use bilinear interpolation and introduce the notation

$$\begin{aligned} {[f(x, k_1, k_2)]}_{ji} &:= (1-t)(1-w) f_{j,i} + t(1-w) f_{j+1,i} \\ &+ tw f_{j+1,i+1} + (1-t) w f_{j,i+1} \end{aligned}$$

where $f_{j,i} := f(x, u_j, v_i)$, $t := \frac{k_1 - u_j}{h_u^j}$, and $w := \frac{k_2 - v_i}{h_v^i}$.

For $x \in \Omega$ define the numerical integration operator by

(4.5) $$\kappa_N \mu(x,s) := \sum_{j=1}^{N-1} \sum_{i=1}^{N-1} \int_{u_j}^{u_{j+1}} \int_{v_i}^{v_{i+1}} H(s,k) [f(x, k_1, k_2)]_{ji} dk_1 dk_2.$$

To obtain a linear system, we choose s to be the nodes of the inner mesh elements $\{s = (u_j, v_i)\}_{j,i=1}^{N}$. Then to form $g(x,s)$ and $\kappa_N \mu(x,s)$ the following integrals must be evaluated for $j, i = 0, \ldots, N$ and $s = (u_m, v_n), m, n = 1, \ldots, N$.

(4.6) $$J_{\alpha\beta}^{ji}(s) := \int_{u_j}^{u_{j+1}} \int_{v_i}^{v_{i+1}} \frac{k_1^\alpha k_2^\beta}{(s-k)} dk_1 dk_2, \quad \alpha, \beta \in \{0, 1\}.$$

Note that when s lies on a corner of the mesh element over which we are integrating, an integrable singularity will be present in the integrand. When s does not coincide with a corner of the mesh element over which we are integrating, the above integrals are not singular, and they can be computed using a numerical quadrature method such as 2-D Gauss-Legendre quadrature. The singular integrals can be evaluated analytically using residue calculus.

Denote $\mu_{ji}(x) := \mu(x,(u_j,v_i))$. By regrouping terms, the numerical integration operator in (4.5) can now be written as

$$(4.7) \qquad \kappa_N \mu(x,s) = \sum_{j=1}^{N-1} \sum_{i=1}^{N-1} A^{ji}(x) \overline{\mu_{ji}(x)}$$

where

$$(4.8) \qquad A^{ji}(x) = a_{ji}(x) J_{00}^{ji}(s) + b_{ji}(x) J_{10}^{ji}(s) + c_{ji}(x) J_{01}^{ji}(s) + d_{ji}(x) J_{11}^{ji}(s).$$

Choose $\{s = (u_k, v_l)\}_{k,l=1}^N$ defining an N^2 by N^2 matrix $\mathbf{A}(x) = (A^{ji}(x))$ where $A^{ji}(x)$ is the linear combination of the $J_{\alpha\beta}^{ji}((u_k, v_l))$ above, $\alpha, \beta \in \{0, 1\}$. This results in the linear system

$$(4.9) \qquad \mathbf{I}\mu(x) - \mathbf{A}\bar{\mu}(x) = \mathbf{g}(x)$$

where $\mathbf{I}$ is the N^2 by N^2 identity matrix. This system can be solved by equating the real and imaginary parts to obtain two linear systems in real variables with two vectors of unknowns. Namely,

$$(4.10) \qquad (\mathbf{I} - Re(\mathbf{A}))\mathbf{a} - Im(\mathbf{A})\mathbf{b} = Re(\mathbf{g})$$

$$(4.11) \qquad (\mathbf{I} + Re(\mathbf{A}))\mathbf{b} - Im(\mathbf{A})\mathbf{a} = Im(\mathbf{g})$$

where $\mu = \mathbf{a} + i\mathbf{b}$. Solving the linear system gives $\{\mu(x,(u_j,v_i))\}_{j,i=1}^N$.

Note that the factors $J_\alpha^{ji}(s)$ in the matrix $\mathbf{A}$ are independent of x, so they need only be computed once and stored. Then in parallel, the matrix $\mathbf{A}$ is assembled by forming the linear combination (4.8) and the resulting systems (4.9) are solved.

5. Numerical Examples

In this section we test the algorithm on three high-contrast conductivities in $C^3(\Omega)$ and a high contrast conductivity in $C^4(\Omega)$ which contains a dip near the boundary of Ω. The first three examples increase in magnitude from 4 to approximately 6 and 8. We compare the corresponding scattering transforms and the reconstructions. The conductivities for Examples 1, 2, and 3 are defined by the following formula. Fix $0 < \rho < 1$ and let $F_\rho \in C_0^3(\mathbb{R})$ be given by

$$(5.1) \qquad F_\rho(x) := (x^2 - \rho^2)^4, \quad -\rho \leq x \leq \rho$$

and $F_\rho(x) \equiv 0$ for $|x| > \rho$. Let

$$(5.2) \qquad \gamma(x) := (\alpha F_\rho(|x|) + 1)^2,$$

Then the support of $\gamma(x) - 1$ is the interval $[-\rho, \rho]$.

Example 1: $\rho = 1/4$ and $\alpha = 16\rho^{-6}$.
Example 2: $\rho = 1/4$ and $\alpha = 23\rho^{-6}$.
Example 3: $\rho = 1/4$ and $\alpha = 29\rho^{-6}$.

The plots of γ_1, γ_2, and γ_3 are found with the reconstructions in Figure 2. In each of these examples, 49 eigenvalues were computed using the method described in Section 3. The upper and lower bounds were in very close agreement. The approximate scattering transform $\mathbf{t}^{\exp}(k)$ was then computed using the series (3.1) with $N = 49$ and the lower bounds for the eigenvalues. The scattering tranforms $\mathbf{t}^{\exp}(k)$ corresponding to γ_1, γ_2, and γ_3 are plotted in Figure 1. One observes that the scattering transform increases in amplitude with γ, and the divergence of the series (3.1) becomes more marked as the amplitude of γ increases. Reconstructions of γ_1, γ_2, and γ_3 were obtained by solving the $\bar{\partial}$ equation as described in Section 4. The $\bar{\partial}$ equation was solved on the k mesh $[-30, 30]^2$ with stepsize $h \approx 1.5$ and for the 41 x values $x = (-0.4, 0), (-0.38, 0), \ldots, (0.38, 0), (0.4, 0)$. Since

$$(5.3) \qquad \gamma^{1/2}(x) = \lim_{k \to 0} \mu(x, k),$$

the value of $\gamma^{1/2}(x)$ was approximated by $\Re(\mu(x, k))$ at a node k with minimum norm. The γ_i reconstructed thusly are plotted in Figure 2. The relative error in the reconstruction increases as the amplitude of γ increases. The relative errors for γ_1, γ_2, and γ_3 were .127, .190, and .251 respectively.

Example 4: Here, the conductivity function under consideration is defined by the following formula. Fix $0 < \rho < 1$ and let $F_\rho \in C_0^4(\mathbb{R})$ be given by

$$(5.4) \qquad F_\rho(x) := (x^2 - \rho^2)^4 \cos \frac{3\pi x}{2\rho}, \quad -\rho \le x \le \rho$$

and $F_\rho(x) \equiv 0$ for $|x| > \rho$. Let

$$(5.5) \qquad \gamma(x) := (10 F_{3/4}(|x|) + 1)^2,$$

Then the support of $\gamma(x) - 1$ is the interval $[-3/4, 3/4]$. See Figure 3 for a plot of the conductivity and the corresponding potential.

Using the method described in Section 3, 75 eigenvalues of the Dirichlet-to-Neumann map were computed. Again, the upper and lower bounds were in very close agreement. The approximate scattering transform $\mathbf{t}^{\text{exp}}(k)$ was then computed using the series (3.1) with $N = 75$. To validate the function $\mathbf{t}^{\text{exp}}(k)$ computed from the series (3.1), the scattering transform was also computed from the definition by solving the forward problem. First, the functions $\mu_{LS}(x, k)$ were computed on the k mesh $[-30, 30]^2$ by solving the Lippmann-Schwinger equation

$$\mu - 1 = -g_k \star (q\mu)$$

using a projection method described in [21]. Recall g_k is the Fadeev's Green's function defined by (2.9). The scattering transform was then computed by applying Gauss-Legendre quadrature to the definition using the computed values μ_{LS}:

$$\mathbf{t}_{LS}(k) := \int_{\mathbb{R}^2} e_k(x) \mu_{LS}(x, k) q(x) dx, \quad k \in \mathbb{C} \setminus 0.$$

Plots of $\mathbf{t}^{\text{exp}}(k)$, $\mathbf{t}_{LS}(k)$, and $\hat{q}(2|k|)$ are found in Figure 4. Note that except for small $|k|$ values, the function $\mathbf{t}_{LS}(k)$ is very similar to $\hat{q}(2|k|)$

Reconstructions of γ were obtained by solving the $\bar{\partial}$ equation using the method described in Section 4 and using both $\mathbf{t}^{\text{exp}}(k)$ and $\mathbf{t}_{LS}(k)$ as data in the equation. The $\bar{\partial}$ equation was solved on the uniform k-mesh $[-20, 20]^2$ with $h \approx 1$. Again, the value of $\gamma^{1/2}(x)$ was approximated by $\Re(\mu(x, k))$ at a node k with minimum norm. The $\bar{\partial}$ equation was solved for the x values $(-0.8, 0), (-.75, 0), \ldots, (.75, 0), (0.8, 0)$, and the γ's reconstructed from $\mathbf{t}^{\text{exp}}(k)$ and $\mathbf{t}_{LS}(k)$ are plotted in Figure 5. The relative error for γ reconstructed from $\mathbf{t}^{\text{exp}}$ is .316 while the relative error for γ reconstructed from $\mathbf{t}_{LS}$ is .082. Although the reconstructed conductivity profiles are not perfectly symmetric, they are nearly so. Note that some incorrect oscillation is present in the reconstruction from $\mathbf{t}^{\text{exp}}$. This may be due to the divergence of the series for large $|k|$, although such oscillations were not present in reconstructions of simpler radially symmetric examples that did not contain a dip such as those above and those in [21]. Figures 6 contains a 2-D plot of the reconstructed γ from $\mathbf{t}^{\text{exp}}(k)$

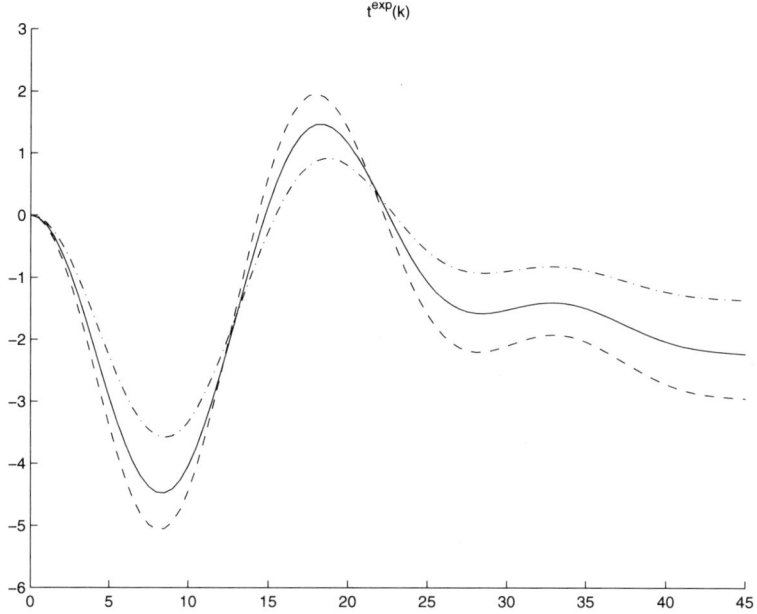

FIGURE 1. The functions $\mathbf{t}^{exp}(k)$ for $\gamma_1(x)$ (dot-dashed line), $\gamma_2(x)$ (solid line), and $\gamma_3(x)$ (dashed line).

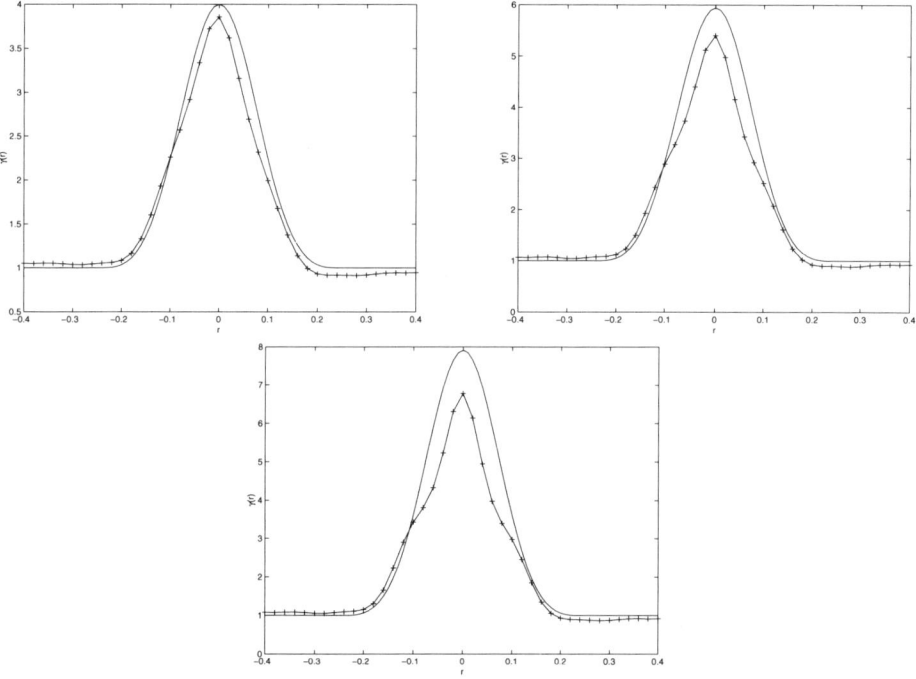

FIGURE 2. Cross-sectional plots of the reconstructed γ_i (-+) and the actual γ (-) for $i = 1$ (upper left), $i = 2$ (upper right), and $i = 3$ (bottom center).

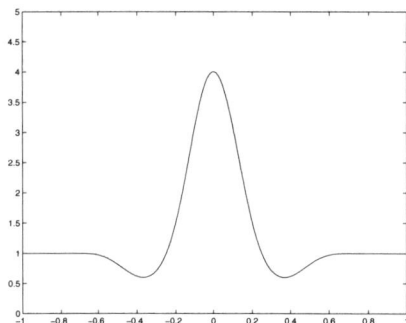 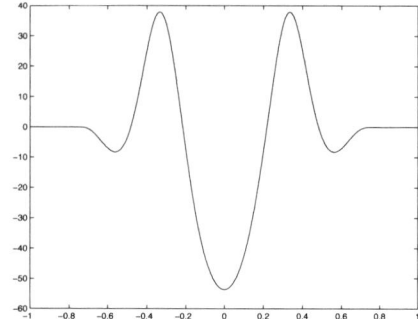

FIGURE 3. Profile of the radially symmetric conductivity γ (left) and the potential q (right) for Example 4.

on the x mesh $[-0.6, 0.6]^2$ with mesh stepsize 0.1. The functions $\Re(\mu_{exp}(x,k))$, $\Re(\mu_{LS}(x,k))$, $\Im(\mu_{exp}(x,k))$, and $\Im(\mu_{LS}(x,k))$ are found in Figures 7 and 8.

6. Acknowledgements

The authors thank A. Nachman for helpful discussions. The numerical calculations in this paper were performed using computing equipment supported by NSF cooperative agreement ACI-9619020 through computing resources provided by the National Partnership for Advanced Computational Infrastructure at the San Diego Supercomputer Center.

References

[1] K. Atkinson. *An Introduction to Numerical Analysis*. Wiley, second edition, 1989.
[2] D. C. Barber and B. H. Brown, *Applied potential tomography*, J. Phys. E. Sci. Instrum. 17 (1984), pp.723–733.
[3] R. Beals and R. R. Coifman, The D-bar approach to inverse scattering and nonlinear evolution equations, *Physica*, 18D (1986) pp.242–249.
[4] R. Beals and R. R. Coifman, Multidimensional inverse scattering, *Proceedings of symposia in pure mathematics*, 43 (1985) pp.45–70.
[5] R. M. Brown and G. Uhlmann, *Uniqueness in the inverse conductivity problem for nonsmooth conductivities in two dimensions*, Communications in partial differential equations, 22 (1997), pp.1009–1027.
[6] L. Borcea, J. G. Berryman and G. Papanicolaou, *High contrast impedance tomography*, Inverse Problems, 12 (1996), pp.835–858.
[7] A. P. Calderón On an inverse boundary value problem In *Seminar on Numerical Analysis and its Applications to Continuum Physics*, Soc. Brasileira de Matemàtica (1980), pp.65–73.
[8] M. Cheney et al, *Noser: An algorithm for solving the inverse conductivity problem*, Int'l J. Imaging Systems and Technology, 2 (1990), pp.66–75.
[9] M. Cheney, D. Isaacson and J. C. Newell *Electrical Impedance Tomography* SIAM Review, 41 (1999), pp.85–101.
[10] D. C. Dobson, *Convergence of a reconstruction method for the inverse conductivity problem*, SIAM J. Appl. Math. 52 (1992), pp.442–458.
[11] L. D. Faddeev. *Increasing solutions of the Schrödinger equation*, Sov. Phys. Dokl, 10 (1966), pp.1033–1035.
[12] D. G. Gisser, D. Isaacson, and J. C. Newell. *Electric current tomography and eigenvalues*, SIAM J. Appl. Math., 50 (1990), pp.1623–1634.
[13] M. Ikehata and S. Siltanen *Numerical method for finding the convex hull of an inclusion in conductivity from boundary measurements*, Inverse Problems, 16 (2000), pp.1043–1052.

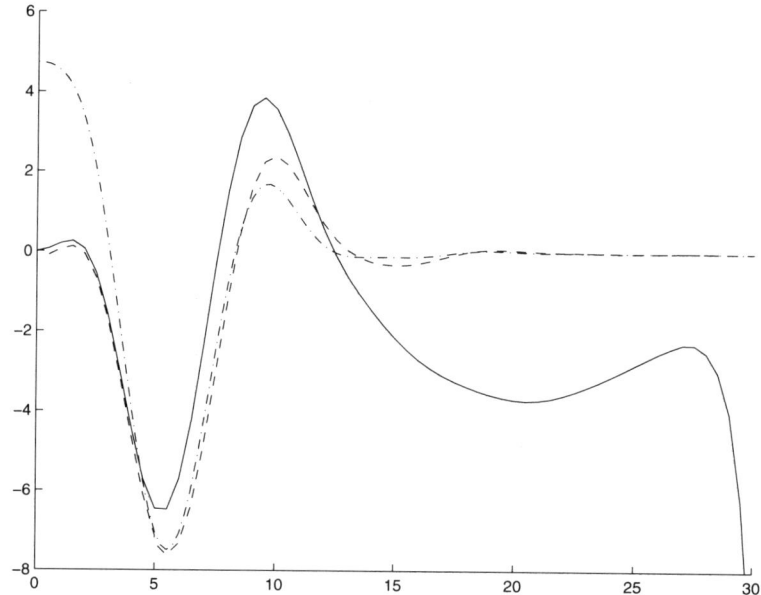

FIGURE 4. The functions $\mathbf{t}^{exp}(k)$ (solid line), $\mathbf{t}^{LS}(k)$ (dashed line), and $\hat{q}(2|k|)$ (dot-dashed line) for Example 4.

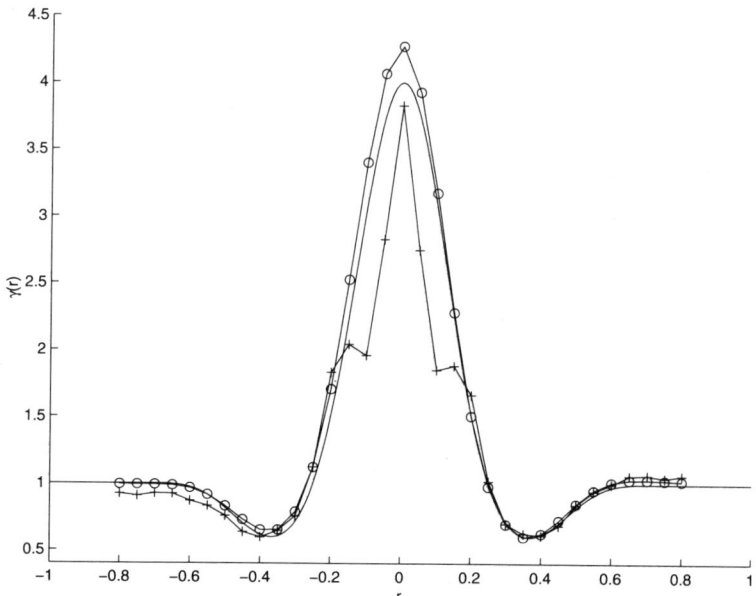

FIGURE 5. Cross-sectional plot of the reconstructed γ from $\mathbf{t}^{exp}$ (-+), the reconstructed γ from $\mathbf{t}^{LS}$ (-o), and the actual γ (solid line) for Example 4.

[14] D. Isaacson and M. Cheney. *Effects of measurement precision and finite numbers of electrodes on linear impedance imaging algorithms*, SIAM J. Appl. Math., 15 (1991), pp.1705–1731.

[15] V. Isakov, *Inverse Problems for Partial Differential Equations*, Springer, 1998.

[16] R. V. Kohn and A. McKenney: *Numerical implementation of a variational method for electrical impedance imaging*, Inverse Problems, 9 (1990), pp.389–414.

[17] R. V. Kohn and M. Vogelius: *Determining conductivity by boundary measurements II. interior results*, Communications on Pure and Applied Mathematics, 38 (1985), pp.643–667.

[18] L. Liu, *Stability Estimates for the Two-Dimensional Inverse Conductivity Problem*, PhD thesis, University of Rochester, 1997.

[19] A. I. Nachman, *Reconstructions from boundary measurements*, Annals of Mathematics, 128 (1988), 531–576.

[20] A. I. Nachman, *Global uniqueness for a two-dimensional inverse boundary value problem*, Annals of Mathematics 143 (1996), pp.71–96.

[21] S. Siltanen, J. Mueller and D. Isaacson, *An implementation of the reconstruction algorithm of A. Nachman for the 2-D inverse conductivity problem*, Inverse Problems 16 (2000), pp.681–699.

[22] E. Somersalo, M. Cheney, D. Isaacson, and E. Isaacson, *Layer stripping: a direct numerical method for impedance imaging*, Inverse Problems, 7 (1991), pp.899–926.

[23] J. Sylvester, *A convergent layer stripping algorithm for the radially symmetric impedance tomography problem*, Communications in partial differential equations, 17 (1992), pp.1955–1994.

[24] J. Sylvester and G. Uhlmann, *A global uniqueness theorem for an inverse boundary value problem*, Annals of Mathematics, 125 (1987), pp.153–169.

[25] J. Sylvester and G. Uhlmann, The Dirichlet to Neumann map and applications, In *Inverse Problems in Partial Differential Equations, Philadelphia*, SIAM, 1990.

HELSINKI UNIVERSITY OF TECHNOLOGY, ESPOO, FINLAND
Current address: Instrumentarium Corporation, Imaging Division, P.O.Box 20, FIN-04301, Tuusula, Finland
E-mail address: `samuli.siltanen@fi.instrumentarium.com`

DEPARTMENT OF MATHEMATICAL SCIENCES, RENSSELAER POLYTECHNIC INSTITUTE, TROY, NY 12180
Current address: Department of Mathematics, Colorado State University, Fort Collins, CO 80523
E-mail address: `mueller@math.colostate.edu`

DEPARTMENT OF MATHEMATICAL SCIENCES, RENSSELAER POLYTECHNIC INSTITUTE, TROY, NY 12180
E-mail address: `isaacd@rpi.edu`

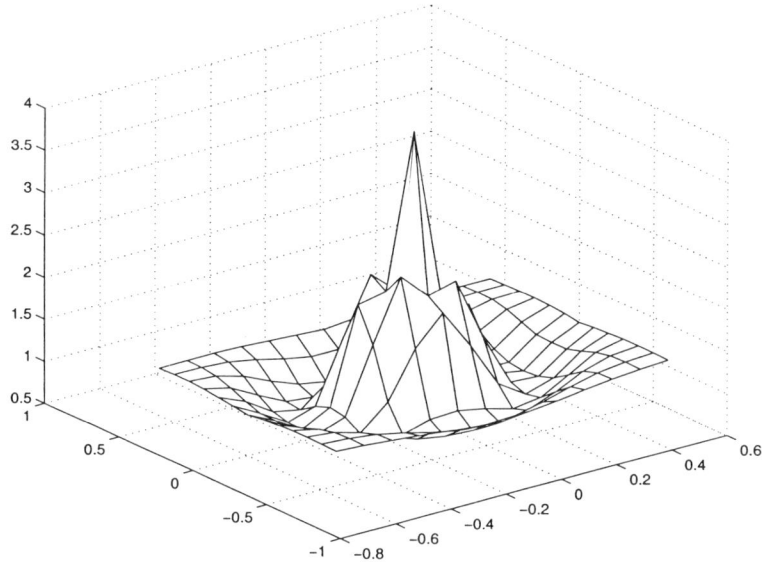

FIGURE 6. Plot of the reconstructed γ from $\mathbf{t}^{exp}$ for Example 4.

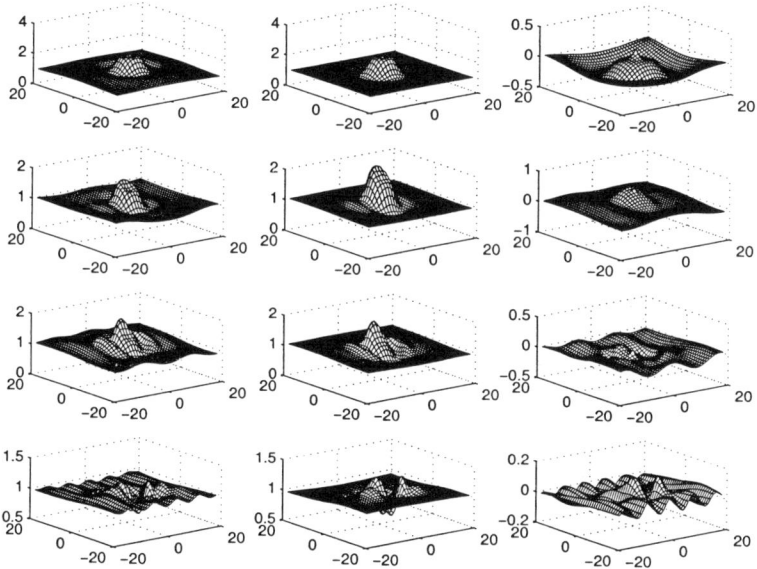

FIGURE 7. The reconstructed $\Re(\mu_{exp}(x,k))$ (left column), $\Re(\mu_{LS}(x,k))$ (center column) and $\Re(\mu_{LS}(x,k)) - \Re(\mu_{exp}(x,k))$ (right column) for $x = 0, .1, .2, .4$ (top to bottom).

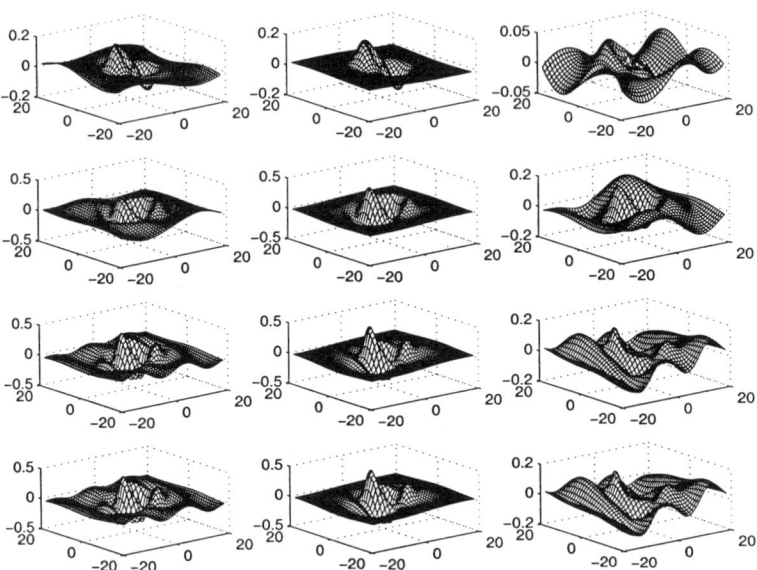

FIGURE 8. The reconstructed $\Im(\mu_{exp}(x,k))$ (left column), $\Im(\mu_{LS}(x,k))$ (center column) and $\Im(\mu_{LS}(x,k)) - \Re(\mu_{exp}(x,k))$ (right column) for $x = 0, .1, .2, .4$ (top to bottom).

Integral Geometry Problem with Incomplete Data for Tensor Fields in a Complex Space.

L. B. Vertgeim

ABSTRACT. The paper is devoted to the problem of reconstructing a tensor field in C^n, if its ray transform is known along all complex lines intersecting a given complex curve. A procedure for recovering the solenoidal part of the tensor field is given.

1. Introduction and some theory of tensor fields in a complex space

For a major reference to integral geometry of tensor fields we refer the reader to the book [2]. In the paper [3] the author considered an integral geometry problem with incomplete data for symmetric tensor fields in a real space. (See [1] for the references to other papers on the integral geometry problems with incomplete data.) In the current article we are going to study a similar problem for tensors in a complex space. The problem for the complete collection of data was considered in the author's dissertation [4], as well as in [5]. We will need to recall some theory from these papers.

Let $p, q \geq 0$ be integers and T_p^q be the space of bidegree (p, q) tensors on C^n, i.e., the functions $f : \underbrace{C^n \times \cdots \times C^n}_{p} \times \underbrace{C^n \times \cdots \times C^n}_{q} \to C$, which are C-linear with respect to each of the first p variables and C-antilinear with respect to the last q. Let S_p^q be a space of tensors, symmetric with respect to the collections of the first p and the last q variables separately. There is a canonical projection $\sigma : T_p^q \to S_p^q$:

$$\sigma f(z_1, \ldots, z_p, w_1, \ldots, w_q) = \frac{1}{p!q!} \sum_{\pi \in \Pi_p} \sum_{\delta \in \Pi_q} f(z_{\pi 1}, \ldots, z_{\pi p}, w_{\delta 1}, \ldots, w_{\delta q}),$$

where Π_p, Π_q are permutation groups. We write each tensor $f \in T_p^q$ in the form

$$f = f_{i_1 \ldots i_p}^{j_1 \ldots j_q} \, dz^{i_1} \otimes \cdots \otimes dz^{i_p} \otimes d\bar{z}^{j_1} \otimes \cdots \otimes d\bar{z}^{j_q}.$$

2000 *Mathematics Subject Classification.* 44A12.

The research described in this publication was made possible in part by Award No. RM2-2242 of the U.S. Civilian Research and Development Foundation for the Independent States of the Former Soviet Union (CRDF). The author is also supported by the Integrated Grant, SB of the Russian Academy of Sciences 2000, N43.

© 2001 American Mathematical Society

Henceforth we will use the Einstein summation convention — summation with respect to the pairs of repeated indices, independently running from 1 to n. The numbers $f_{i_1\ldots i_p}^{j_1\ldots j_q}$ are called the coordinates (or the components) of the tensor f. A map $C^n \to T_p^q$ is called a *tensor field on C^n*. By $C^\infty(T_p^q)$ and $\mathcal{S}(T_p^q)$ we denote the spaces of tensor fields on C^n with smooth and rapidly decreasing components respectively. We will need the following operators, defined in coordinates:

$$(d_l f)_{i_1\ldots i_{p+1}}^{j_1\ldots j_q} = \sigma\big(\frac{\partial}{\partial z^{i_{p+1}}} f_{i_1\ldots i_p}^{j_1\ldots j_q}\big), \quad (\bar{d}_u f)_{i_1\ldots i_p}^{j_1\ldots j_{q+1}} = \sigma\big(\frac{\partial}{\partial \bar{z}^{j_{q+1}}} f_{i_1\ldots i_p}^{j_1\ldots j_q}\big),$$

$$(\bar{\delta}_l f)_{i_1\ldots i_{p-1}}^{j_1\ldots j_q} = \sum_{i=1}^n \frac{\partial}{\partial \bar{z}^i} f_{i_1\ldots i_{p-1}i}^{j_1\ldots j_q}, \quad (\delta_u f)_{i_1\ldots i_p}^{j_1\ldots j_{q-1}} = \sum_{j=1}^n \frac{\partial}{\partial z^j} f_{i_1\ldots i_p}^{j_1\ldots j_{q-1}j}.$$

The operators d are the operators of inner differentiation of the different kinds ("l" - lower, "u" - upper), δ — the divergence operators.

Here, as usual,

$$\frac{\partial}{\partial z^k} = \frac{1}{2}\big(\frac{\partial}{\partial x^k} - \sqrt{-1}\frac{\partial}{\partial y^k}\big) \,,\quad \frac{\partial}{\partial \bar{z}^k} = \frac{1}{2}\big(\frac{\partial}{\partial x^k} + \sqrt{-1}\frac{\partial}{\partial y^k}\big).$$

Let $C_0^n = C^n \setminus \{0\}$.

DEFINITION 1.1. *The ray transform of a tensor field $g \in C^\infty(S_p^q)$ is the function Ig, defined on $C^n \times C_0^n$ by the expression:*

$$Ig(z,\xi) = \int_C g_{i_1\ldots i_p}^{j_1\ldots j_q}(z+t\xi)\, \xi^{i_1}\ldots\xi^{i_p}\, \bar{\xi}^{j_1}\ldots\bar{\xi}^{j_q}\, dS(t),$$

where $dS(t)$ is the area form on C, and where we assume the absolute convergence of all integrals involved.

The problem we will be dealing with is to reconstruct g from Ig. It turns out that the operator I has a nontrivial kernel. In $\mathcal{S}(S_p^q)$ it consists exactly of the tensor fields of the form $d_l v + \bar{d}_u w$, where $v \in C^\infty(S_{p-1}^q), w \in C^\infty(S_p^{q-1})$ and v, w vanish sufficiently fast at infinity.

We need the following statement.

THEOREM 1.2. *For a tensor field $g \in \mathcal{S}(S_p^q)$ there exists a unique tensor field $f \in C^\infty(S_p^q)$ such that for some tensor fields $v \in C^\infty(S_{p-1}^q), w \in C^\infty(S_p^{q-1})$ one has*

$$g = f + d_l v + \bar{d}_u w\,,\ \bar{\delta}_l f = 0\,, \delta_u f = 0\,;\ \ f, v, w \to 0\ as\ |z| \to \infty\,.$$

The field f is called the *solenoidal part* of g, we denote $f = {}^s g$. It turns out that by knowing Ig we can reconstruct $f = {}^s g$, and there is in [4], [5] an explicit inversion formula.

We will be using the following version of the Fourier transform for tensor fields: in coordinates,

$$\hat{g}_{i_1\ldots i_p}^{j_1\ldots j_q}(\zeta) = (2\pi)^{-n} \int_{C^n} e^{-\frac{\sqrt{-1}}{2}(\langle z,\zeta\rangle + \langle \zeta, z\rangle)}\, g_{i_1\ldots i_p}^{j_1\ldots j_q}(z)\, dV_{2n}(z),$$

where $\langle\cdot,\cdot\rangle$ is the standard Hermitian form on C^n and $dV_{2n}(z)$ is the volume form on C^n.

We have the following expression for $\hat{f}$ in terms of $\hat{g}$ ($f = {}^s g$), which will be useful later:

(1.1) $$\hat{f}^{j_1...j_q}_{i_1...i_p}(\zeta) = \varepsilon^{k_1}_{i_1}(\zeta)\ldots\varepsilon^{k_p}_{i_p}(\zeta)\,\varepsilon^{j_1}_{l_1}(\zeta)\ldots\varepsilon^{j_q}_{l_q}(\zeta)\,\hat{g}^{l_1...l_q}_{k_1...k_p}(\zeta),$$

where
$$\varepsilon^j_i(\zeta) = \delta^j_i - \frac{\bar{\zeta}^i \zeta^j}{|\zeta|^2},$$

δ^j_i — the Kroneker symbol.

2. Integral geometry problem with incomplete data, reconstruction of the solenoidal part

For $n \geq 3$ let $\gamma \subset C^n$ be a C^1-smooth complex, but not necessarily holomorphic curve, parameterized as follows:

$$x = \phi(\lambda),\ \lambda \in \Lambda \subset C,\ \phi \in C^1(\Lambda).$$

Problem. Let $g \in \mathcal{S}(S^q_p)$. Reconstruct its solenoidal part $f = {}^s g$ by the known values $Ig(z,\xi)$ for all $z \in \gamma$, $\xi \in C^n_0$.

To formulate a condition on γ we need to consider the following algebraic setting. Let $P(z)$ be an arbitrary degree m polynomial on C^N, which is not necessarily holomorphic:

$$P(z) = \sum_{l+r \leq m} p^{(l,r)j_1...j_r}_{i_1...i_l} z^{i_1}\ldots z^{i_l} \bar{z}^{j_1}\ldots \bar{z}^{j_r},\ p^{(l,r)} \in S^r_l.$$

Altogether there are $\mathcal{L}_{N,m} := \binom{2N+m}{m}$ independent coefficients (taking into account symmetries).

DEFINITION 2.1. *A collection of $\mathcal{L}_{N,m}$ points in C^N: $b_1,\ldots,b_{\mathcal{L}_{N,m}}$ is called defining of order m, if any degree m polynomial $P(z)$ is determined uniquely by its values $P(b_j)$, $j = 1,\ldots,\mathcal{L}_{N,m}$.*

Almost all collections are defining in the sense that they form in $(C^n)^{\mathcal{L}_{N,m}}$ the complement of an algebraic hypersurface.

DEFINITION 2.2. *We say that a complex curve γ satisfies the complex Kirillov-Tuy condition of order $m \geq 1$, if for every $z \in C^n$, $\eta \in S^{2n-1}$ ($|\eta|=1$) we can find a defining collection of order m: $a_1(z,\eta),\ldots,a_{\mathcal{L}_{n-1,m}}(z,\eta)$ in the intersection of the complex hyperplane $\langle a,\eta\rangle = \langle z,\eta\rangle$ with γ. (Defining, that is, for the polynomials on this hyperplane.)*

THEOREM 2.3. *Let $\gamma \subset C^n$ ($n \geq 3$) be a C^1-smooth complex curve, satisfying the complex Kirillov-Tuy condition of order $(p+q)$. If $g \in \mathcal{S}(S^q_p)$, then its solenoidal part $f = {}^s g$ can be uniquely reconstructed from the known values $Ig(z,\xi)$ for all $z \in \gamma$, $\xi \in C^n_0$.*

PROOF. We notice the following homogeneity property of Ig with respect to the second variable:

$$Ig(z,\tau\xi) = \frac{\tau^p \bar{\tau}^q}{|\tau|^2} Ig(z,\xi).$$

Thus for a fixed z we can treat $Ig(z,\cdot)$ as a tempered distribution from $\mathcal{S}'(C^n)$ and consider its Fourier transform.

LEMMA 2.4. *We have the following formula in $\mathcal{S}'(C^n)$:*

$$(\widehat{Ig})(a,\eta) = \lim_{H\to\infty} \sum_{l=0}^{p}\sum_{r=0}^{q} \sum_{1\le\alpha_1<\cdots<\alpha_l\le p} \sum_{1\le\gamma_1<\cdots<\gamma_r\le q} (-1)^{l+r} a^{i\alpha_1}\ldots a^{i\alpha_l} \times$$

$$\times \bar{a}^{j\gamma_1}\ldots\bar{a}^{j\gamma_r} \int_{|\rho|\le H} |\rho|^{2n-4}\rho^p\bar\rho^q e^{\frac{\sqrt{-1}}{2}(\langle\rho a,\eta\rangle + \langle\eta,\rho a\rangle)} \times$$

(2.1) $$\times \bigl(z^{i\beta_1}\ldots z^{i\beta_{p-l}}\bar z^{j\delta_1}\ldots \bar z^{j\delta_{q-r}}\, g^{j_1\ldots j_q}_{i_1\ldots i_p}(z)\bigr)^{\wedge}(\bar\rho\eta)\, dS(\rho).$$

Here we set $\{\beta_1\ldots\beta_{p-l}\} = \{1\ldots p\}\setminus\{\alpha_1\ldots\alpha_l\}$, $1\le\beta_1<\cdots<\beta_{p-l}\le p$; $\{\delta_1\ldots\delta_{q-r}\} = \{1\ldots q\}\setminus\{\gamma_1\ldots\gamma_r\}$, $1\le\delta_1<\cdots<\delta_{q-r}\le q$. The limit is taken in the weak sense in $\mathcal{S}'(C^n)$.

PROOF OF LEMMA 2.4. We need to apply both parts of (2.1) to a test function $\psi(\eta)\in\mathcal{S}(C^n)$. The left-hand side will then be

(2.2) $$\langle (\widehat{Ig})(a,\eta), \psi(\eta)\rangle = \langle Ig(a,y), \widehat\psi(y)\rangle.$$

Consider the right-hand side before taking the limit:

$$\int_{C^n} \sum_{l=0}^{p}\sum_{r=0}^{q}\sum_{1\le\alpha_1<\cdots<\alpha_l\le p}\sum_{1\le\gamma_1<\cdots<\gamma_r\le q} (-1)^{l+r} a^{i\alpha_1}\ldots a^{i\alpha_l}\bar a^{j\gamma_1}\ldots\bar a^{j\gamma_r} \times$$

$$\times \int_{|\rho|\le H} |\rho|^{2n-4}\rho^p\bar\rho^q e^{\frac{\sqrt{-1}}{2}(\langle\rho a,\eta\rangle+\langle\eta,\rho a\rangle)} \times$$

$$\times \bigl(z^{i\beta_1}\ldots z^{i\beta_{p-l}}\bar z^{j\delta_1}\ldots \bar z^{j\delta_{q-r}}\, g^{j_1\ldots j_q}_{i_1\ldots i_p}(z)\bigr)^{\wedge}(\bar\rho\eta)\, dS(\rho)\,\psi(\eta)\, dV_{2n}(\eta) =$$

$$= \int_{|\rho|\le H} |\rho|^{2n-4}\rho^p\bar\rho^q \int_{C^n} \sum_{l=0}^{p}\sum_{r=0}^{q}\sum_{1\le\alpha_1<\cdots<\alpha_l\le p}\sum_{1\le\gamma_1<\cdots<\gamma_r\le q} (-1)^{l+r} a^{i\alpha_1}\ldots a^{i\alpha_l} \times$$

$$\times \bar a^{j\gamma_1}\ldots\bar a^{j\gamma_r} z^{i\beta_1}\ldots z^{i\beta_{p-l}}\bar z^{j\delta_1}\ldots\bar z^{j\delta_{q-r}} g^{j_1\ldots j_q}_{i_1\ldots i_p}(z)\times$$

$$\times (2\pi)^{-n}\int_{C^n} e^{\frac{\sqrt{-1}}{2}(\langle\rho(z-a),\eta\rangle+\langle\eta,\rho(z-a)\rangle)}\psi(\eta)\, dV_{2n}(\eta)\, dV_{2n}(z)\, dS(\rho).$$

We can change the order of integration, because ψ and all the components of g are from the Schwartz space.

The last expression above equals

$$\int_{|\rho|\le H} |\rho|^{2n-4}\rho^p\bar\rho^q \int_{C^n} (z-a)^{i_1}\ldots(z-a)^{i_p}\overline{(z-a)}^{j_1}\ldots\overline{(z-a)}^{j_q} g^{j_1\ldots j_q}_{i_1\ldots i_p}(z)\times$$

$$\times \widehat\psi(\rho(z-a))\, dV_{2n}(z)\, dS(\rho).$$

Introducing variable change $y = \rho(z-a)$ and $t = 1/\rho$, we obtain

$$\int_{|\rho|\le H}\int_{C^n} g^{j_1\ldots j_q}_{i_1\ldots i_p}(a+\frac{1}{\rho}y) y^{i_1}\ldots y^{i_p}\bar y^{j_1}\ldots \bar y^{j_q}\widehat\psi(y)\cdot dV_{2n}(y)\frac{1}{|\rho|^4}\, dS(\rho) =$$

$$
\text{(2.3)} \qquad = \int\limits_{|t|\geq H^{-1}} \int\limits_{C^n} g^{j_1\ldots j_q}_{i_1\ldots i_p}(a+ty) y^{i_1}\ldots y^{i_p} \bar{y}^{j_1}\ldots \bar{y}^{j_q} \widehat{\psi}(y)\, dV_{2n}(y)\, dS(t).
$$

The integral above converges absolutely, i.e.

$$
\int\limits_{C^n}\int\limits_{C} |g^{j_1\ldots j_q}_{i_1\ldots i_p}(a+ty)||y^{i_1}|\ldots |y^{i_p}||\bar{y}^{j_1}|\ldots |\bar{y}^{j_q}||\widehat{\psi}(y)|\, dS(t)\, dV_{2n}(y) < \infty,
$$

because the function

$$
y \to \int\limits_{C} |g^{j_1\ldots j_q}_{i_1\ldots i_p}(a+ty)||y^{i_1}|\ldots |y^{i_p}||\bar{y}^{j_1}|\ldots |\bar{y}^{j_q}|\, dS(t)
$$

is positively homogeneous of the degree $(p+q-2)$ and $\widehat{\psi} \in \mathcal{S}(C^n)$.

Thus in (2.3) we can take the limit as $H \to \infty$ and obtain

$$
\int\limits_{C^n}\int\limits_{C} g^{j_1\ldots j_q}_{i_1\ldots i_p}(a+ty) y^{i_1}\ldots y^{i_p} \bar{y}^{j_1}\ldots \bar{y}^{j_q}\, dS(t)\, \widehat{\psi}(y)\, dV_{2n}(y) =
$$

$$
= \int\limits_{C^n} Ig(a,y)\widehat{\psi}(y)\, dV_{2n}(y) = \langle Ig(a,y), \widehat{\psi}(y)\rangle,
$$

which is the same as in (2.2).

This proves Lemma 2.4. $\qquad\square$

We notice that in the right-hand side of (2.1) we have a pointwise limit as $H \to \infty$ in the domain $\{\eta \in C^n \mid \eta \neq 0\}$, because the corresponding Fourier transform is rapidly decreasing (if $\eta = 0$, then the limit does not exist). We will need to show that the restriction of the distribution $\widehat{(Ig)}(a,\eta)$ to this domain coincides with the regular distribution, defined by the pointwise limit.

Each term in (2.1) up to a coefficient has the form

$$
\text{(2.4)} \qquad \lim_{H\to\infty} \int\limits_{|\rho|\leq H} |\rho|^{2n-4} \rho^p \bar{\rho}^q\, e^{\frac{\sqrt{-1}}{2}(\langle \rho a,\eta\rangle + \langle \eta,\rho a\rangle)} \widehat{G}(\bar{\rho}\eta)\, dS(\rho),
$$

$$
G(z) = z^{i\beta_1}\ldots z^{i\beta_{p-l}} \bar{z}^{j\delta_1}\ldots \bar{z}^{j\delta_{q-r}}\, g^{j_1\ldots j_q}_{i_1\ldots i_p}(z).
$$

The components of g are from the Schwartz space, therefore $G(z)$ and $\widehat{G}(z)$ are rapidly decreasing and

$$
|\widehat{G}(\zeta)| \leq \frac{C_M}{1+|\zeta|^M},
$$

for every M.

So, in (2.4) we have for each $\eta \neq 0$ the following value

$$
\int\limits_{C} |\rho|^{2n-4} \rho^p \bar{\rho}^q\, e^{\frac{\sqrt{-1}}{2}(\langle \rho a,\eta\rangle + \langle \eta,\rho a\rangle)} \widehat{G}(\bar{\rho}\eta)\, dS(\rho).
$$

Take a test function $\psi \in \mathcal{S}(C^n)$ with $\operatorname{supp}\psi \subset C_0^n$. Then for some $r > 0$ we have $|\eta| \geq r$ on $\operatorname{supp}\psi$. Consider the following expression

$$
\text{(2.5)} \qquad \int\limits_{C^n}\int\limits_{|\rho|\leq H} |\rho|^{2n-4}\rho^p \bar{\rho}^q\, e^{\frac{\sqrt{-1}}{2}(\langle \rho a,\eta\rangle + \langle \eta,\rho a\rangle)} \widehat{G}(\bar{\rho}\eta) dS(\rho) \cdot \psi(\eta)\, dV_{2n}(\eta).
$$

We have the following estimate for each $\eta \in \operatorname{supp} \psi$:

$$\left| \int_{|\rho| \leq H} |\rho|^{2n-4} \rho^p \bar{\rho}^q \, e^{\frac{\sqrt{-1}}{2}(\langle \rho a, \eta \rangle + \langle \eta, \rho a \rangle)} \widehat{G}(\bar{\rho}\eta) \, dS(\rho) \right| \leq$$

$$\leq \int_{|\rho| \leq H} |\rho|^{2n+p+q-4} \frac{C_M}{1 + |\rho|^M |\eta|^M} \, dS(\rho) \leq$$

$$\leq \int_C |\rho|^{2n+p+q-4} \frac{C_M}{1 + |\rho|^M r^M} \, dS(\rho) = C(M) < \infty,$$

if M is sufficiently large for the last integral to converge.

Since ψ belongs to the Schwartz space and because of the Lebesgue dominated convergence theorem, we can take the pointwise limit under the integral sign over C^n in (2.5) and get

$$\int_{C^n} \int_C |\rho|^{2n-4} \rho^p \bar{\rho}^q e^{\frac{\sqrt{-1}}{2}(\langle \rho a, \eta \rangle + \langle \eta, \rho a \rangle)} \widehat{G}(\bar{\rho}\eta) \, dS(\rho) \cdot \psi(\eta) \, dV_{2n}(\eta).$$

By the hypothesis of the theorem, we therefore know the following expression for every $a \in \gamma$ and $\eta \in S^{2n-1}$:

$$\sum_{l=0}^{p} \sum_{r=0}^{q} \sum_{1 \leq \alpha_1 < \cdots < \alpha_l \leq p} \sum_{1 \leq \gamma_1 < \cdots < \gamma_r \leq q} (-1)^{l+r} a^{i_{\alpha_1}} \ldots a^{i_{\alpha_l}} \bar{a}^{j_{\gamma_1}} \ldots \bar{a}^{j_{\gamma_r}} \int_C |\rho|^{2n-4} \rho^p \bar{\rho}^q \times$$

$$(2.6) \qquad \times e^{\frac{\sqrt{-1}}{2}(\langle \rho a, \eta \rangle + \langle \eta, \rho a \rangle)} \widehat{\left(z^{i_{\beta_1}} \ldots z^{i_{\beta_{p-l}}} \bar{z}^{j_{\delta_1}} \ldots \bar{z}^{j_{\delta_{q-r}}} g_{i_1 \ldots i_p}^{j_1 \ldots j_q}(z) \right)} (\bar{\rho}\eta) \, dS(\rho).$$

We fix an arbitrary $z_0 \in C^n$ and $\eta \in S^{2n-1}$. By the hypothesis we can find a defining collection of points $a_1(z_0, \eta), \ldots, a_{\mathcal{L}_{n-1,m}}(z_0, \eta)$ in the intersection of the hyperplane $\langle a, \eta \rangle = \langle z_0, \eta \rangle$ with γ. Note that the restriction of the expression in (2.6) to this hyperplane is a polynomial $P(a)$ on it (because there we have $\langle \rho a, \eta \rangle = \rho \langle a, \eta \rangle = \rho \langle z_0, \eta \rangle$ and the dependence on a is purely polynomial).

The values $P(a_j(z_0, \eta))$ are known, because $a_j(z_0, \eta) \in \gamma$. Therefore $P(a)$ is known on the whole hyperplane.

We introduce the following polynomial $\widetilde{P}(a)$, defined everywhere on C^n:

$$\widetilde{P}(a) = P(z_0 + \pi_\eta(a - z_0)),$$

where $\pi_\eta(z) = z - \langle z, \eta \rangle \eta$ is the orthogonal projection to the complement $\eta^\perp$ of η with respect to the Hermitian form. It is clear that $\widetilde{P}$ is known on C^n. Its homogeneous part of the highest degree $(p+q)$ has the form

$$(-1)^{p+q} \int_C |\rho|^{2n-4} \rho^p \bar{\rho}^q \, e^{\frac{\sqrt{-1}}{2}(\langle \rho z_0, \eta \rangle + \langle \eta, \rho z_0 \rangle)} \hat{g}_{i_1 \ldots i_p}^{j_1 \ldots j_q}(\bar{\rho}\eta) \, dS(\rho) \times$$

$$\times (a^{i_1} - \langle a, \eta \rangle \eta^{i_1}) \ldots (a^{i_p} - \langle a, \eta \rangle \eta^{i_p}) (\bar{a}^{j_1} - \langle \eta, a \rangle \bar{\eta}^{j_1}) \ldots (\bar{a}^{j_q} - \langle \eta, a \rangle \bar{\eta}^{j_q}) =$$

$$= (-1)^{p+q} \int_C |\rho|^{2n-4} \rho^p \bar{\rho}^q e^{\frac{\sqrt{-1}}{2}(\langle \rho z_0, \eta \rangle + \langle \eta, \rho z_0 \rangle)} \hat{g}_{i_1 \ldots i_p}^{j_1 \ldots j_q}(\bar{\rho}\eta) \, dS(\rho) \times$$

$$\times \varepsilon_{k_1}^{i_1}(\eta) \ldots \varepsilon_{k_p}^{i_p}(\eta) \, \varepsilon_{j_1}^{l_1}(\eta) \ldots \varepsilon_{j_q}^{l_q}(\eta) a^{k_1} \ldots a^{k_p} \bar{a}^{l_1} \ldots \bar{a}^{l_q} =$$

$$= (-1)^{p+q} \int_C |\rho|^{2n-4} \rho^p \bar{\rho}^q \, e^{\frac{\sqrt{-1}}{2}(\langle \rho z_0, \eta \rangle + \langle \eta, \rho z_0 \rangle)} \, \hat{f}^{l_1 \ldots l_q}_{k_1 \ldots k_p}(\bar{\rho}\eta) \, dS(\rho) \times$$

(2.7)
$$\times a^{k_1} \ldots a^{k_p} \bar{a}^{l_1} \ldots \bar{a}^{l_q},$$

where $f = {}^s g$ is the solenoidal part of g. (See the formula (1.1) and use $|\eta| = 1$ and $\varepsilon_i^j(\bar{\rho}\eta) = \varepsilon_i^j(\eta)$.)

Thus, we know all the coefficients in (2.7):

$$\int_C |\rho|^{2n-4} \rho^p \bar{\rho}^q \, e^{\frac{\sqrt{-1}}{2}(\langle \rho z_0, \eta \rangle + \langle \eta, \rho z_0 \rangle)} \, \hat{f}^{l_1 \ldots l_q}_{k_1 \ldots k_p}(\bar{\rho}\eta) \, dS(\rho).$$

Consider now a fixed $\eta_0 \in S^{2n-1}$ and introduce the variable $\mu = \bar{\rho}$. If we take $z_0 = \lambda \eta_0$, $\lambda \in C$, we therefore obtain

$$\int_C |\mu|^{2n-4} \bar{\mu}^p \mu^q \, e^{\frac{\sqrt{-1}}{2}(\langle z_0, \mu \eta_0 \rangle + \langle \mu \eta_0, z_0 \rangle)} \, \hat{f}^{l_1 \ldots l_q}_{k_1 \ldots k_p}(\mu \eta_0) \, dS(\mu) =$$

$$= \int_C e^{\frac{\sqrt{-1}}{2}(\lambda \bar{\mu} + \mu \bar{\lambda})} |\mu|^{2n-4} \bar{\mu}^p \mu^q \, \hat{f}^{l_1 \ldots l_q}_{k_1 \ldots k_p}(\mu \eta_0) \, dS(\mu).$$

Noting that $\frac{\sqrt{-1}}{2}(\lambda \bar{\mu} + \mu \bar{\lambda}) = \sqrt{-1} \operatorname{Re}(\lambda \bar{\mu})$, we recognize here the 2-dimensional Fourier transform (up to a coefficient) of the function

$$\mu \to |\mu|^{2n-4} \bar{\mu}^p \mu^q \hat{f}^{l_1 \ldots l_q}_{k_1 \ldots k_p}(\mu \eta_0).$$

The value $\lambda \in C$ can be taken arbitrary, therefore this Fourier transform is known on C. Applying the inversion formula for it, we find $\hat{f}^{l_1 \ldots l_q}_{k_1 \ldots k_p}(\mu \eta_0)$ for all $\mu \in C$ (and all $\eta_0 \in S^{2n-1}$). Then, applying Fourier inversion in C^n, we obtain all the components $f^{l_1 \ldots l_q}_{k_1 \ldots k_p}$ of the solenoidal part f. This completes the proof of Theorem 2.3. □

Acknowledgments

The author is grateful to the organizers of the 2000 Summer AMS/IMS/SIAM Research Conference "Radon transform and Tomography" for their hospitality.

References

[1] *Natterer F.* The Mathematics of Computerized Tomography. B.G.Teubner, Stuttgart, and John Wiley & Sons Ltd, 1986.
[2] *Sharafutdinov V. A.* Integral Geometry of Tensor Fields. VSP, Utrecht, the Netherlands, 1994.
[3] *Vertgeim L. B.* Integral geometry problems for symmetric tensor fields with incomplete data. J.Inv. Ill-Posed Problems, Vol. 8, No. 3 (2000), pp. 353–362.
[4] *Vertgeim L. B.* Integral geometry of symmetric tensor fields in a complex space and integral geometry of matrices. Candidate dissertation. Sobolev Institute of Math., Siberian Branch of the Russian Acad. of Sciences, Novosibirsk. 1999. (in Russian)
[5] *Vertgeim L. B.* Integral geometry of symmetric tensor fields in a complex space. Solution methods for the conditionally-correct problems. Novosibirsk. Institute of Math., Sib. Branch of the USSR Acad. of Sciences. 1991. pp. 29–46. (in Russian)

Sobolev Institute of Mathematics, Siberian Branch of the Russian Academy of Sciences, Acad. Koptyug prosp., 4, Novosibirsk, 630090, Russia.
E-mail address: vert@math.nsc.ru

Selected Titles in This Series

(Continued from the front of this publication)

253 **Charles N. Delzell and James J. Madden, Editors,** Real algebraic geometry and ordered structures, 2000

252 **Nathaniel Dean, Cassandra M. McZeal, and Pamela J. Williams, Editors,** African Americans in Mathematics II, 1999

251 **Eric L. Grinberg, Shiferaw Berhanu, Marvin I. Knopp, Gerardo A. Mendoza, and Eric Todd Quinto, Editors,** Analysis, geometry, number theory: The Mathematics of Leon Ehrenpreis, 2000

250 **Robert H. Gilman, Editor,** Groups, languages and geometry, 1999

249 **Myung-Hwan Kim, John S. Hsia, Yoshiyuki Kitaoka, and Rainer Schulze-Pillot, Editors,** Integral quadratic forms and lattices, 1999

248 **Naihuan Jing and Kailash C. Misra, Editors,** Recent developments in quantum affine algebras and related topics, 1999

247 **Lawrence Wasson Baggett and David Royal Larson, Editors,** The functional and harmonic analysis of wavelets and frames, 1999

246 **Marcy Barge and Krystyna Kuperberg, Editors,** Geometry and topology in dynamics, 1999

245 **Michael D. Fried, Editor,** Applications of curves over finite fields, 1999

244 **Leovigildo Alonso Tarrío, Ana Jeremías López, and Joseph Lipman,** Studies in duality on noetherian formal schemes and non-noetherian ordinary schemes, 1999

243 **Tsit Yuan Lam and Andy R. Magid, Editors,** Algebra, K-theory, groups, and education, 1999

242 **Bernhelm Booss-Bavnbek and Krzysztof Wojciechowski, Editors,** Geometric aspects of partial differential equations, 1999

241 **Piotr Pragacz, Michał Szurek, and Jarosław Wiśniewski, Editors,** Algebraic geometry: Hirzebruch 70, 1999

240 **Angel Carocca, Víctor González-Aguilera, and Rubí E. Rodríguez, Editors,** Complex geometry of groups, 1999

239 **Jean-Pierre Meyer, Jack Morava, and W. Stephen Wilson, Editors,** Homotopy invariant algebraic structures, 1999

238 **Gui-Qiang Chen and Emmanuele DiBenedetto, Editors,** Nonlinear partial differential equations, 1999

237 **Thomas Branson, Editor,** Spectral problems in geometry and arithmetic, 1999

236 **Bruce C. Berndt and Fritz Gesztesy, Editors,** Continued fractions: From analytic number theory to constructive approximation, 1999

235 **Walter A. Carnielli and Itala M. L. D'Ottaviano, Editors,** Advances in contemporary logic and computer science, 1999

234 **Theodore P. Hill and Christian Houdré, Editors,** Advances in stochastic inequalities, 1999

233 **Hanna Nencka, Editor,** Low dimensional topology, 1999

232 **Krzysztof Jarosz, Editor,** Function spaces, 1999

231 **Michael Farber, Wolfgang Lück, and Shmuel Weinberger, Editors,** Tel Aviv topology conference: Rothenberg Festschrift, 1999

230 **Ezra Getzler and Mikhail Kapranov, Editors,** Higher category theory, 1998

229 **Edward L. Green and Birge Huisgen-Zimmermann, Editors,** Trends in the representation theory of finite dimensional algebras, 1998

For a complete list of titles in this series, visit the
AMS Bookstore at **www.ams.org/bookstore/**.